DESALINATION, TRENDS AND TECHNOLOGIES

Edited by **Michael Schorr**

Desalination, Trends and Technologies
Edited by Michael Schorr

Published by InTech
Janeza Trdine 9, 51000 Rijeka, Croatia

Publishing Process Manager Iva Lipovic

Technical Editor Teodora Smiljanic

Cover Designer InTech Design Team

Additional hard copies can be obtained from orders@intechopen.com

Desalination, Trends and Technologies, Edited by Michael Schorr
p. cm.
ISBN 978-953-307-311-8

For EU product safety concerns:
IN TECH d.o.o., Prolaz Marije Krucifikse Kozulić 3, 51000 Rijeka, Croatia,
info@intechopen.com or visit our website at intechopen.com.

Meet the editor

Michael Schorr is a Professor and Doctor Honoris Causa at the Institute of Engineering of the University of Baja California in Mexico, and a member of the Mexican National System of Investigators, Level II. He was granted the B.Sc. degree in Chemistry and the M.Sc. degree in Materials Engineering, both by the Technion – Israel Institute of Technology and, during the period of 1986-2004, served as Editor of the Corrosion Reviews journal. He was actively involved in the development of desalination processes in Israel, focusing particularly on the selection and testing of corrosion-resistant materials required for fabrication of the pertinent equipment and plants. Professor Schorr has fifty years of experience in controlling environmental and industrial corrosion in Latin America, Europe, USA, South Africa and Israel. He authored and co-authored 240 scientific and technical publications on corrosion science, engineering and technology, as well as several books in English, Spanish and Hebrew.

Contents

Preface

The sustainability and prosperity of the ancient civilizations of China, Egypt, Babylonia, Phoenicia, Persia and Roma were based on the extensive use of water for human consumption, crop irrigation, canal navigation and energy generation. Today, the worldwide scarcity of water and clean energy constitutes a central and critical problem for the whole humankind. This situation is aggravated as industrial, agricultural and municipal effluents reach the water bodies, or the coastal seawater that is used as feed for desalination plants. All these problems are linked to the actual, natural and anthropogenic changes of climate, global warming and greenhouse-gas emissions, all interrelated phenomena that affect our planet.

In order to avoid damage to its facilities and equipment, the desalination industry invests considerable efforts to deal with these changes, in particular with extreme events such as torrential rains, devastating floods, dry seasons with devouring fires, as well as with extended spells of cold weather with freezing temperatures.

The book chapters are arranged in an hierarchical sequence, starting with conventional and novel desalination processes and following with energy, environmental, economic and ecological issues, all affecting the desalination industry image and profitability.

Leading experts from academia and industry, as well as environment researchers, distinguished teachers and experienced engineers have written special chapters for this impressive collection. The contributing authors offer a large amount of practical information, presenting it in a highly condensed yet coherent body of useful knowledge and practical expertise. Moreover, the multi-authored characteristic of this volume offers a wide spectrum of knowledge and experience, as the authors are specialists in different fields and express diverse approaches and orientations. The intended multifacet content of this publication certainly contributes to enrich it.

This compendium provides valuable, encyclopedic knowledge on research, development, processes, plants and technologies of this industry, from the fundamental concepts up to many practical cases collected from around the world. Hence, it provides a useful insight into the world of water, energy and desalination, easy to follow and to apply.

This volume is an essential companion to chemists, as well as to civil and chemical engineers who design, build and operate desalination plants. It is also highly relevant to maintenance personnel, corrosion specialists, material- and mechanical engineers.

Also, university lecturers and researchers will find it useful for their students while preparing their thesis on subjects related to desalination processes and plants. Not less so, desalination industry executives should make sure that their field managers and engineers in charge of running their plants will have access to it, and apply the built-in know-how in their daily work routine.

Another strong part of this book is the wealth of references listed for each chapter, amounting to hundreds of sources of detailed information from the modern scientific and technical literature. Anyone interested in desalination will be thrilled by their diverse content.

All in all, this volume enables the reader to gain a deeper understanding of the state of the art of the desalination industry and to become acquainted with the most recent developments and technologies in this area.

Finally, it is my pleasant duty to acknowledge with thanks each of the learned authors for contributing their chapters to this volume.

December 2010

Prof. Michael Schorr
Institute of Engineering
University of Baja California
Mexicali, Mexico

Part 1

Desalination Processes and Plants

1

Electrodialysis Technology - Theory and Applications

Fernando Valero, Angel Barceló and Ramón Arbós
Aigues Ter Llobregat (ATLL). Spain

1. Introduction

First commercial equipment based on Electrodialysis (ED) technology was developed in the 1950s to demineralize brackish water (Juda & McRae, 1950; Winger et al. 1953). Since then ED has advanced rapidly because of improved ion exchange membrane properties, better materials of construction and advances in technology. In the 1960s, Electrodialysis Reversal (EDR) was introduced, to avoid organic fouling problems (Mihara & Kato, 1969). Over the past twenty years EDR has earned a reputation as a membrane desalination process that works economically and reliably on surface water supplies, reuse water and some specific industrial applications when designed and operated properly.
Some applications of ED/EDR were its use to reduce inorganics like radium (Hays, 2000), perchlorate (Roquebert et al., 2000), bromide (Valero & Arbós, 2010), fluoride (GE W&P, 2010), iron and manganese (Heshka, 1992) and nitrate (Menkouchi Sahlia et al., 2008) in drinking water. In addition the technology can be used to recycle municipal and industrial wastewater (Broens et al,. 2004; Chao & Liang, 2008), recovering reverse osmosis reject (Reahl, 1990; Korngold, 2009), desalting wells (Harries et al., 1991), surface waters (Lozier et al. 1992), final effluent treatment for reuse in cooling towers (De barros, 2008), whey and soy purification (MEGA a.s.,2010), table salt production (Kawahara, 1994) and many other industrial uses (Schoeman & Stein, 2000; Dalla Costa et al., 2002; Pilat, 2003). For this kind of applications, this technology had shown best hydraulic recovery and cost effective in front of other membrane technologies, specially compared with Reversal Osmosis (RO). In these sense, the lower residues produced during ED/EDR process, is another important advantatge of this technique (AWWA, 2004). Moreover, electrodialysis is not always a cost effective option for seawater desalination and does not have a barrier effect against microbiological contamination.
This chapter reviews some aspects related with the theory of the technology, design, operation and maintenance (O&M), manufacturers, applications, operational costs and finally shows two cases studies involving the two world's biggest EDR systems, both located near to Barcelona (Spain). The first of them is located in Abrera (Valero et al., 2007) with a capacity of treatment of 220.000 m^3/d (576 stacks in two stages, provided by GE Water & Process) and it is related with desalting brackish water to improve the quality of the produced drinking water. The second one is located in Sant Boi del Llobregat (Segarra et al., 2009) with a capacity of treatment of 57,000 m^3/d (96 stacks in two stages, provided by MEGA a.s.) and represents a tertiary treatment of a wastewater treatment plant (WWTP) for agricultural reuse.

2. Theory

ED is an electrochemical separation process in which ions are transferred through ion exchange membranes by means of a direct current (DC) voltage. The process uses a driving force to transfer ionic species from the source water through cathode (positively charged ions) and anode (negatively charged ions) to a concentrate wastewater stream, creating a more dilute stream (Figure 1).

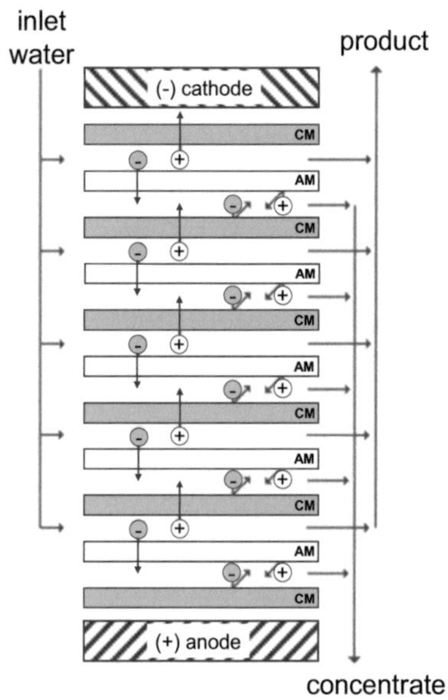

Fig. 1. Principles of ED

ED selectively removes dissolved solids, based on their electrical charge, by transferring the brackish water ions through a semi permeable ion exchange membrane charged with an electrical potential. It points out that the feed water becomes separated into the following three types of water (AWWA, 1995):

- product water, which has an acceptably low conductivity and TDS level;
- brine, or concentrate, which is the water that receives the brackish water ions; and
- electrode feed water, which is the water that passes directly over the electrodes that create the electrical potential.

EDR is a variation on the ED process, which uses electrode polarity reversal to automatically clean membrane surfaces. EDR works the same way as ED, except that the polarity of the DC power is reversed two to four times per hour. When the polarity is reversed, the source water dilute and concentrate compartments are also reversed and so are the chemical reactions at the electrodes. This polarity reversal helps prevent the formation of scale on the membranes. The setup is very similar to an ED system except for the presence of reversal valves (Ionics Inc., 1984).

2.1 Membrane stacks

All ED and EDR systems are designed specifically for a particular application. The amount of ions to be removed is determined by the configuration of the membrane stack. A membrane stack may be oriented in either a horizontal or vertical position.

Cell pairs form the basic building blocks of an EDR membrane stack (Figure 1). Each stack assembled has the two electrodes and groups of cell pairs. The number of cell pairs necessary to achieve a given product water quality is primarily determined by source water quality, and can design stacks with more than 600 cell pairs for industrial applications (Strathmann, 2004).

A cell pair consists of the following:

- Anion permeable membrane
- Concentrate spacer
- Cation permeable membrane
- Dilute stream spacer

In each stack, we can observe different flows (Figure 2):

1. Source water (feed) flows parallel only through demineralizing compartments, whereas the concentrate stream flows parallel only through concentrating compartments.
2. As feed water flows along the membranes, ions are electrically transferred through membranes from the demineralized stream to the concentrate stream.
3. Flows from the two electrode compartments do not mix with other streams. A degasifier vents reaction gases from the electrode waste stream.
4. Top and bottom plates are steel blocks that compress the membranes and spacers to prevent leakage inside the stack.

Effluent from these compartments may contain oxygen, hydrogen, and chlorine gas. Concentrate from the electrode stream is sent to a degasifier to remove and safely dispose of any reaction gases.

The first type of commercial ED system was the batch system. In this type of ED system, source water is recirculated from a holding tank through the demineralizing spacers of a single membrane stack and back to the holding tank until the final purity is obtained. The production rate is dependent on the dissolved minerals concentration in the source water

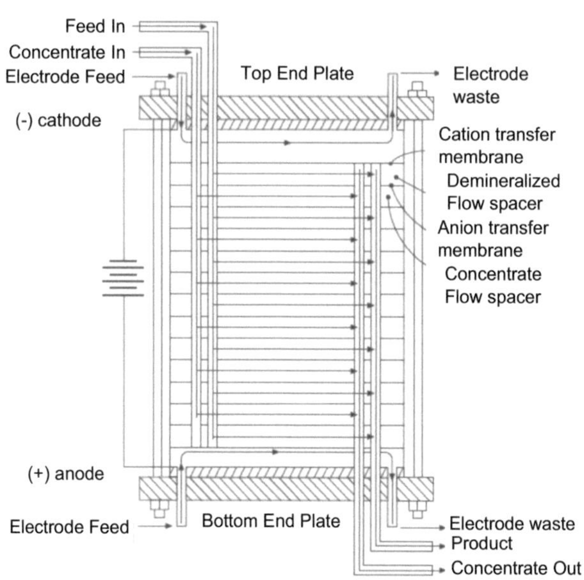

Fig. 2. Stack description (Ionics Inc., 1984)

and on the degree of demineralization required. The concentrate stream is also recirculated to reduce wastewater volume, and continuous addition of acid is required to prevent membrane stack scaling.

The second type of commercially available system was the unidirectional continuous-type ED. In this type of system, the membrane stack contains two stages in series; each stage helps demineralize the water. The demineralized stream makes a single pass through the stack and exits as product water. The concentrate stream is partially recycled to reduce wastewater volume and is injected with acid to prevent scaling. EDR was patented in 1969 (Mihara & Kato, 1969) and is a variation of this system which uses electrode polarity reversal to automatically clean membrane surfaces.

2.2 Membranes

The membranes are produced in the form of foils composed of fine polymer particles with ion exchange groups anchored by polymer matrix. Impermeable to water under pressure, membranes are reinforced with synthetic fiber which improves the mechanical properties of the membrane (AWWA, 1995).

The two types of ion exchange membranes used in electrodialysis are:

- Cation transfer membranes which are electrically conductive membranes that allow only positively charged ions to pass through. Commercial cation membranes generally consists of crosslinked polystyrene that has been sulfonated to produce $^{-}SO_3H$ groups attached to the polymer, in water this group ionizes producing a mobile counter ion (H^+) and a fixed charge ($-SO_3^-$).
- Anion transfer membranes, which are electrically conductive membranes that allow only negatively, charged ions to pass through. Usually, the membrane matrix has fixed positive charges from quaternary ammonium groups ($^{-}NR_3^+OH^-$) which repel positive ions.

Both types of membranes shows common properties: low electrical resistance, insoluble in aqueous solutions, semi-rigid for ease of handling during stack assembly, resistant to change in pH from 1 to 10, operate temperatures in excess of 46°C, resistant to osmotic swelling, long life expectancies, resistant to fouling and hand washable.

The membranes are permselective (or ion selective) that refers to their ability to discriminate between different ions to allow passage or permeation through the membrane. In these sense membranes can be tailored to inhibit the passage of divalent anions or cations, such as sulfates, calcium, and magnesium. For example, some membranes show good permeation or high transport numbers for monovalent anions, such as Cl^- or NO_3^-, but have low transport numbers and show very low permeation rates for divalent or trivalent ions, such as SO_4^{-2}, PO_4^{-3}, or similar anions. This is achieved by specially treating the anion membrane, and the effect can be exploited to separate various ions. The relative specificities vary, with the monovalent anion membrane showing the greatest specificity, for example, the ratio of chloride to sulfate ion transport numbers. (Xu, 2005).

It depends on the manufacturer by usually each membrane is 0.1 to 0.6 mm thick and is either homogeneous or heterogeneous, according to the connection way of charge groups to the matrix or their chemical structure (Xu, 2005). In the case of homogeneous membranes, charged groups are chemically bonded and for heterogeneous they are physically mixed with the membrane matrix. Different manufacturers of ion exchange membranes are available in the market (Table 1). Each one offers membranes for specific applications, and they have different properties involving, size, thickness, area resistance and composition.

Manufacturer/Reference	Country	Commercial brand
Asahi Chemical Industry Co.	Japan	Aciplex
Asahi Glass Col. Ltd	Japan	Selemion
DuPont Co.	USA	Nafion
FuMA-Tech GmbH	Germany	Fumasep
GE Water & Process	USA	AR, CR,..
LanXess Sybron Chemicals	Germany	Ionac
MEGA a.s.	Czech Republic	Ralex
PCA GmbH	Germany	PC
Tianwei Membrane Co.Ltd.	China	TWAED
Tokuyama Co-Astom	Japan	Neosepta

Table 1. Main manufacturers of ion exchange membranes.

2.3 Spacers

The spaces between the membranes represent the flow paths of the demineralized and concentrated streams formed by plastic separators which are called demineralized and concentrate water flow spacers respectively. These spacers are made of polypropylene or low density polyethylene and are alternately positioned between membranes in the stack to create independent flow paths, so that all the demineralized streams are manifolded together and all the concentrate streams are manifolded together too.

Demineralizing and concentrating spacers are created by rotating an identical spacer 180°. Demineralizing spacers allow water to flow across membrane surfaces where ions are removed, whereas concentrating spacers prevent the concentrate stream from contaminating the demineralized stream.

There is a spacer design with a "tortuous path" in which the spacer is folded back upon it self and the liquid flow path is much longer than the linear dimensions or the unit. Another kind of spacers is a "sheet flow" that consists of an open frame with a plastic screen separating the membranes. These spacers are operated at lower flow velocities, to achieve a degree of desalting in each pass through the stack, comparable to the tortuous path or sheet flow spacers. In general the increase of turbulence promotes mixing of the water, use of the membrane area, and the transfer of ions. Turbulence resulting from spacers also breaks up particles or slime on the membrane surface and attracts ions to the membrane surface. Flow velocity ranges from (18 to 35 cm/s, creating a pressure drop between the inlet and outlet. A velocity less than 18 cm/s promotes polarization, or the point of limiting density of water (AWWA, 1995).

Maximum pressure for ED and EDR systems is generally limited to 50 psi (345 kPa), and pressure is lost at each stage of the system. Since pressure must be maintained throughout the system, the impact of spacers on pressure is an important design consideration.

Different models and sizes of spacers satisfy specific design applications. The main difference in spacer models is the number of flow paths, which determines water velocity across the membrane stack and contact time of the source water with the membrane. Since water velocity is responsible for the degree of mixing and the amount of desalting that occurs across membranes, velocity is an important design parameter for spacer choice. Because the same spacers are used for both demineralized and concentrated water in EDR systems, the flow rates of both these streams should be equalized to prevent high differential pressures across the membranes.

2.4 Electrodes

A metal electrode at each end of the membrane stack conducts DC into the stack. Electrode compartments consist of an electrode, an electrode water-flow spacer, and a heavy cation membrane. The electrode spacer is thicker than a normal spacer, which increases water velocity to prevent scaling. This spacer also prevents the electrode waste from entering the main flow paths of the stack (Ionics, 1984).

Because of the corrosive nature of the anode compartments, electrodes are usually made of titanium and plated with platinum. Its life span is dependent on the ionic composition of the source water and the amperage applied to the electrode. Large amounts of chlorides in the source water and high amperages reduce electrode life. Polarity reversal (as in EDR) also results in significantly shorter electrode lifetimes than for nonreversing systems (AWWA, 1995).

2.5 Operation

When DC potential is applied across the electrodes, the following take place (AWWA, 1995):
At the cathode, or negative electrode (-):

- Cations (Na^+) attraction
- Pairs of water molecules break down (dissociate) at the cathode to produce two hydroxyl (OH^-) ions plus hydrogen gas (H_2). Hydroxide raises the pH of the water, causing calcium carbonate ($CaCO_3$) precipitation.

And at the anode, or positive electrode (+):

- Anions (Cl^-) attraction
- Pairs of water molecules dissociate at the anode to produce four hydrogen ions (H^+), one molecule of oxygen (O_2), and four electrons (e^-). The acid tends to dissolve any calcium carbonate present to inhibit scaling.
- Chlorine gas (Cl_2) may be formed.

Colloidal particles or slimes that are slightly electronegative may accumulate on the anion membrane and cause membrane fouling. This problem is common to all classes of ED systems. These fouling agents are removed by flushing with cleaning systems. In EDR systems, the polarity of the electrodes is reversed two to four times each hour. When polarity is reversed, chemical reactions at the electrodes are reversed. Valves in the electrode streams automatically switch flows in the two types of compartments. Streams that were in demineralizing compartments become concentrate streams, and concentrate streams become demineralizing streams. The alternating exposure of membrane surfaces to the product dilute and brine concentrate streams provides a self-cleaning capability that enables purification and recovery higher than 90% of source water, reducing the burden on water sources, and minimizing the volume of waste that requires disposal (AWWA, 2004).

2.6 Design

In commercial practice, the basic apparatus for ED/EDR is a stack of rectangular membranes terminated on each end by an electrode. Flow of the process streams is contained and directed by spacers that alternate with the membranes. The membranes are arranged alternately cation and anion. The assembly of membrane spacers and electrodes is held in compression by a pair of end plates. The apparatus thus resembles a plate-and-frame filter press. Stack is completed with pumps, piping and an electrical subsystem trhat includes: adjustable DC power supply, rectifiers to convert alternating current (AC) power to DC power, internal control system with controls, reversal timing (only for EDR), and alarms.

The design of an ED/EDR plant is based on the product water requeriments of the application and the characteristic of the inlet water to be treated (Tsiakis & Papageorgiou, 2005). The parameters, which characterize the working optimum of an EDR, are the values of the applied voltage used in the electrical stages and the feed water pressure corresponding to the maximum separation percentage and minimum energy consumption. These values were obtained from the surfaces corresponding to separation percentage and consumed power versus applied voltage and pressure. It could be important to carry out a pilot study before the industrial design of the system (Valerdi-Pérez et al., 2001). Along the pilot study, operators had to been check the quality of the product in different conditions, focused in the behaviour of several limiting parameters characterizing an ED/EDR system (Ionics, 1984):

- Limiting Current Density (Polarization)
- Current Leakage
- Back Diffusion
- Langelier Saturation Index
- Calcium Sulfate Saturation
- Pressure Drop
- Differential Pressure
- Water Transfer
- Temperature Limits

With these data, ED and EDR plants can be designed to remove from 50 to 99 percent of source water contaminants or dissolved solids. Source water salinities of less than 100 mg/L up to 12,000 mg/L TDS can be successfully treated to produce finished water of less than 10 mg/L TDS.

To calculate the efficiency of the process during the design step, we have to take into account the Faraday's Law. In these sense, the passage of 96,500 amperes of electric current for one second will transfer one gram equivalent of salt. One Faraday is equal to 96,500 ampere-second and that is equal to 26,8 amperes, of current passing for one hour. Thus, when 26.8 ampere-hours, one gram equivalent of salt will be transferred in each cell pair, we have a process 100 percent efficient. To determine the voltage requirements for a given system, the current is determined from Faraday's Law and the resistance is determined by the components of the membrane stack and the solution under treatment, according to the Ohm's Law (Lee et al, 2006; Valerdi-Pérez & Ibañez-Mengual, 2001).

Typically, maximum salt removal for any hydraulic stage is 55-60 percent with normal design values at 40-50 percent. To increase the amount of salt removal in a EDR system, additional hydraulic stages must be incorporated. Then in systems where high capacities are required, additional hydraulic stages are made by simply adding more stacks in series to achieve the desired water purity (Larchet et al, 2008).

In addition, to increase water recovery some functions can be incorporated into the EDR system to take advantage of a substantial increase in recovery and production rates at minimal costs. In this sense, three main flows can be recycled: concentrate stream, off-specification product and electrode stream. With those products, water can be recycled back to the system feed, eliminating the need to send it to waste. To achieve this water recovery, it could be necessary to dose some chemical into the system. In this sense, an antiscalant (1-5mg/L) can be added into the concentrate stream to control calcium scaling, and acid (HCl) is continuously feed into the electrode flow and into the concentrate stream to maintain the Langelier Saturation Index (LSI) at +1.8 for calcium carbonate control.

3. Maintenance

Scheduled maintenance depends on the use and application of the system and varies with each manufacturer. Nowadays, ED and EDR systems are designed with fully automated control systems. Thus O&M procedures are scheduled to check control settings and operating parameters, supported with detection systems that recognize operation levels or critical conditions.

Several setpoints are implemented to operate the system. In these sense, operators check the values and alarms related with temperature, conductivitiy, pH, current voltage, intensity, flows and pressures. Data are collected or logged for the different streams and, in the case of EDR, during positive or negative polarity.

It could be necessary to clean the membranes periodically. Cleaning is a means of removing mineral scale, organic matter, biological growth (slime), colloidal particles, or insoluble constituents which build up on the surface of the membrane. To prevent scaling and fouling, ED and EDR units are equipped with a clean-in-place (CIP) system to allow periodic flushing of the membrane stack and piping with an acid solution. A chemical feed pump and storage tank form the main components of the CIP system.

In ED systems, acid is continuously fed into the electrode stream of the cathode to prevent scaling. In EDR systems, on the other hand, electrode clean-in-place (ECIP) is a routine preventive-maintenance procedure to remove scale or fouling from the electrode system. In addition during the preventive maintenance a CIP process is a required procedure that flushes scale or reversible fouling from the membrane stack and hydraulic piping. The chemical solution circulated through the stack depends on the type of contamination. The following chemical solutions used in the CIP process are the only chemicals that should be used for stack cleaning (AWWA, 1995):

- Hydrochloric acid solution. Periodic cleaning with a 2 to 5 percent hydrochloric acid (HCl) solution is the most frequently used method to remove scale and biofouling.
- Sodium chloride solution. A 3 to 5 percent NaCl solution removes organic foulants, which are present in many surface waters, from the membranes. The solution should be at least 3 percent NaCl (0.55N chloride) and have a pH between 8.0 and 10.0, adjusted with NaOH. A pH greater than 11 can damage the anion membrane. This solution should then be circulated through the system. After the NaCl application, the operator should flush the membranes with HCl to remove excess salt.
- Chlorine solution. A 10- to 50-mg/L chlorine solution disinfects the membranes and hydraulic piping.

The use of chlorine is one of the advantatges of ED/EDR membranes in front of other membranes technologies. They can operate on waters with up to 0.5 mg/L chlorine to control the biological nature of feed water, and can be shock-chlorinated up to 30 mg/L for maximum cleaning efficiency if required. Additionally, ED/EDR presents some other advantages, for example it is possible to repair and disassembly the stack and if it is necessary, it easy to manually clean the membranes or replaced them. In these cases, stacks should not be allowed to remain dry for long periods of time because membranes may become damaged. Generally, disassembly requires that each piece be removed separately, with the exception of the top electrode, which can be replaced without the removal of any of the membranes. It is important to maintain correct component orientation and to store membranes in water. The stack should be rebuilt in the order it was disassembled; incorrect assembly can reduce performance or cause scaling.

Besides the cleaning procedure, the most frequent manual operation is to check the intermembrane voltage to prevent "hot spots" or current leakage. The intermembrane voltage has to be similar along the entire stack, and operators had to check it frequently. Excess current can melt or "burn" the membranes and spacers. Normal design practices limit this voltage to 80% of the current that would cause burning. The limits is determined by water temperature, conductivity of the source water, membrane stack size, and the internal manifold that splits flow into concentrate and dilute streams. When operators find increases of current in a located point, they had to check the voltage along some days to prevent a "hot spot".

4. EDR vs RO

Most of desalting processes are related with membrane pressure technologies in general and especially with RO. In this way, RO represents a worldwide solution for many desalination problems, but EDR could be a cost effective solution for many industrial applications of different size (Strathmann, 2010). The RO process requires extensive pretreatment, higher pumping power and more chemicals. RO also has a lower water recovery rate if the water has positive scaling tendencies.

Only EDR could be innapropiate in two cases: desalting sea water directly as drinking water (because is not cost effective) and if the problem to solve is a microbiological contamination (because EDR not provides a barrier effect). In the case of the emerging contaminants, it will depend of the chemical status of each compound, and it needs to check it.

Then it is possible to point out some advantatges in favour of EDR in comparison with RO. Most of these advantages are included in O&M tasks, and are listed below (Strathmann, 2004, GE W&P, 2010):

1. The EDR system does not require high feedwater quality and is less sensitive to pre-treatment problem in comparison with an RO system. EDR system is able to operate with Silt Density Index (SDI) average of 12 compared to 3 for the RO system. High SiO_2 content water can be treated without forming precipitation on ion exchange membranes
2. The EDR system is capable of operating with a continuous free chlorine residual of up to 1 ppm. The RO system will require a dechlorination process to protect RO membrane from degradation by free chlorine oxidation. The EDR ability to operate with chlorine residual minimises biological fouling of the membrane in a more reliable system.
3. The EDR system has a nominal initial brackish water recovery in the range of 80%-90%. The RO system normally has a water recovery in a much lower range, 65%-75%. The high EDR water recovery reduces this project's feedwater usage and wastewater discharge cost.
4. The EDR membrane is not attacked by bacteria or affected by high temperatures. Therefore no special storage solutions are necessary for long term storage. The RO system requires special storage solutions and controlled storage temperatures. The EDR

membrane can be cleaned with acid and brine/caustic flush while the RO membrane requires special and expensive cleaning chemicals. It is important to determine if these chemicals can be discharged to the environment without further treatment.

5. The EDR rugged thick membrane technology has ensured membrane life of 7 to 10 years. The RO membrane is designed for 5-7 years due to the membrane sensitivity to various operating factors.
6. The reversal feature of the EDR system controls membrane scaling with no chemical addition. The RO system requires the addition of acid and a sequestering agent. The resulting waste from the RO system is highly acidic requiring caustic neutralisation and may not be discharged to the environment.
7. The EDR membrane can be manually cleaned without damaging the membrane properties. This is due to the "plate and frame" configuration. The RO membrane has a spiralwound configuration, it can not be cleaned manually, and therefore it must be replaced.

5. Applications

Over the last ten to fifteen years, numerous advances in membrane and system technology have made EDR an especially attractive technology, both in terms of performance and cost-effectiveness.

Many applications of EDR technology can be founded worldwide. From small installations that have only one stack to the biggest one equipped with 576 stacks. Diferent suppliers and applications are involved. Desalting process is applied mainly to brackish water process, tertiary wastewater production and specific industrial applications, ranging from mining to pharmaceutical and food & beverages industries.

Table 2, shows a list of some of the worldwide industrial installations. Diferent suppliers and membrane manufacturers are available. The biggest supplier is General Electric Water

LOCATION	COUNTRY	APPLICATION		Production m^3/d	YEAR
EURODIA					
Montefano	Italy	Groundwater	Nitrate removal	1.000	1991
Munchenbuschsee	Switzerland	Groundwater	Nitrate removal	1.200	1996
Kleylehof	Austria	Groundwater	Nitrate removal	3.500	1997
GENERAL ELECTRIC WATER & PROCESS (fomerly ionics Inc)					
Abrera, BCN	SPAIN	Surface water	bromide reduction	200.000	2008
Magna, Utah	USA	Groundwater	As, Perchlorate reduction	22.728	2008
Sherman, Texas	USA	Surface water	salinty reduction	27.700	1993-96-98
Suffolk, Virginia	USA	Groundwater	Fluoride reduction	56.000	1990
Sarasota, Or	USA	Groundwater	Hardness & salts reduction	45.420	1995
Maspalomas	SPAIN	Groundwater	salinty reduction	37.000	1986
Barranco Seco, Canary Is.	SPAIN	Waste Water	Reuse	26.000	2002
Bermuda WaterWorks	Bermudas	Groundwater	Hardness & Nitrate reduction	2.300	1989
Falconera, Valencia	SPAIN	Groundwater	Nitrate reduction	16.000	2007
MEGA a.s.					
Sant Boi, BCN	SPAIN	Waste Water	salinty reduction	55.296	2010
Dolni Rozinka	Czech Rep.	Uranium mining	Desalination of sludge	1.752	2007
ZIAR nad HRONOM	Slovakia	Waste water	Desalination of sludge	350	2003
Arak	Iran	Waste water	cooling tower	4.800	2008 -10
Alberta	Canada	Well water	Gas well water desalination	40	2008

Table 2. Some worldwide EDR systems

& Process (USA), that adquired Ionics Inc. in 2004. Ionics Inc. developed the first ED commercial system in the 1950s (Juda and Mc Rae, 1950). Now is it possible to use differents suppliers of this technology, involving the whole system, or associated with membrane manufacturers as internal suppliers. In this sense, could cite: MEGA (Czech Republic), Eurodia (France), Hydrodex (Brasil) and Tecnoimpanti (Italy).

6. Case studies

We show two technical cases involved the biggest EDR plants for drinking water and reuse water for agricultural applications, respectively. Both are located near Barcelona (Spain). The first one is the Abrera Drinking Water Treatment Plant (DWTP) and the second the Depurbaix Waste Water Treatment Plant (WWTP).

6.1 Case study 1: The Abrera DWTP.

The Llobregat and the Ter rivers, typical Mediterranean catchments in Northeast Spain, supply water to more than 4.5 millions inhabitants residing in the metropolitan area of Barcelona. Aigües Ter Llobregat (*ATLL*) a public company of the Autonomous Government of Catalonia has been appointed to manage the system that includes two DWTP and one sea water reverse osmosis (SWRO) plant, with a whole capacity of continuous treatment of 14 m^3/s.

The Abrera DWTP takes raw water directly from the Llobregat River. It presents a low and irregular flow and some quality problems such as high salinity with a significant presence of sulphates, Ba^{2+}, Sr^{2+}, Na^+, Ca^{2+}, K^+, Cl^- and specially Br^- (Fernández-Turiel et al., 2000). Furthermore, many problems are associated with the increase in micropollutant and microbiological levels due to both urban and industrial sewage. These problems produce many interruptions of the process, some of them lasting many hours. Consequently, the high levels of bromide (ranging between 0.5-1.2 mg/L), NOM and T, produce high concentrations of trihalomethanes (THMs) after chlorination (Chang et al., 2001; Rook, 1977), showing a high brominated profile.

The total THMs represent the sum of the concentrations of four THMs; chloroform ($CHCl_3$), bromoform ($CHBr_3$), bromodichloromethane ($CHBrCl_2$), and dibromochloromethane ($CHBr_2Cl$). They have been regulated since 1998 in the European Union (Council Directive 98/83/CE), and in Spain since 2003, with a parametric value of 100 μg/L established in 2009 (Real Decreto 140/2003).

To minimize the THMs problem, *ATLL* searched for a new technology based on a membrane process. In this sense, the concern about THMs at the Llobregat DWTP has to do with changes in the treatment process, starting with enhanced coagulation in 1994, followed by the inclusion of a new step of GAC filtration in 1995, after sand filters. Subsequently, several changes have been carried out to improve the process. In 1999 *ATLL* carried out some trials using a pilot plant of Reverse Osmosis (RO) during a period of 6 months. The study showed that RO technology had a low recovery and certain instability due to the frequent shutdowns of the process because of the poor quality of raw water (floods, high turbidity, high fouling potential, chemical pollution…). Additional problems were the sensibility to high concentrations of sulphates, barium, calcium, alumina, and the disinfectant chlorine, that can be used at different points of the process. Later, to assess the reduction of the salts concentration (Kimbrough & Suffet, 2005) and consequently the THMs-Formation Potential (THMs-FP, 25ºC, 48h), a pilot study was carried out during 28 months using a pilot plant of

EDR technology. Results showed that the EDR step improved the chemical and aesthetic quality of drinking water (Devesa et al., 2009, García et al., 2010) and allows a THMs-FP after 48h that is lower than the regulated level of 100 μg/L (Valero et al., 2007).
The final decision was the enlargement of the plant production from 3 m^3/s to 4m^3/s and the inclusion of a new EDR step after Granular Activated Carbon (GAC) filtration, with a production capacity of 2.3 m^3/s. EDR takes feedwater from GAC step by means of a derivation of filtered water pipeline.
In addition, EDR permeates are aggressive showing a pH ranged between 6.5 and 7.3 and a LSI that varies between -1 and -2. Thus, a remineralization step is necessary, to supply EDR product water without blending with GAC filtered water. In this sense remineralization of EDR produced water was applied using lime contactors and CO_2 dosing. Only if the quality of raw water makes it possible, conventional treatment will be blended to produce up to 4 m^3/s.
This plant is the world's largest desalination plant using this technology, and a new example of a large scale application of a desalting technology to improve the quality of drinking water. The work was carried out by the Spanish temporary consortium SACYR-SADYT using EDR technology provided by General Electric Water&Process.
The main characteristics of the DWTP are:

- Conventional process: pre-oxidation with potassium permanganate, coagulation, flocculation, oxidation with chlorine dioxide, sand filtration, GAC filtration and final chlorination using chlorine gas.
- Average current flow supplied by the DWTP: 2.3 m^3/s. Maximum extended flow of the DWTP: 4 m^3/s

Design of EDR's Stage:

- Maximum flow treatment : 2.3 m^3/s (58 MGD)
- Range conductivity inlet water: 900-3000 μS/cm.
- Temperature range inlet water: 5-29 ºC
- Pump station : 9+3 pumps of 1030m^3/h to 60 mca
- Cartridge filters: 18 filters with 170 cartridges each of 50 inches and 5 μm.
- 9 modules with 576 stacks wit 600 cell pairs each one, in double stage.
- Homogeneous membranes: AR204 (anionic) and CR67 (cationic)
- Wet technology.
- Voltage range: 340-450 V 1st stage, 320-390 V 2nd stage.
- Bromides reduction: 60-80 %
- Conductivity reduction: 60-80 %
- Maximum volume of brines: 154 Tm/d, sent via a pipeline to the sea at the mouth of the Llobregat River.
- Water recovery>90% (including off-spec and concentrate recycle).
- Remineralization (when necessary) with $Ca(OH)_2$ up to 7 Tm/d and CO_2.

Every module is provided with reversing systems of flow for the changes of polarity, automatic valves and pumps equipped with electronic frequency variators that allow a full automated system. EDR process is operated according with the levels of THMs expected in the final drinking water. Then 1 to 9 modules were worked when necessary to blend with conventional treatment product to get the THMs levels at the lower cost.
The plant started operating on a trial basis in June 2008, and came into the normal operation from April 2009. Along the period April, 2009 to August, 2010, more than 20 hm^3 had been

Fig. 3. Details of the EDR step at the Abrera DWTP.

produced through the EDR line. THMs's average values in the water product of the DWTP ranged between 40 and 60 µg/L. The energetic average consumption for the EDR process (stacks and pumps) has been lower than 0.6 kWh/m^3. During the indicated period the hydraulic performance has been higher than 90%, with a reduction of salts (measures like conductivity) higher than 80% in summer. Specifics consumptions of HCl were of 0.08 Kg HCl/m^3 and for antiscalant in the rejection of brine 0,002 Kg/m^3 (Valero et al., 2010)

Due to the large size of the industrial plant, additional R&D studies will be focused on O&M procedures. Maintenance related to cleaning membranes and spacers, the measure of the inter-membranes voltages and "hot spots" detection, would be simplified using specific tools designed by the technical staff.

The cost of the new enlargement project was 61,218,478€. Given the considerable interest of these works, their repercussion on the quality of the supply and the technology used, a subsidy of 85% of the budget of the works was obtained from European Union funds.

6.2 Case study 2: The Depurbaix WWTP.

The project is located in Sant Boi de Llobregat, near Barcelona. It is a brackish water desalination facility for some of the effluent treated in the Depurbaix WWTP, which produces more than 57,000 m^3/d using EDR technology (Segarra et al., 2009).

The facility is one of the largest in the world that treats wastewater for agricultural use. The work was carried out by the Spanish temporary consortium BEFESA-ACSA using EDR technology provided by MEGA a.s.

The main characteristics of the EDR system are:

- Inlet water: tertiary treatment of the WWTP + anthracite/sand filters. Average conductivity 3.040 µS/cm
- Expected EDR product water: 55,296 m^3/d.
- Expected plant product water after blending: 57,024 m^3/d.
- Pump station : 2+1 pumps

- Cartridge filters: 4 filters with 300 cartridges each one (20 µm).
- 4 modules with 96 stacks with 600 cell pairs each one, in double stage.
- Heterogeneous ion-exchange membranes: RALEX AM(H) (anionic) and CM(H) (cationic)
- Dry technologie
- Conductivity reduction: 60-80 %
- Water recovery>85%.

The plant started operating on a trial basis in January 2010 and came into the normal operation from September 2010. The full automatic modular system allows the operation according to the expected use of the product water.

Fig. 4. EDR stacks at the Depurbaix WWTP.

7. Discussion

In recent years membrane technology has become an important useful tool for the desalination of seawater, the use of brackish water and polluted water resources which were not suitable for producing drinking water, and for the physicochemical and microbiological improvement of the water obtained by conventional treatment.

Based in the important advantatges of ion-exchange membranes (rugged, resistant to organic fouling, chlorine stable, broad range for pH and Temperature,..) compared with other membranes technologies, the improvement of EDR allows to use it for many applications that are cost effective than other technologies with a better commercial marketing like UF or RO. Maybe the use of EDR still has a label of a technology to solve local problems involving small communities or specific industrial applications. However, during last years big systems are in operation showing good performances and cost effective results. In this sense the T. Maybry Carlton WTP located at Sarasota (FL, USA) was pioneer in operate a big system since 1995. In that case, EDR was selected due to its ability to

maximize recovery of freshwater and minimize wastewater volume. The plant produces 45.420 m^3/d and is equipped with 320 stacks. Later, improvement of EDR allows installing more systems worldwide, some of them in Spain related with drinking water and water reuse. EDR was introduced in the Canary Islands during the 80's, but during lasts years some big facilities were building in the Spanish Mediterranean area: two plants (16,000 m^3/d each) in Valencia to reduce nitrate levels and two more in Barcelona: the first to reduce bromide levels and then the THMs formation (200.000 m^3/d, 576 stacks) and the last to reduce salinity for reuse water for irrigation (55.296 m^3/d, 96 stacks).
In addition, desalination of brackish water using membranes technologies like ED and specially EDR it is a cost effective method to supply good quality drinking water water and could be a good solution for some industrial water utilities. Besides, EDR systems now are simpler and more reliable, which means that the demineralization of difficult-to-treat water is easier for municipalities to handle. In addition, the costs are becoming easier to swallow.
Some aspects could be improved in a near future: spacer configuration, membranes chemistry, materials and configuration of electrodes, specific antiscalants for EDR, elimination of degasifiers and the increase of the production of the stacks.
Finally, there are some interesting works related with the use of hybrid systems to get synergies between technologies (Turek, 2002; Kahraman, 2004), and some innovations are under study to improving the EDR technology (Balster et al., 2009; Charcosset, 2009; Ortiz et al., 2008; Turek et al., 2008; Veerman et al., 2009).

8. Conclusions

- EDR should be effectively applied for water and salt recovery from an industrial effluent for pollution prevention and for resource recovery.
- The growing popularity among municipalities of the EDR systems is related with its capacity to reduce TDS and some inorganics elements like nitrates, sulphates, radon, bromides and others, with high water recovery and easily operation and control by adjusting amount of electricity applied to membrane stack.
- The correct operation of big EDR systems, compared with classical membrane pressure systems like RO, allows extending EDR to new cost effective applications.
- Future steps of EDR systems could improve the design of membranes and spacers as well as a more compact design, lowering the capital and O&M costs.
- EDR could be in a near future the technology of choice for many applications because its efficiency to desalt water needed in differents fields like drinking water, reuse water and many industrial applications, like food, beverages and mining among others.
- Hybrid systems between different membranes technologies including EDR, could be useful solutions for specific applications, and could improve recovery and reduce waste.

9. References

Asahi Chemical Industry Co. (October, 2010).www.asahi-kasei.co.jp
Asahi Glass Col. Ltd (October, 2010).www.agc.com
AWWA (1995). AWWA M38. Electrodialysis and Electrodialysis Reversal, American Water Works Association, Denver, CO.

AWWA (2004). Committee report: current perspectives on residual management for desalting membranes. J. AWWA, 96: 73-87.

Balster, J., D., Stamatialis D.F. & Wessling, M. (2009). Towards spacer free electrodialysis. J. Memb. Sci. 341: 131-138.

Broens, L., Liebrand, N., Futselaar, H. & de Armas, J.C. (2004). Effluent reuse at Barranco Seco (Spain): a 1,000 m^3/h case study. Desalination 167: 13-16

Chang, E.E., Lin, Y.P. & Chiang, P.C. (2001). Effects of bromide on the formation of THMs and HAAs. Chemosphere 43: 1029-1034.

Chao, Y.M. & Liang, T.M. (2008). A feasibility study of industrial wastewater recovery using electrodialysis reversal. Desalination 221:433–439

Charcosset, C. (2009). A review of membrane processes and renewable energies for desalination. Desalination 245: 214–231

Council Directive 98/83/EC of 3 November 1998 on the quality of water intended for human consumption Official Journal L 330 , 05/12/1998 P. 0032 - 0054.

Dalla Costa R.F., Klein, C.W., Bernades A.M. & Ferreira, J.Z. (2002). Evaluation of the Electro dialysis Process for the treatment of metal finishing wastewater. J. Braz.Chem. Soc. Vol 13, Nº 4: 540-547.

De Barros, M. (2008). Avaliaçâo do processo de electrodiálise reversa no tratamento de efluentes de refinaria de petróleo. PhD. Thesis, Escola de Engenharia, Universidade federal do Rio Grande do Soul.

Devesa R., García V. & Matía, L. (2010). Water flavour improvement by membrane (RO and EDR) treatment. Desalination, 250:113-117.

DuPont Co.(October, 2010).www.2dupont.com

Eurodia (October, 2010).www.eurodia.com

Fernandez-Turiel, J.L., Roig, A., Llorens, Antich, N., Carnicero, M. & Valero, F. (2000). Monitorig of drinking water treatment plants of Ter and Llobregat (Barcelona NE Spain) using ICP-MS, Toxicological and Environmental Chemistry 74: 87-103.

FuMA-Tech GmbH (October, 2010).www.fumatech.com

García, V., Fernández, A., Ferrer, O., Cortina, J.L., Valero, F. & Devesa, R. Aesthetic assessment of blends between desalinated waters and conventional resources. Proceedings of the IWA World Water Congress & Exhibition, Montreal 2010.

GE Water W&P. (October, 2010). www.ge.com

Harries R.C, Elyanow D., Heshka D.N. & Fischer K.L. (1991). Desalination of brackish groundwater for a prairie community using electrodialysis reverersal. Desalination, 84: 109-121.

Hays, J. (2000).Iowa's first electrodialysis reversal water treatment plant. Desalination, 132: 161-165.

Heshka, D. (1992). EDR water treatment desalination on the prairies. Desalination, 88: 109-121.

Hidrodex. (October, 2010). www.hidrodex.com.br

Ionics Inc., (1984). Electrodialysis-Electrodialysis Reversal Technology, Ionics Inc., Watertown, MA.

Juda, W. & Mc Rae. W.A. (1950). Coherent ion-exchange gels and membranes. J. Am. Chem. Soc., 72:1044.

Kahraman N., Cengel Y.A., Wood B, Cerci Y.(2004). Exergy analysis of a combined RO, NF, and EDR desalination plant. Desalination, 171: 217-232.

Kawahara, T. (1994). Construction and operation experience of a large-sacale electrodialysis water desalination plant. Desalination, 96: 341-348.

Kimbrough, D.E. & Suffet, I.H. (2002) Electrochemical removal of bromide and reduction of THM formation potential in drinking water. Water Res. 36: 4902-4906.

Korngold, E., Aronov, L. & Daltrophe, N. (2009). Electrodialysis of brine solutions discharged from an RO plant. Desalination 242: 215–227

LanXess Sybron Chemicals (October, 2010).www.ionexchange.com

Larchet, C., Zabolotsky, V.I., Pismenskaya, N., Nikonenko, V.V., Tskhay, A., Tastanov, K. & Pourcelly, G. (2008) Comparison of different ED stack conceptions when applied for drinking water production from brackish waters. Desalination 222: 489–496.

Lee, H-J., Strathmann, H. & Moon, S-H. (2006). Determination of the limiting current density in electrodialysis desalination as an empirical function of linear velocity. Desalination 190: 43–50.

Lozier, J.C., Smith G., Chapman J.W. & Gattis D.E. (1992) Selection, design, and procurement of a demineraliztion system for a surface water treatment plant. Desalination 88: 3-31.

MEGA a.s. (October, 2010).www.mega.cz

Melnyk L. & Goncharuk V. (2009). Electrodialysis of solutions containing Mn (II) ions. Desalination 241: 49-56.

Menkouchi Sahlia, M.A., Annouarb, S., Mountadarb, M., Soufianec, A. & Elmidaouia, A. (2008). Nitrate removal of brackish underground water by chemical adsorption and by electrodialysis. Desalination 227: 327–333

Mihara K. & Kato, M. (1969). Polarity reversing electrode units and electrical switching means therefore, U.S.Patent 3,453,201.

Ortiz, J.M., Expósito, E., Gallud, F., García-García, V., Montiel, V. & Aldaz, A. (2008). Desalination of underground brackishwaters using an electrodialysis system powered directly by photovoltaic energy. Solar Energy Materials & Solar Cells 92:1677–1688.

PCA GmbH (October,2010).www.pccell.de

Pilat B.V. (2003). Industrial application of electrodialysis reversal systems. Desalination 158: 87-89.

Reahl E.R. Reclaiming reverse osmosis blowdown with electrodialysis reversal. (1990). Desalination: 78, 1: 77-89.

Real Decreto 140/2003 de 7 de febrero de 2003 por el que se establecen los criterios sanitarios de la calidad del agua de consumo humano. BOE 45, 7228-7245 (In Spanish).

Roquebert, V., Booth, S., Cushing R.S., Crozes G., Hansen, E.(2000). Electrodialysis reversal (EDR) and ion exchange as polishing treatment for perchlorate treatment. Desalination 131: 285-291

Rook, J.J.(1977) Chlorination reactions of fulvic acids in natural waters. Environ. Sci & Tecnol., 11:478-482.

Segarra, J., Iglesias, A., Pérez, J. & Salas, J. (2009). Construcción de la planta con mayor capacidad de producción mundial con tecnología EDR para agues regeneradas. Tecnología del Agua, 309: 56-62.

Schoeman, J.J. & Steyn, A. (2000) Evaluation of electrodialysis for water and salt recovery from an industrial effluent. Proceedings of the WISA 2000 Biennial Conference, Sun City, South Africa, 28 May - 1 June 2000

Strathmann, H. (2004). Ion Exchange membrana separation process. Elsevier, Ámsterdam.

Strathmann, H. (2010). Electrodialysis, a mature technology with a multitude of new applications. Desalination, 264: 268-288.

Tecnoimpianti. (October, 2010). www.tecnoimp.com

Tianwei Membrane Co.Ltd. (October, 2010) www.sdtianwei.com

Tokuyama Co-Astom (October, 2010).www.tokuyama.co.jp

Tsiakis, P. & Papageorgiou, L.G. Optimal design of an electrodialysis brackish water desalination plant. (2005). Desalination, 173: 173-186.

Turek, M. (2002). Dual-purpose desalination–salt production electrodialysis. Desalination 153:377–381

Turek, M., Bandura, B., Dydo, P. (2008). Power production from coal-mine brine utilizing reversed electrodialysis Desalination 221: 462–466.

Valerdi-Pérez, R.M. & Ibáñez-Mengual, J.A. (2001).Current-voltage curves for an electrodialysis reversal pilot plant : determination of limiting currents. Desalination 141 :23-37.

Valerdi-Pérez, R.M., López-Rodríguez, M. & Ibáñez-Mengual, J.A. (2001). Characterizing an electrodialysis reversal pilot plant. Desalination 137 :199-206.

Valero, F., García, J.C., González, S., Medina, M.E., de Armas, J.C., Hernández, M.I. & Rodríguez, J.J. (2007) Control of THMs at the Llobregat DWTP (NE, Spain) using Electrodialisys Reversal (EDR). Abstract book, IDA World Congress-Maspalomas,Gran Canaria –Spain. REF: IDAWC/MP07-207.

Valero, F., & Arbós, R. (2010). Desalination of brackish river water using Electrodialysis reversal (EDR). Desalination, 253: 170-174.

Valero, F., Tous, J.F. & Arbós, R. (2010). Mejora de la calidad salnitaria del agua durante el primer año de explotación de la etapa de electrodialisis reversible (EDR) en la ETAP del Llobregat. Proceedings of the VII Congreso AEDYR, Barcelona octubre 2010.

Veerman J., Saakes M., Metz, S.J. & Harmsen, G.J. (2009).Reverse elecrodialysis: Performance of a stack with 50 cells on the mixing of sea river water. J. Memb. Sci. 327: 136-144.

Winger, A.G., Bodamer, G.W. & Kunin, R. (1953) Some electrochemical properties of new synthetic ion exchange memebranes. J. Electrochem. Soc. 100: 178.

Xu, T. (2005). Ion exchange membranes: state of their development and perspective. J. Memb. Sci. 263:1-29.

2

Water Desalination by Membrane Distillation

Marek Gryta
West Pomeranian University of Technology, Szczecin
Poland

1. Introduction

Water is the most common substance in the world, however, 97% is seawater and only 3% is fresh water. The availability of water for human consumption is decreasing due to increasing the environmental pollution. According to the World Health Organisation (WHO), about 2.4 billion people do not have access to basic sanitation facilities, and more than one billion people do not have access to safe drinking water (Singh, 2006). Moreover, the world's population is expected to rise to nine billion from the current six billion in the next 50 years. Chronic water pollution and growing economies are driving municipalities and companies to consider the desalination as a solution to their water supply problems.

Generally, desalination processes can be categorized into two major types: 1) phase-change/thermal and 2) membrane process separation. Some of the phase-change processes include multi-stage flash, multiple effect boiling, vapour compression, freezing and solar stills. The pressure driven membrane processes, such as reverse osmosis (RO), nanofiltration (NF), ultrafiltration (UF) and microfiltration (MF), have found a wide application in water treatment (Charcosset, 2009).

The energy required to run desalination plants remains a drawback. The energy limitations of traditional separation processes provided the impetus for the development and the commercialisation of membrane processes. Membrane technologies (simple, homogenous in their basic concepts, flexible in application), might contribute to the solution of most of the existing separation problems. Nowadays, membranes are used for the desalination of seawater and brackish water, potable water production, and for treating industrial effluents. RO membrane separation has been traditionally used for sweater desalination (Charcosset, 2009; Schäfer et al., 2005; Singh, 2006).

One of the limitations of membrane processes is severe loss of productivity due to concentration polarisation and fouling or scaling (Baker & Dudley, 1998; Schäfer et al., 2005). Membrane pretreatment processes are designed to minimise the potential problems of scaling resulting from the precipitation of the slightly soluble ions. Membrane (MF or UF) pretreatment of RO desalinations plants is now a viable options for removing suspended solids, fine particles, colloids, and organic compounds (Banat & Jwaied, 2008; Singh, 2006). NF pretreatment of sweater is also being used to soften RO feed water instead of traditional softening (Schäfer et al., 2005).

The industrial development of new membrane processes, such as membrane distillation (MD), is now being observed (Banat & Jwaied, 2008; Gryta, 2007). In MD process feed water is heated to increase its vapour pressure, which generates the difference between the partial pressure at both sides of the membrane (El-Bourawi et al., 2006). Hot water evaporates through non-wetted pores of hydrophobic membranes, which cannot be wetted by liquid water (Gryta & Barancewicz, 2010). The passing vapour is then condensed on a cooler surface to produce fresh water (Alklaibi & Lior, 2005; Charcosset, 2009). In the case of solutions of non-volatile substances only water vapour is transported through the membrane. Thus, MD process has a potential application for the water desalination and the treatment of wastewater (Banat et al., 2007; El-Bourawi et al., 2006; Wang, et al., 2008). The MD has a significantly lower requirements concerning pretreatment of feed water, therefore, it enables the production of pure water from water sources, the quality of which impedes a direct application of the RO for this purpose. However, the feed usually contains various impurities, which in turn lead to the formation of deposit (Gryta, 2008). Deposits both pollute surfaces of membranes and make it easier for water to penetrate membrane pores (Gryta, 2007b; He et al., 2008). Consequently, membranes lose their separation properties and the MD process stops. This is why it is essential to prevent formation of deposits on the membrane surfaces.

2. Principles of membrane distillation

An expanded definition of MD process was created in 1986 at the "Workshop on Membrane Distillation" in Rome (Smolder & Franken, 1989). The term "Membrane Distillation" should be applied for membrane operations having the following characteristics:

- the membrane should be porous and not be wetted by the process liquids;
- no capillary condensation should take place inside the pores of the membrane;
- only vapour should be transported through the pores;
- the membrane must not alter the vapour-liquid equilibrium of the different components in the process liquids;
- at least one side of the membrane should be in direct contact with the process liquid;
- the driving force for each component is a partial pressure gradient in the vapour phase.

In membrane distillation heat is required to evaporate the feed components, therefore, in such context (similarly as in the classical distillation) it can be concluded that MD is a thermal-diffusion driven process. However, it operates at low temperatures (323-363 K), therefore, the feed water can be heated be using renewable energy (Banat & Jwaied, 2008).

The MD is carried out in various modes differing in a way of permeate collection, the mass transfer mechanism through the membrane, and the reason for driving force formation (Gryta, 2005; Smolder & Franken, 1989). These differences were taken into consideration in the nomenclature by the addition to the term "Membrane Distillation" the words, which emphasised a feature of a given variant. Various types of MD are known for several years (Fig.1): direct contact MD (DCMD), air gap MD (AGMD), sweeping gas MD (SGMD) and vacuum MD (VMD). DCMD variant is the most frequently studied and described mode of MD process (Alklaibi & Lior, 2005; El-Bourawi et al., 2006; Gryta, 2010; Wang, et al., 2008).

Several theoretical mass transfer models have been presented to describe membrane distillation. The models of DCMD were based on the assumption that vapour permeates through the porous membrane, as a result of molecular diffusion, Knudsen flow and/or the transition between them (Alklaibi & Lior, 2005; El-Bourawi et al., 2006; Gryta, 2008). Using the Stefan-Maxwell model diffusion of vapour through the air layer, the permeate flux can be described as proportional to the membrane permeability and water partial pressure difference (Alklaibi & Lior, 2007; Gryta et al., 1998):

$$J_V = \frac{\varepsilon D_{WA}}{\chi s_m} \frac{M_W P}{R T_m p_{in}} (p_F - p_D) \qquad (1)$$

where p_F and p_D are the partial pressures of the saturated water vapour at interfacial temperatures T_1 and T_2; ε, χ, s_m, M_W, R, T_m, P, D_{WA} and p_{in} are membrane porosity, pore tortousity, membrane thickness, molecular weight, gas constant, membrane temperature, total pressure, vapour diffusion coefficient and air concentration inside the pores, respectively.

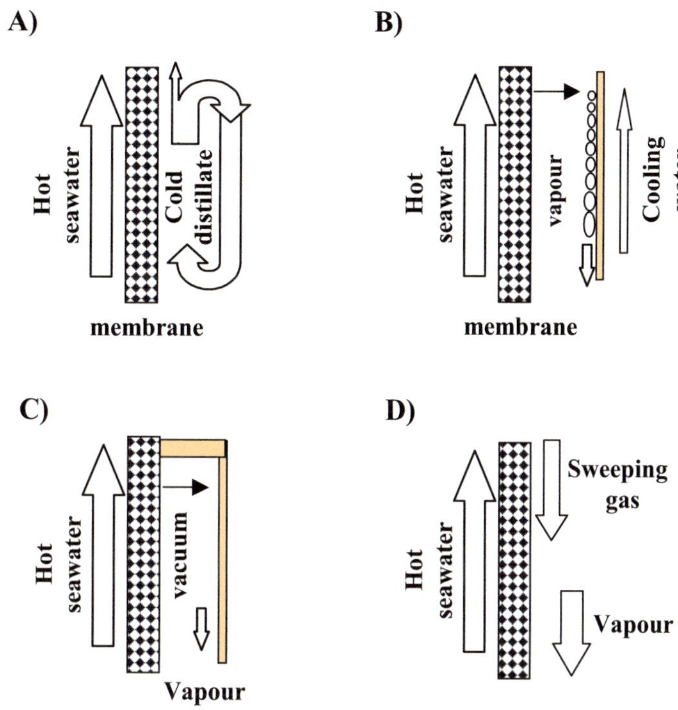

Fig. 1. Types of membrane distillation: A) DCMD, B) AGMD, C) VMD, D) SGMD

In MD process the mass transfer (J_V) occur simultaneously with heat conduction (Q) across the membrane material, and as a results, the temperature of the boundary layer on the feed side is lower, whereas on the distillate side it is higher than that of the bulk (Fig.2). This phenomenon is termed as the temperature polarization (Martínez-Díez & Vázquez-González, 1999). It causes the decrease of vapour pressure difference across the membrane which leads to the reduction of the magnitude of the mass flux (permeate) flowing through the membrane. The interfacial temperatures T_1 and T_2 cannot be measured directly. Several equations used to calculate these temperatures have been presented in the MD literature (Gryta et al., 1998; Khayet et al., 2004; Srisurichan et al., 2006). Their values depend in essential way on the conditions of a heat exchange in the MD module. Thus the correct description of the heat transport across the membrane will determine the accuracy of the mathematical calculation of MD process run (El-Bourawi et al., 2006; Gryta et al., 1998; Gryta, 2008).

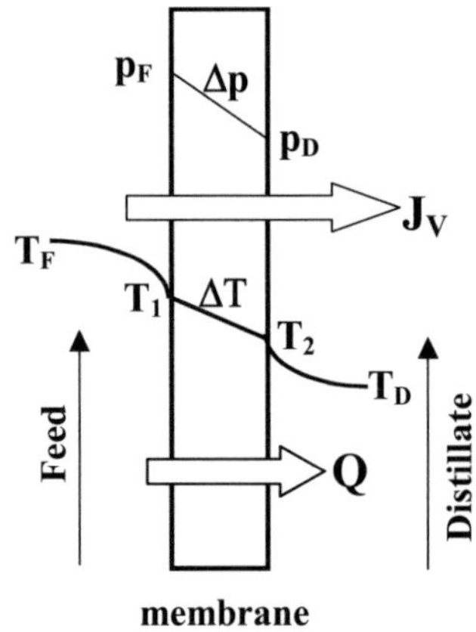

Fig. 2. Principles of DCMD: T_1, T_2, T_F, T_D — temperatures at both sides of the membrane, and temperatures of feed and distillate, respectively; p_F, p_D — water vapor partial pressure at the feed and distillate sides, respectively

2.1 Membranes and modules

The porous and hydrophobic MD membranes are not selective and their pores are filled only by the gas phase. This creates a vapour gap between the feed and the produced distillate, what is necessary for MD process operation. However, during the MD a part of the membrane pores may be wetted, that decreases a thickness of vapour gap inside the membrane wall (Gryta & Barancewicz, 2010). Therefore, the properties of membrane material and membrane porous structure are important for MD process performance (Bonyadi & Chung, 2009; Khayet et al., 2006).

Membrane for MD process should be highly porous, hydrophobic, exhibit a desirable thermal stability and chemical resistance to feed solution (El-Bourawi et al., Gryta et al., 2009). These requirements are mostly fulfilled by the membranes prepared from polymers with a low value of the surface energy such as polytetrafluoroethylene (PTFE), polypropylene (PP) or poly(vinylidene fluoride) (PVDF) (El Fray & Gryta, 2008; Gryta, 2008; Li & Sirkar, 2004; Teoh et al., 2008; Tomaszewska, 1996). Apart from the hydrophobic character of the membrane material, also the liquid surface tension, pores diameter and the hydraulic pressure decide about the possibility of the liquid penetration into the pores. This relation is described by the Laplace - Young (Kelvin law) equation (Schneider et al., 1988):

$$\Delta P = P_F - P_D = \frac{-4\,B\,\sigma\cos\Theta}{d_p} \tag{2}$$

where: ΔP is liquid entry pressure (LEP), B is the pore geometry coefficient (B = 1 for cylindrical pores), σ is the surface tension of the liquid, Θ is the liquid contact angle, d_P is the diameter of the pores, P_F and P_D are the hydraulic pressure on the feed and distillate side, respectively. Water and the solutions of inorganic compounds have high surface tension ($\sigma > 72 \times 10^{-3}$ N/m), however, when the organics are present, its value diminishes rapidly. Thus, taking into consideration the possibility of membrane wetting, it is recommended that for MD the maximum diameter of membrane pores does not exceed the 0.5 μm (Gryta, 2007b; Gryta & Barancewicz, 2010; Schneider et al., 1988).

Hydrophobic polymers are usually low reactive and stable, but the formation of the hydrophilic groups on their surface is sometimes observed (Gryta et al., 2009). The surface reactions usually create a more hydrophilic polymer matrix, which may facilitate the membrane wettability (El Fray & Gryta, 2008; Khayet & Matsuura, 2003). The amount of hydrophilic groups can be also increased during MD process and their presence leads to an increase the membrane wettability (Gryta et al., 2009; Gryta & Barancewicz, 2010).

The application of membranes with improved hydrophobic properties allows to reduce the rate of membrane wettability. Blending of PTFE particles into a spinning solution modified the PVDF membrane, and enhances the hydrophobicity of prepared membranes (Teoh & Chung, 2009). Moreover, the resistance to wetting can be improved by the preparation of MD membranes with the uniform sponge-like membrane structure (Gryta & Barancewicz, 2010).

Apart from membrane properties, the MD performance also depends on the module design. The capillary modules can offer several significant advantages in comparison with the plate modules (flat sheet membranes), such as a simple construction and suppression of the temperature polarization (El-Bourawi et al., 2006; Gryta, 2007; He et al., 2008; Li & Sirkar, 2004; Teoh et al., 2008). The efficiency of the MD capillary module is significantly affected by the mode of the membranes arrangement within the housing (Fig. 3).

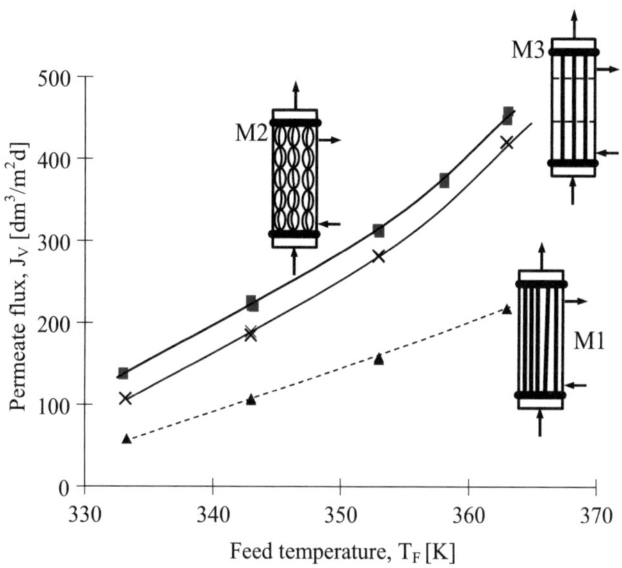

Fig. 3. The influence of feed temperature and the mode of membrane arrangement in a capillary module on the permeate flux. M1 - bundle of parallel membranes; M2 - braided capillaries; and M3 – capillaries mounted inside mesh of sieve baffles

The driving force for the mass transfer increases with increasing the feed temperature, therefore, the permeate flux is also increased at higher feed temperatures. A traditional construction (module M1) based upon the fixation of a bundle of parallel membranes solely at their ends results in that the membranes arrange themselves in a random way. This creates the unfavourable conditions of cooling of the membrane surface by the distillate, which resulted in a decrease of the module efficiency. In module M3 the membranes were positioned in every second mesh of six sieve baffles, arranged across the housing with in 0.1–0.15 m. The most advantageous operating conditions of MD module were obtained with the membranes arranged in a form of braided capillaries (module M2). This membrane arrangement improves the hydrodynamic conditions (shape of braided membranes acted as a static mixer), and as a consequence, the module yield was enhanced.

2.2 MD process efficiency

Although the potentialities of MD process are well recognised, its application on industrial scale is limited by the energy requirements associated. Therefore, high fluxes must be obtained with moderate energy consumption. DCMD has been widely recognised as cost-efficient for desalination operating at higher temperatures, when waste heat is employed to power the process (Alklaibi & Lior, 2005). The performance of membrane distillation mainly depends on the membrane properties, the module design and it operating conditions (Bui et al., 2010; Li & Sirkar, 2004).

Concerning the operating conditions (Figs. 3 and 4), the feed temperature has the most significant influence on the permeate flux, followed by the feed flow rate and the partial pressure established at the permeate side. This last depending on the distillate temperature for DCMD and on the vacuum applied for VMD (Criscuoli et al, 2008; El-Bourawi et al., 2006).

The results presented in Fig. 4 confirmed that the distillate velocities had a minor role in improving the mass transfer, but a distillate velocity below 0.3 m/s would cause a rapid decrease in mass flux (Bui et al., 2010). Moreover, Bui et al. were indicated, that the distillate temperature has had a significant greater influence on DCMD energy efficiency. It is known that decreasing the water temperature from 283 to 273 K results in a very small an increase of mass driving force. Therefore, it is recommended that the DCMD process be operated at a distillate temperature higher than 283 K.

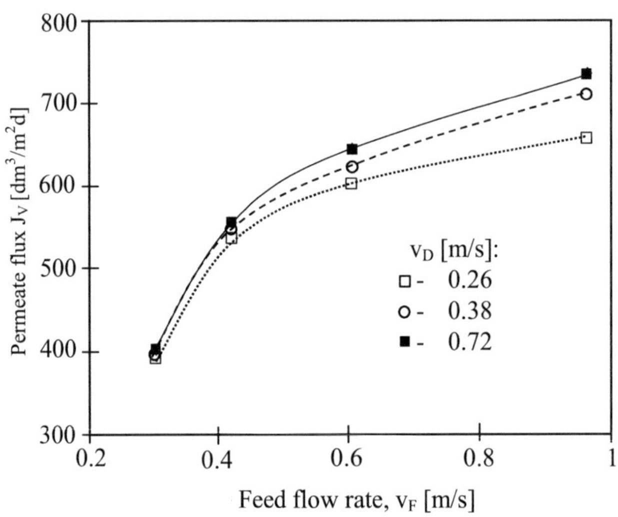

Fig. 4. The effect of the flow rate of streams in a module with braided membranes (module M1) on the permeate flux. T_F = 353 K, T_D = 293 K

The viability of MD process depends on an efficient use of available energy. The heat transfer inside the membrane (Q – total heat) takes place by two possible mechanisms, as conduction across the membrane material (Q_C) and as latent heat associated with vapour flowing through the membrane (Q_V). The heat efficiency (η_T) in the MD process can be defined by Eq. 3.

$$\eta_T = \frac{Q_V}{Q} = \frac{Q_V}{Q_V + Q_C} \tag{3}$$

The heat transfer which occurs in MD module leads to a cooling of the hot feed and to a heating of the distillate. Therefore, in the DCMD process it is necessary to supply heat to the hot stream and to remove heat from the distillate stream. The heating and the cooling steps represent the energy requirements of the DCMD process.

The amount of heat exchanged in the MD module increases along with an increase of the feed temperature (Fig. 5). However, under these conditions the permeate flux also increases, which causes the limitation of heat losses (heat conducted through the membrane material). As a results, an increase in the module yield influences on the enhancement of heat efficiency of the MD process (Fig. 6). For the highest permeate flux the η_T coefficient equal to 0.75 was obtained. It was concluded that energy efficiency of DCMD process could be maximised if the process were operated at the highest allowable feed temperature and velocity (Bui et al., 2010). A nonuniform arrangement of the capillary membranes in the module housing (module M1) caused a decrease in the energy consumption efficiency.

The unitary energy consumption in the MD process decreases along with temperature of feeding solution. This consumption was reduced from 5000 to 3000 kJ per 1 kg of obtained distillate when the feed temperature increased from 333 to 363 K (Gryta, 2006).

A decrease of the membrane wall thickness significantly increases the obtained permeate flux. However, during the MD process the liquid systematically wetted the consecutives pores, which reduced the thickness of the air-layer inside the membrane wall. In this

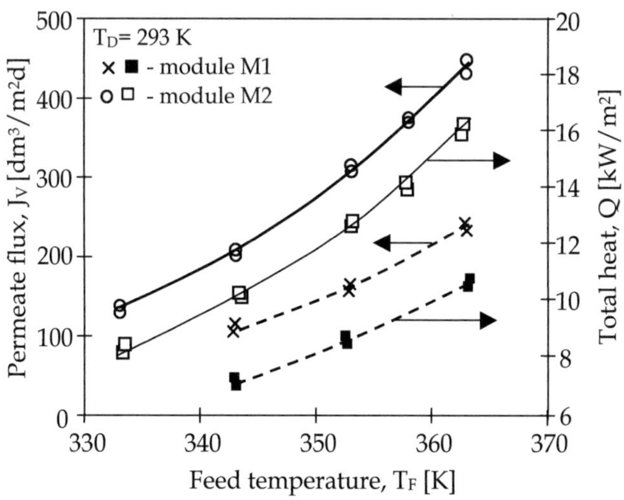

Fig. 5. Effect of feed inlet temperature and mode of membrane arrangement (M1 - parallel, irregular, M2 – braided membranes) on permeate flux and heat transfer in DCMD

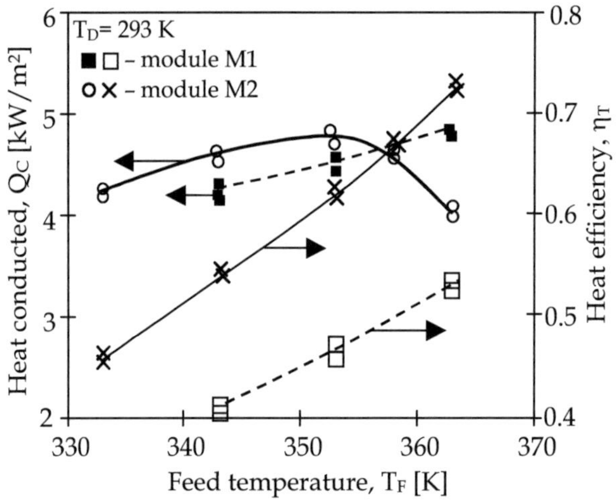

Fig. 6. Effect of feed temperature and mode of membrane arrangement (M1 - parallel, irregular, M2 - braided membranes) on heat conducted and heat efficiency in DCMD

situation, the membranes having a thin wall will be wetted in a relatively short time. Therefore, the hydrophobic membranes with thicker walls are recommended for commercial DCMD applications (Gryta & Barancewicz, 2010).

3. Membranes fouling

Fouling is identified as a decrease of the membrane permeability (permeate flux) due to deposition of suspended or dissolved substances on the membrane surface and/or within its pores (Schäfer et al., 2005). Several types of fouling can occur in the membrane systems, e.g. inorganic fouling or scaling, particulate and colloidal fouling, organic fouling and biological fouling (Baker & Dudley, 1998; Singh, 2006; Srisurichan et al., 2005). Scaling occurs in a membrane process when the ionic product of sparingly soluble salt in the concentrate feed exceeds its equilibrium solubility product. The term scaling is commonly used when the hard scales are formed (e.g. $CaCO_3$, $CaSO_4$) (He et al., 2008; Lee & Lee, 2000). Fouling is also one of the major obstacles in MD process because the deposit layer formed on the membrane surface may cause membrane wetting. This phenomenon will certainly be accelerated if the salt crystals were formed inside the pores (Alklaibi & Lior, 2005; Gryta, 2002; Gryta, 2007; Tun et al., 2005).

The possible origins of fouling in MD process as follows: chemical reaction of solutes at the membrane boundary layer (e.g. formation of ferric hydroxides from soluble forms of iron), precipitation of compounds which solubility product was exceeded (scaling), adsorption of organic compounds by membrane-forming polymer, irreversible gel formation of macromolecular substances and colonization by bacteria and fungi (Gryta, 2002; Gryta, 2005b; Gryta, 2007; Gryta, 2008). The operating conditions of membrane distillation restricted the microbial growth in the MD installation; therefore, one should not expect the problems associated with biofouling in the degree encountered in other membrane processes such as UF, NF or RO (Gryta, 2002b).

A large influence on the fouling intensity has a level of feed temperature. During concentration of bovine serum albumin aqueous solution by DCMD was found that fouling was practically absent in the process operated at low temperature (i.e. 293–311 K) (Ortiz de Zárate et al., 1998). On the contrary, a severe fouling by proteins was observed at higher feed temperatures (Gryta et al., 2001; Gryta et al., 2006c). The $CaCO_3$ scaling is also increased with an increase of the feed temperature. As a result of feed heating the HCO_3^- ions, present in the water, undergo the decomposition and a significant amount of $CaCO_3$ precipitates on the membrane surface (Drioli et al., 2004; Karakulski & Gryta, 2005; Gryta, 2005b; Schneider, et al., 1988). Although the acidification of feed water to pH 4 limited $CaCO_3$ scaling in the MD process, a slight fouling caused by other compounds (such as silicates), was still observed (Karakulski & Gryta, 2005). The foulants concentration may be reduced in the pretreatment stage, e.g. by using the NF or RO processes (Karakulski et al, 2002; Gryta, 2005b).

The deposit layers can be divided into two basic categories: porous and homogenous (non-porous) - Fig. 7. The deposit covered a part of the membrane surfaces, which reduced the membrane permeability and changed the temperature polarisation (Gryta, 2007). The values of heat transfer coefficients in both liquid phases and the membrane have a dominant influence on the values of T_1 and T_2 temperature of surfaces adjacent to the membrane (Fig. 2). The deposit layer creates an additional thermal resistance, thus decreasing the heat transfer coefficient from the feed bulk to the evaporation and condensation surfaces, and the temperature polarisation increased. As a result, the driving force for mass transfer is reduced and a significant decline of the permeate flux was observed (Gryta, 2008). The formation of non-porous layer causes a significant increase in the mass transfer resistance and the value of the permeate flux approach zero in an exponential way (Gryta, 2008).

Fig. 7. SEM image of deposit on the MD membranes (Accurel PP S6/2). A) porous ($CaCO_3$); B) non-porous (proteins)

The supersaturation state enables the nucleation and crystal growth, what in MD is mainly caused by water evaporation and temperature changes (Alklaibi & Lior, 2005; Gryta, 2002; He et al., 2008; Yun et al., 2006). In the case when the solute solubility decreases along with a temperature drop, deposit can be formed as a result of the temperature polarization (He et al., 2008; Gryta, 2002).

The formation of deposit on the MD membrane surface begins in the largest pores (Fig. 8), because they undergo wettability the most rapidly (Alklaibi & Lior, 2005; Schneider et al., 1988). The wetted pores are filled by the feed, what facilitates the oversaturation and formation of deposits. The salt crystallization inside the pores was limited through a reduction of the surface porosity (Gryta, 2007b; He et al., 2008).

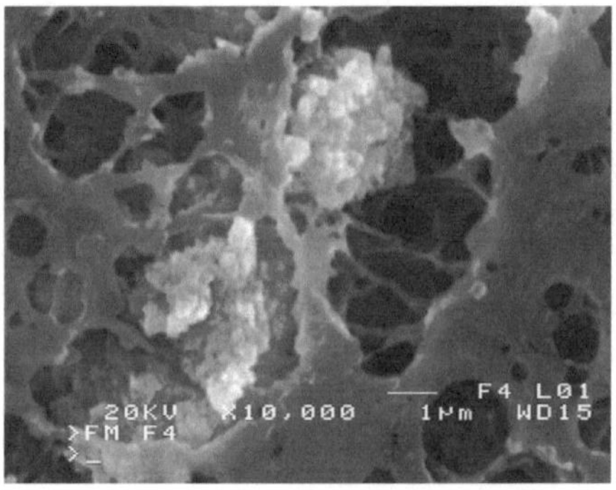

Fig. 8. SEM images of deposits formed inside the large pores (3-5 μm of diameter)

The adherence of the deposit to the membrane surface is a critical factor for MD performance, as well as for other membrane processes (Gryta, 2008; Gryta, 2009). It was found, that the deposit of $CaCO_3$ on the membrane surface can easily be removed by rinsing the module with a 2–5 wt.% solution of HCl, what allowed to restore the initial permeate flux (Fig. 9). However, the repetitions of module cleaning procedure by this method resulted in a gradual decline of the maximum permeate flux (Gryta, 2008).

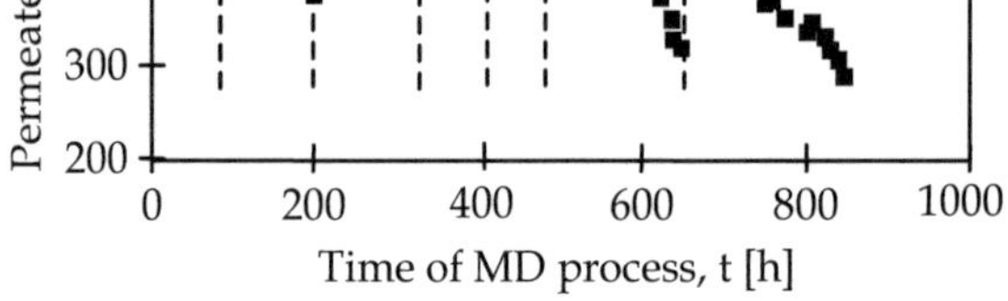

Fig. 9. Changes of the permeate flux during MD process of tap water

The SEM investigation of the membrane cross-sections revealed that the deposit covered not only the membrane surfaces but also penetrated into the pore interior (Fig. 10). The SEM-EDS line analysis of a change of the calcium content located into the membrane wall demonstrated that the deposit occurred up to the depth of 20–30 μm. Although, a rinsing acid solution dissolves the crystals, the wettability of the pores filled by deposit was accompanied to this operation. Therefore, the elimination of the scaling phenomenon is very important for MD process. The application of chemical water softening and the net filters (surface crystallization) allows to limit the amounts of precipitates deposited on the membrane surface during water desalination by MD process (Gryta, 2008c).

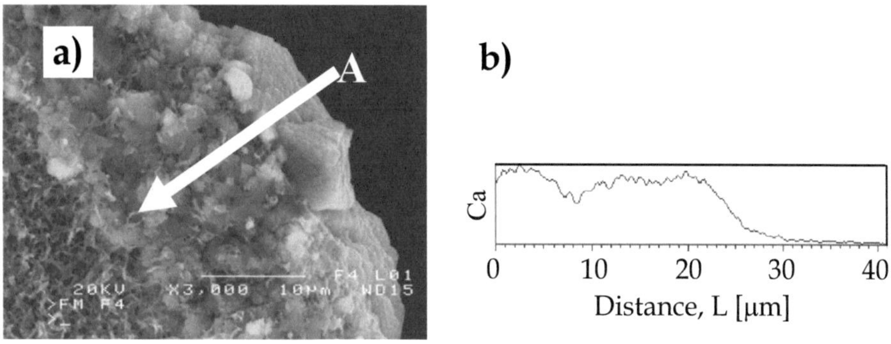

Fig. 10. $CaCO_3$ deposit on the membrane surface. a) membrane cross section, b) SEM-EDS line analysis (direction A)

4. Water pretreatment and membrane cleaning

The main techniques currently used to control fouling are feed pretreatment and membrane cleaning (Baker & Dudley, 1998; Schäfer et al., 2005, Gryta, 2008). The degree of pretreatment depends on the nature of the feeding water, the kind of membrane, the water recovery level and frequency of membrane cleaning (Karakulski et al., 2006; Schäfer et al., 2005). It was found that a significant amount of foulants from effluents obtained during ion-exchangers regeneration was successfully removed by the addition of the $Ca(OH)_2$ to treated wastewater (Gryta et al., 2005c). The fouling intensity can be also limited by combining the MD with other membrane processes (Drioli et al., 2004; Jiao, 2004; Karakulski et al., 2006). The UF/MD integrated processes enables the concentration of solutions polluted by significant amounts of petroleum derivatives (Karakulski et al., 2002; Gryta et al., 2001b). On the other hand, an excessively advanced pretreatment system significantly increases the installation costs (Karakulski et al., 2006), which may render the application of MD process as unprofitable. Moreover, an effective water pretreatment by NF and RO processes did not allow to completely eliminate fouling (Karakulski et al., 2002; Karakulski & Gryta, 2005), therefore, its negative consequences should also be limited through the development of appropriate procedure of installation operation.

The majority of problems encountered during the water desalination by MD process are associated with water hardness. As the water is heated, CO_2 content decreases and the precipitation of $CaCO_3$ takes place due to the decomposition of bicarbonate ions (Figs. 7-11). For this reason, the feed water has to be pretreated before feeding the MD installation (Singh, 2006; Karakulski et al., 2006; Gryta, 2006b). Several operations such as coagulation, softening and filtration are used during the production of technological water. The possibility of such pretreated water utilization as a feed for the MD process is an attractive option (Gryta, 2008b). Contact clarifiers (accelators) are usually applied to the chemical pretreatment of feed water in power stations (Powell, 1954, Singh, 2006). The chemicals (e.g. lime, aluminum or ferric sulphate) are added directly to the accelator containing a relatively high concentration of precipitated sludge near the bottom of the tank, and raw water is treated with this mixture. Inside the accelator, water flowing downward from the mixing and reaction zone passes the outer section of a much larger diameter, which is free of turbulence. Subsequently, the water flows upward, and the removal of flocks by settling takes place. A larger portion of this water passes through the return zone to the primary mixing and to the reaction zone. This recirculation improves the quality of the treated water.

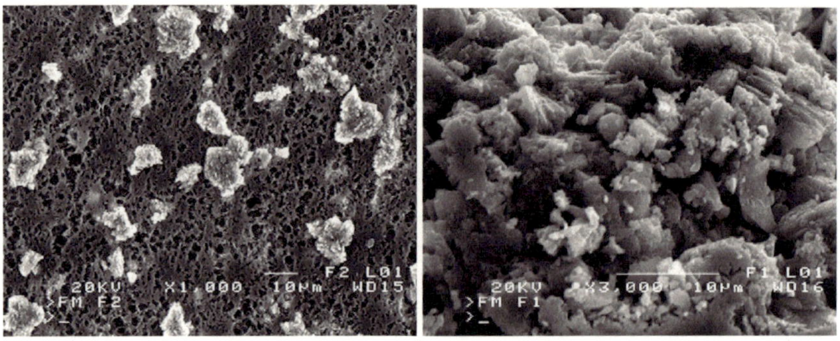

Fig. 11. SEM images of $CaCO_3$ deposit on membrane surface after: A) 10 h, and B) 50 h desalination of surface water by MD process

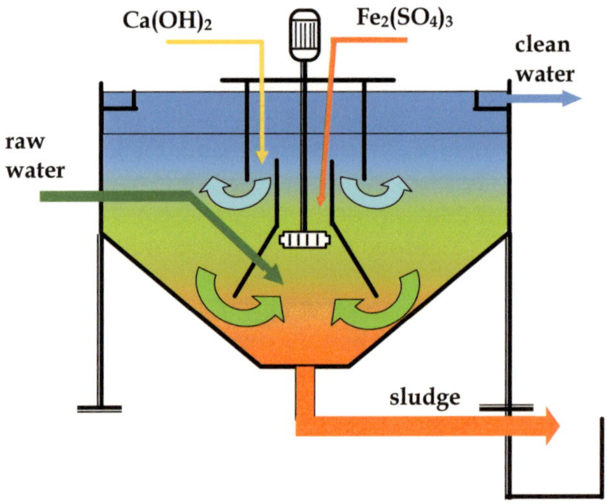

Fig. 12. Water treatment using the contact clarifiers (accelator)

The chemical pretreatment of ground water caused a significant decrease of the concentration of compounds responsible for the formation of a deposit on the membrane surface during the MD process (Gryta, 2008). However, the treatment of water carried out in an accelator, employed in the power station for production of demineralized water by the ion exchange process, was found to be insufficient for the MD process (Fig. 13). The formation of crystallites on the membrane surface was confirmed by SEM observations. Thus, a further purification of water produced by accelator is required in order to use it as a feed for the MD process.

A very efficient method for preventing $CaCO_3$ precipitation is dosing an acid (Karakulski & Gryta, 2005). In this case HCO_3^- ions are converted into CO_2 according to the following reaction:

$$HCO_3^- + H^+ \rightarrow CO_2 + H_2O \quad (4)$$

A major disadvantage of this method is an increase of concentration of chloride (HCl) or sulphates (H_2SO_4) in the retentate. The later anions (SO_4^{-2}) are particularly hazardous for the membrane (Fig. 14).

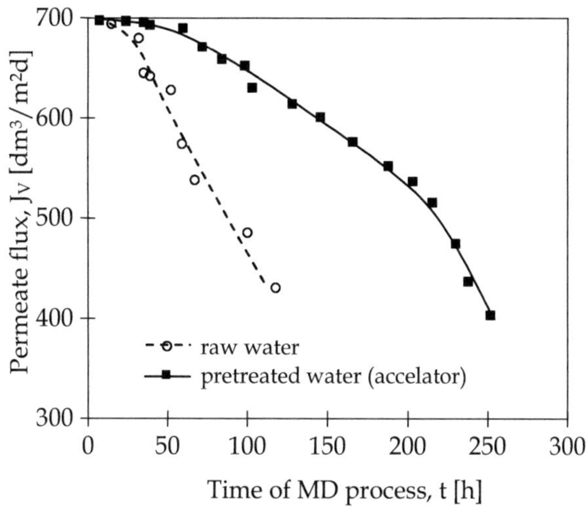

Fig. 13. Effect of the feed pretreatment (accelator) on the MD permeate flux

Fig. 14. SEM image of $CaSO_4$ deposit on the MD membrane surface

Sulphates comprise the second type of fouling components, the scaling of which can be encountered during water desalination by MD. The $CaSO_4$ solubility often determines the maximum recovery rate of demineralised water from feeding water (Gryta, 2009b).

The feed water before flowing into MD modules is heated in heat exchangers. In this case, a thermal softening of water can also be performed (Gryta, 2006b). As the water is heated, CO_2 content decreases and the precipitation of $CaCO_3$ takes place due to the decomposition of bicarbonate ions. A precipitated deposit may also cause substantial fouling of membranes; therefore, this deposit should be removed by using an additional filtration (Karakulski & Gryta, 2005). Other option is the application of heat exchanger, the design of which allows to remove the deposit of carbonates formed during water heating (Gryta, 2004).

Thermal pretreatment allows to remove most bicarbonates from water, which in turn reduces the amount of precipitate forming during MD process. However, the degree of water purification sometimes is too low and precipitate is still forming on the membrane surface. The SEM-EDS analysis revealed that apart a large amount of Ca, this deposit also contained Mg, Si, S, Fe, Ni, Al and Na. When the majority of HCO_3^- ions was removed from water, the carbonates formed an amorphous deposit with increased content of silicon (Gryta, 2010b). Such a nonporous form of deposit increases the rate of decline of the MD process efficiency (Fig. 15). For this reason an additional operation of the feed treatment was required to prevent the formation of deposit. The residual of HCO_3^- ions, from the thermally softened water, were removed by acidifying the boiled water down to pH = 4. This operation retained the formation of precipitate and as a result the MD process proceeded without the flux decrease.

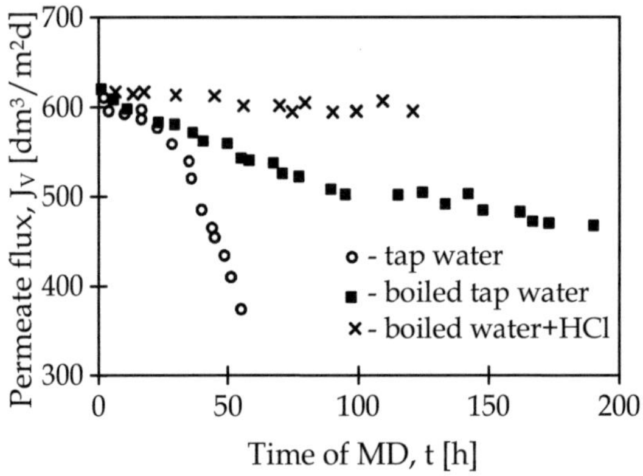

Fig. 15. The dependence of permeate flux as a function of the mode of feed pretreatment

Increasing the speed of the feed flow can reduce the negative influence deposit formation on MD process efficiency. SEM investigations demonstrated that the layer of the deposit was in this case more porous (Gryta, 2008c).

The induction period of $CaCO_3$ nucleation decreases as the supersaturation increases, but for the low saturation ratios (5-20) the induction period was higher than 30 min. It was reported that the induction time decreased from 12.9 to 1.1 min when the saturation ratio increased from 4 to 16 (Qu et al.; 2009). The elimination of membrane scaling is possible when the induction time will be longer than the residue time of feed inside the MD module.

A heterogeneous crystallization performed inside a net filter may decrease the saturation ratio and as a result, the amount of deposit formed on the membrane surface will be reduced (Gryta, 2006b). The application of pre-filter element assembled directly to the MD module inlet allows to significantly limit the amounts of precipitates deposited on the membrane surface during the desalination of natural water by MD process (Gryta, 2009c). The removal of formed deposit from this element (rinsing by HCl solutions) would not result in the membrane wettability. The period between consecutive rinsing operations of the pre-filter is dependent on the several factors, such as a water hardness level, parameters of MD process and the residence time of the feed inside the MD installation. On the basis of the obtained results it can be assumed, that this period would be in the range of 2–5 h. The efficiency of this system was found to decrease along with an increase of distance of pre-filter element from the module inlet.

5. Practical aspects of MD process

The MD separation mechanism is based on vapour/liquid equilibrium of a liquid mixture. For solutions containing non-volatile solutes only water vapour is transferred through the membrane; hence, the obtained distillate comprises demineralized water (Alklaibi & Lior, 2005; Karakulski & Gryta, 2005; Schneider et al., 1988). However, when the feed contains several volatile components, they are also transferred through the membranes to the distillate (El-Bourawi et al., 2006; Gryta, 2010c). Based on this separation mechanism, the major application areas of membrane distillation include water treatment technology, seawater desalination, production of high purity water and the concentration of aqueous solutions (El-Bourawi et al., 2006; Drioli et al. 2004; Gryta et al., 2005c; He et al., 2008; Karakulski et al., 2006, Li & Sirkar, 2005; Srisurichan et al., 2005; Teoh et al., 2008).

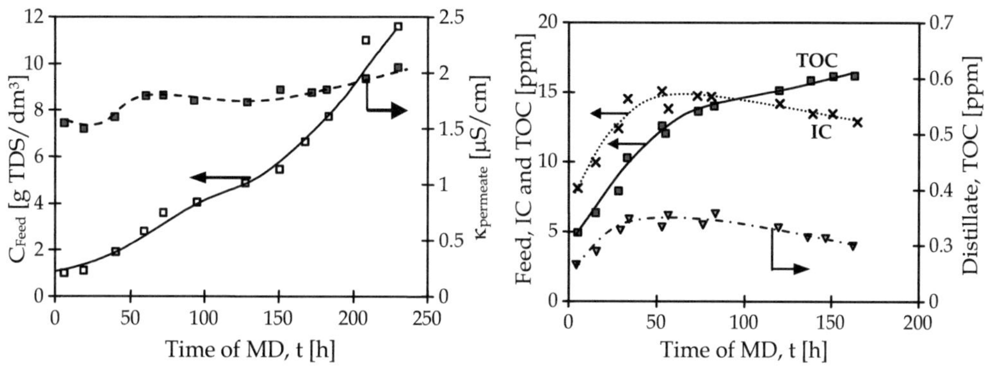

Fig. 16. Desalination of surface water by MD process

The results shown in Fig. 16 indicate that an increase in the feed concentration had a negligible effect on the quality of produced distillate. Despite the increasing value of the feed concentration the content of inorganic carbon (IC) in the distillate was close to the analytic zero. Only a slight amount of total organic carbon (TOC), below 0.5 mg TOC/dm³, was detected in the distillate, which can be associated with the transport of the volatile compounds through the MD membranes. It was found that volatile organic compounds (VOCs) diffuse through the pores of hydrophobic membranes, similarly to water vapour, hence, they are not completely rejected in the MD process (Gryta, 2010c; Karakulski & Gryta, 2005; Lawson & Loyd, 1997).

The produced MD distillate usually has the electrical conductivity in the range 0.5–5 µS/cm and contained below 0.5 ppm of inorganic carbon. It confirms the fact that regardless of the time of the process duration, the MD membranes demonstrated a high retention of inorganic solutes (Alklaibi et al., 2005; Gryta, 2006b).

The possibility of application of the MD process for the treatment of saline effluents generated during the regeneration of ion exchangers was investigated. The feasibility studies were also performed in the MD pilot plant (Gryta, 2007). A corrosion phenomenon was noticed in this installation during a long-term operation of process. The pilot plant was constructed using a typical heat exchanger made of stainless steel, however, the employed construction material was found to undergo the corrosion in studied solutions. A more appropriate heat exchangers for this process should be made of tantalum, but their price is 2-times higher than the cost of constructed MD installation. Therefore, the treatment of the effluents from ion exchangers regeneration would be unprofitable due to a high investment cost. Moreover, the fouling caused by iron oxides does not always result from the corrosion of installation, but also from the reactions proceeding in the feed. Therefore, the utilization of plastics for the construction of the entire MD installation will not prevent the formation of iron oxides that subsequently will precipitate onto the membrane surface. Such a phenomenon has been observed in the hybrid MD/absorber system utilised for gas purification by the absorption of SO_2 in a solution of Fe (II) sulphate (VI) proceeding simultaneous with the catalytic oxidation of SO_2 to sulphuric acid (Lewicki & Gryta, 2004).

Membrane processes associated with renewable energy for water desalination offer alternative solutions to decrease the dependence on fossil fuels (Charcosset, 2009). The potential use of solar thermal-driven MD process for water desalination has been studied extensively. Although the desalted water was produced using free energy, it was stated that this technology is still expensive compared to other desalination processes (Banat & Jwaied, 2008). However, it was found that increasing the reliability of the MD technology and plant life-time could reduce the cost of the produced water significantly.

6. Conclusion

In comparison with other desalination processes, the main advantages of membrane distillation are: (1) 100% separation (in theory) of ions, macromolecules, colloids, cells etc., (2) lower operating pressures, (3) lower requirements concerning the mechanical properties of the membrane, and (4) less space requirement compared to conventional distillation processes. However, besides these advantages, membrane distillation still faces difficulties for commercialization.

The availability of the industrial MD modules is currently one of the limitations for MD process implementation. Flat-sheet membranes in plate and frame modules or spiral wound modules and capillary membranes in tubular modules have been used in various MD studies. The design of the MD modules should provide not only good flow conditions, but also has to improve the heat transfer and thermal stability. Several advantages offer the capillary MD modules. The efficiency of these modules was significantly improved when the cross flow or a devices with membranes arranged in a twisted or braided form in the housing were used.

The major difficulties are basically associated with a phenomenon of membrane wetting and the formation of the deposit on its surface. The use of an adapted pretreatment minimizes the fouling problems and can provide good protection of the membranes. Moreover, the module scaling may be reduced using the appropriate MD process conditions. The $CaCO_3$ precipitation was limited by lowering the feed temperature and by increasing the feed flow rate. The HCO_3^- ions concentration may be reduced by chemical water softening or by using pressure driven membrane processes. An effective solution would be the complete removal of the HCO_3^- ions from feed water, which can be achieved by the acidification of water to pH 4. However, the significant amounts of acids are required for feed acidification and as a result, the amount of salt increased in the retentate discharged to the environment.

The fouling and scaling accelerated the membrane wetting; therefore, more work will have to be done for a thorough evaluation of these phenomena.

7. References

Alklaibi, A.M. & Lior, N. (2005). Membrane-distillation desalination: status and potential, *Desalination*, Vol. 171, No. 2 (January 2005) 111–131, ISSN 0011-9164

Alklaibi, A.M. & Lior, N. (2007). Comparative study of direct-contact and air-gap membrane distillation processes. *Ind. Eng. Chem. Res.* Vol.46, No.2 (January 2007) 584–590, ISSN

Baker, J.S. & Dudley L.Y. (1998). Biofouling in membrane system - A review. *Desalination*, Vol. 18, No.1-3 (September 1998) 81–90, ISSN 0011-9164

Banat, F.; Jwaied, N.; Rommel, M.; Koschikowski. J. & Wieghaus, M. (2007). Performance evaluation of the "large SMADES" autonomous desalination solar-driven membrane distillation plant in Aqaba, Jordan. *Desalination,* Vol.217, No.1-3, (November 2007) 17-28, ISSN 0011-9164

Banat, F. & Jwaied, N. (2008). Economic evaluation of desalination by small-scale autonomous solar-powered membrane distillation units. *Desalination,* Vol.220, No.1-3, (March 2008) 566–573, ISSN 0011-9164

Bonyadi, S. & Chung, T.S. (2009). Highly porous and macrovoid-free PVDF hollow fiber membranes for membrane distillation by a solvent-dope solution co-extrusion approach, *J. Membr. Sci.*, Vol. 331, No.1-2 (April 2009) 66–74, ISSN 0376-7388

Bui, V.A.; Vu, L.T.T. & Nguyen, M.H. (2010). Simulation and optimization of direct contact membrane distillation for energy efficiency. *Desalination,* Vol.259, No.1-3, (September 2010) 29–37, ISSN 0011-9164

Charcosset, C. (2009). A review of membrane processes and renewable energies for desalinastion. *Desalination,* Vol.245, No.1-3, (September 2009) 214-231, ISSN 0011-9164

Criscuoli, A.; Carnevale, M.C. & Drioli, E. (2008). Evaluation of energy requirements in membrane distillation, *Chem. Eng. Proc.,* Vol.47, No.7, (July 2008) 1098-1105, ISSN 0009-2509

Drioli, E.; Curcio, E.; Criscuoli, A. & Di Profio, G. (2004). Integrated system for recovery of $CaCO_3$, NaCl, $MgSO_4 \cdot 7H_2O$ from nanofiltration retentate, *J. Membr. Sci.*, Vol.239, No.1, (August 2004) 27–38, ISSN 0376-7388

El-Bourawi, M.S.; Ding, Z.; Ma, R. & Khayet, M. (2006). A framework for better understanding membrane distillation separation process. *J. Membr. Sci.,* Vol.285, No.1-2, (November 2006) 4–29, ISSN 0376-7388

El Fray, M. & Gryta, M. (2008). Environmental fracture of polypropylene membranes used in membrane distillation process, *Polimery*, Vol.53, No.11-12, (November 2008) 865–870, ISSN 0032-2725

Gryta, M.; Tomaszewska, M. &. Morawski, W. (1998). Heat transport in the membrane distillation process, *J. Membr. Sci.*, Vol.144, No.1-2, (June 1998) 211–222, ISSN 0376-7388

Gryta, M.; Tomaszewska, M.; Morawski, A.W. & Grzechulska J., (2001). Membrane distillation of NaCl solution containing natural organic matter, *J. Membr. Sci.*, Vol.181, No.2, (January2001) 279–287, ISSN 0376-7388

Gryta, M.; Karakulski, K. & Morawski, A.W. (2001b). Purification of oily wastewater by hybrid UF/MD. *Water Res.*, Vol. 35, No.15, (October 2001) 3665–3669, ISSN 0043-1354

Gryta, M. (2002). Direct contact membrane distillation with crystallization applied to NaCl solutions, *Chem. Pap.*, Vol. 56, No.1, (January 2002) 14–19, ISSN 0366-6352

Gryta, M. (2002b). The assessment of microorganism growth in the membrane distillation system, *Desalination*, Vol.42, No.1 (January 2002) 79–88, ISSN 0011-9164

Gryta, M. (2004). Water membrane distiller, *Inż. Chem. Proc.,* Vol. 25, No.2 (April 2004), 381–391, ISSN 0208-6425

Gryta, M. (2005). Osmotic MD and other membrane distillation variants. *J. Membr. Sci.*, Vol. 246, No.2 (January 2005), 45–56, ISSN 0376-7388

Gryta, M. (2005b). Long-term performance of membrane distillation process, *J. Membr. Sci.*, Vol. 265, No.1-2, (November 2005) 153–159, ISSN 0376-7388

Gryta, M.; Karakulski, K.; Tomaszewska, M. & Morawski, A. (2005c). Treatment of effluents from the regeneration of ion exchangers by MD process, *Desalination*, Vol.180, No.1-3, (August 2005) 173–180, ISSN 0011-9164

Gryta, M. (2006). Heat efficiency of the capillary modules for membrane distillation process, *Inż. Chem. Proc.*, Vol. 27, No.1 (January 2006) 305-314, ISSN 0208-6425

Gryta, M. (2006b). Water purification by membrane distillation process, *Sep. Sci. Technol.* Vol. 41, No.9 (September 2006) 1789–1798, ISSN 0149-6395

Gryta, M.; Tomaszewska, M. & Karakulski, K. (2006c). Wastewater treatment by membrane distillation, *Desalination*, Vol. 98, No.1-3 (October 2006) 67–73, ISSN 0011-9164

Gryta, M. (2007). Effect of iron oxides scaling on the MD process performance, *Desalination*, Vol. 216, No.1-3 (October 2007) 88–102, ISSN 0011-9164

Gryta, M. (2007b). Influence of polypropylene membrane surface porosity on the performance of membrane distillation process, *J. Membr. Sci.*, Vol.287, No. 1 (January 2007) 67–78, ISSN 0376-7388

Gryta, M. (2008). Fouling in direct contact membrane distillation process, *J. Membr. Sci.*, Vol. 325, No.1, (November 2008) 383–394, ISSN 0376-7388

Gryta, M. (2008b). Chemical pretreatment of feed water for membrane distillation. *Chem. Pap.* Vol.62, No.1, (January 2008) 100–105, ISSN 0366-6352

Gryta, M. (2008c). Alkaline scaling in the membrane distillation process, *Desalination*, Vol.228, No.1-3 (August 2008) 128–134, ISSN 0011-9164

Gryta, M.; Grzechulska-Damszel, J.; Markowska, A. & Karakulski, K. (2009). The influence of polypropylene degradation on the membrane wettability during membrane distillation, *J. Membr. Sci.*, Vol. 326, No.2 (January 2009) 493–502, ISSN 0376-7388

Gryta, M. (2009b). Calcium sulphate scaling in membrane distillation process. *Chem. Pap.*, Vol. 63, No. 2 (March 2009) 146–151, ISSN 0366-6352

Gryta, M, (2009c). Scaling diminution by heterogeneous crystallization in a filtration element integrated with membrane distillation module, *Pol. J. Chem. Tech.*, Vol. 11, No. 1 (February 2009) 59–64, ISSN 1509-8117

Gryta, M. & Barancewicz, M. (2010). Influence of morphology of PVDF capillary membranes on the performance of direct contact membrane distillation, *J. Membr. Sci.*, Vol. 358, No.1-2 9August 2010) 158–167, ISSN 0376-7388

Gryta, M. (2010b). Desalination of thermally softened water by membrane distillation process. *Desalination*, Vol. 257, No.1-3 (July 2010) 30–35, ISSN 0011-9164

Gryta, M. (2010c). Application of membrane distillation process for tap water purification. *Membrane Water Treatment*, Vol. 1, No. 1 (January 2010), 1-12, ISSN 2005-8624

He, F.; Gilron, J.; Lee, H.; Song, L. & Sirkar, K. (2008). Potential for scaling by sparingly soluble salts in crossflow DCMD. *J. Membr. Sci.*, Vol. 311, No.1-2 (March 2008) 68–80, ISSN 0376-7388

Jiao, B.; Cassano, A. & Drioli, E. (2004). Recent advanced on membrane processes for the concentration of fruit juices: a review, *J. Food Eng.* Vol. 63, No.3 (August 2004) 303–324, ISSN 0260-8774

Karakulski, K.; Gryta, M. & Morawski, A. (2002). Membrane processes used for potable water quality improvement, *Desalination*, Vol. 145, No.1-3 (September 2002) 315–319, ISSN 0011-9164

Karakulski, K. & Gryta, M. (2005). Water demineralisation by NF/MD integrated processes, *Desalination*, Vol.177, No.1-3 (June 2005) 109–119, ISSN 0011-9164

Karakulski, K.; Gryta, M. & Sasim,M. (2006). Production of process water using integrated membrane processes, *Chem. Pap.*, Vol.60, No.6 (November 2006) 416–421, ISSN 0366-6352

Khayet, M. & Matsuura, T. (2003). Application of surface modifying macromolecules for the preparation of membranes for membrane distillation, *Desalination*, Vol. 158, No.1-3, (August 2003) 51–56, ISSN 0011-9164

Khayet, M.; Godino, M.P. & Mengual, J.I. (2004). Study of asymmetric polarization in direct contact membrane distillation, *Sep. Sci. Technol.*, Vol. 39, No. 1 (January 2004) 125–147, ISSN 0149-6395

Khayet, M.; Matsuura, T.;. Mengual, J.I. & Qtaishat, M. (2006). Design of novel direct contact membrane distillation membranes, *Desalination*, Vol.192, No.1-3 (May 2006) 105–111, ISSN 0011-9164

Lawson, K.W. & Lloyd, D. R. (1997). Membrane distillation. *J. Membr. Sci.*, Vol.124, No.1 (February 1997) 1–25, ISNN 0376-7388

Lee, S. & Lee, Ch. (2000). Effect of operating conditions on $CaSO_4$ scale formation mechanism in nanofiltration for water softening, *Water Res.*, Vol.34, No.15, (October 2000) 3854–3866, ISSN 0043-1354

Li, B. & Sirkar, K.K. (2004). Novel membrane and device for direct contact membrane distillation-based desalination process, *Ind. Eng. Chem. Res.*, Vol. 43, No.17 (August 2004) 5300–5309, ISSN 0888-5885

Martínez-Díez, L. & Vázquez-González, M.I. (1999). Temperature and concentration polarization in membrane distillation of aqueous salt solutions, *J. Membr. Sci.*, Vol. 156, No.2 (April 1999) 265–273, ISSN 0376-7388

Ortiz de Zárate, J.M.; Rincón, C. & Mengual, J.I. (1998) Concentration of bovine serum albumin aqueous solutions by membrane distillation, *Sep. Sci. Technol.*, Vol. 33, No., 283–296, ISSN 0149-6395

Schäfer, A.I.; Fane, A.G. & Wait, T.D., (Eds). (2005). *Nanofiltration: Principles and Application*, Elsevier, ISBN 1-85617-405-0, Oxford

Schneider, K.; Hölz, W. & Wollbeck, R. (1988). Membranes and modules for transmembrane distillation, *J. Membr. Sci.*, Vol. 39, No.1 (October 1988) 25–42, ISSN 0376-7388

Singh, R (2006). *Hybrid Membrane Systems for Water Purification*, Elsevier, ISBN 1-856-17442-5, Kidlington

Smolders, K. & Franken, A.C.M. (1989). Terminology for membrane distillation, *Desalination*, Vol. 72, No.1-2 (April/May 1989) 249–262, ISSN: 0011-9164

Srisurichan, S.; Jiraratananon, R. & Fane, A.G. (2005). Humic acid fouling in the membrane distillation, *Desalination*, Vol.174, No.1 (April 2005) 63–72, ISSN 0011-9164

Srisurichan, S.; Jiraratananon, R. & Fane, A.G. (2006). Mass transfer mechanisms and transport resistances in direct contact membrane distillation process, *J. Membr. Sci.*, Vol. 277, No.1-2 (June 2006) 186–194, ISSN 0376-7388

Teoh, M.M.; Bonyadi, S. & Chung, T.S. (2008). Investigation of different hollow fiber module designs or flux enhancement in the membrane distillation process, *J. Membr. Sci.*, Vol. 311, No.1-2 (March 2008) 371–379, ISSN 0376-7388

Teoh, M.M. & Chung, T.S. (2009). Membrane distillation with hydrophobic macrovoid free PVDF–PTFE hollow fiber membranes, *Sep. Purif. Technol.*, Vol.66, No.2 (April 2009), 229–236, ISSN 1383-5866

Tomaszewska, M. (1996). Preparation and properties of flat-sheet membranes from poly(vinylidene fluoride) for membrane distillation, *Desalination*, Vol.104, No.1-2 (April 1996) 1–11, ISSN 0011-9164

Tomaszewska, M. (1993). Concentration of the extraction fluid from sulfuric acid treatment of phosphogypsum by membrane distillation, *J. Membr. Sci.*, Vol. 78, No.3 (April 1993) 277–282, ISSN 0376-7388

Tun, C.M.; Fane, A.G.; Matheickal, J.T. & Sheikholeslami, R. (2005) Membrane distillation crystallization of concentrated salts-flux and crystal formation, *J. Membr. Sci.*, Vol.257, No.1-2 (July2005) 144–155, ISSN 0376-7388

Qu, D.; Wang, J.; Wang, L., Hou, D.; Lian, Z. & Wang, B. (2009). Integration of accelerated precipitation softening with membrane distillation for high-recovery desalination of primary reverse osmosis concentrate, *Sep. Purif. Technol.*, Vol.67, No.1 (May 2009) 21–25, ISSN 1383-5866

Wang, K.Y.; Chung, T.S. & Gryta, M. (2008). Hydrophobic PVDF hollow fiber membranes with narrow pore size distribution and ultra-thin skin for the freshwater production through membrane distillation, *Chem. Eng. Sci.*, Vol. 63, No.9 (May 2008) 2587–2594, ISSN 0009-2509

Yun, Y.; Ma, R.; Zhang, W.; Fane, A.G. & Li, J. (2006). Direct contact membrane distillation mechanism for high concentration NaCl solutions, *Desalination*, Vol.188, No.1-3 (February 2006) 251–262, ISSN 0011-9164

3

Desalination of Coastal Karst Springs by Hydro-geologic, Hydro-technical and Adaptable Methods

Marko Breznik and Franci Steinman
University of Ljubljana, Faculty of Civil and Geodetic Engineering, Slovenia

1. Introduction

The karst landscape consists of rocks such as limestone, dolomite, gypsum and various salts which are, to a greater or lesser extent, soluble in water, and through which underground water flows. According to this, the latest definition of the karst, 10 % of the world's land surface, and as much as 40 % of Slovenia's surface, is covered by carbonate karst rocks, which are the only kind of karst rocks that are important from the point of view of the exploitation of their water resources. On typical carbonate karst, only short lengths of rivers flowing through karst poljes are to be found. Elsewhere, due to the fact that water flows underground, the karst is a dry area, with a lack of drinking water; next to the sea, brackish karst springs are found.

This paper is concerned with the successes and failures of engineering -works which have attempted to improve natural conditions through the construction of various structures for the desalination of brackish springs. The fact that many completed works have been successful should encourage engineers to design and build new hydro-technical structures in karst environments (Breznik, 1998).

In the Ice ages were the differences between the lowest mean temperatures of the cold periods and the highest ones of the warm periods 5 degrees Celsius. These differences between the lowest ones of the Ice ages and the present highest ones are 7 degrees Celsius. During the last 30 years, 10 warmest were between 1990 and 2006. We are living probably in the warmest period during the last 150.000 years (Rošker, 2007). The yearly air temperatures in Ljubljana have increased by 1,7 degrees Celsius in the last 50 years (Kajfež-Bogataj, 2006). Sixty years ago, we had to walk for 1km over the Triglav mountains glacier, after climbing over the Triglav's north wall, 800 m high. This glacier has nearly melted till the present.

Precipitations have decreased from 1100 to 1000 mm/year in the Trieste town during last 100 years. Italian scientists believe that the Azores Island's anticyclone with sunny weather has extended towards the Mediterranean. Precipitations in the Portorož town have decreased by 14%, during the last 50 years (Kajfež-Bogataj, 2006). An about 1000 km large belt of severe drought hazards event extends along the southern Spain, Italy, Greece, Turkey and northern Africa to Syria and Iraq.

Lučka Kajfež-Bogataj, Professor of the Ljubljana University and the Vice-chair of the Working Group II: Impacts, Adaptation and Vulnerability of the Intergovernmental Panel on Climate Change (IPCC), warns that the climate changes will continue and threaten the world population with shortages of water, energy and food. The adaptive measures have to be taken quickly. The proposed desalination of larger karstic coastal springs could provide fresh water for drinking and irrigation (Breznik, 1998; Breznik & Steinman, 2008).

2. Exploitation of karst ground water in coastal areas. Theory with examples

2.1 Sea water intrusion

Brackish karst springs are a regular phenomenon of any seashore consisting of limestone or dolomite. Fresh water from a calcareous karst aquifer is contaminated by the intrusion of sea water, which renders spring water useless. The development of brackish springs, therefore, is of great human and economic importance for areas which are short of fresh water. The first developments were made by the ancient Phoenicians, who covered submarine springs with lead funnels and fed fresh water into leather bags (Kohout, 1966).

2.2 Springs in karst aquifers of isotropic permeability

The porosity and ground water movements in an isotropically permeable karst aquifer, and in an aquifer in granular sediments, are similar. The flow of ground water is of a diffused type. The mechanism of contamination of fresh ground water with sea water in aquifers in sand and gravel has been explained by Ghyben (1888), Herzberg (1901) and Hubbert (1940). Fresh water floats on denser sea water. A 40 m high column of sea water exerts the same pressure at the bottom as a fresh water column about 41 m high. This is known as the Ghyben-Herzberg law. The plane that separates the fresh and sea water in the aquifer is called the interface and is at a depth of about 40 times the height of the fresh water table above sea level.

In areas in which ground water flows towards the sea, some sea water mixes with flowing fresh water and creates a zone-of-mixing some meters high, which replaces the interface. In this zone, ground water is brackish, while above it is fresh water and beneath it unchanged sea water. The mixing process is partly the result of diffusion, but mostly of hydraulic mixing due to the different velocities of fresh and sea water. The thickness of the zone-of-mixing depends on the velocity of ground water movement and the fluctuations of the sea. Ghyben-Herzberg rules can be used for the calculation. Numerous small, brackish springs at small heights of 0.1 to 1 m above or some meters below sea level are typical of such a system. Relevant examples are the lower part of the Postire and Marina Stupin valleys in Croatia and a coastal aquifer in karstic sandstone in Israel. This paper does not discuss such aquifers (Breznik, 1998).

2.3 Springs in karst aquifers of anisotropic permeability

2.3.1 Principle, cases, theory

In the depths of the karst, ground water circulation tends to concentrate along a limited number of well-karstified zones. This is demonstrated by the concentration of drainage in the direction of a few large springs. The karst of the Central Dinaric Alps, with an area of 17,500 km^2, has only 55 large springs. Each spring, with a discharge from 7 to 9 m^3/s, drains a surface of 320 km^2 (Komatina, 1968). A similar situation is found on the island of Crete. Each of three separated karst regions, Dikti, Psiloritis and Lefka Ori, with areas of 150, 300 and 400 km^2, is drained by a single large spring, with respective discharges of 2, 6 and 8 m^3/s. The water collecting galleries, Postire II, Dubrava, Zaton, Gustirne and Blaž, all in Croatia, have also shown a concentration of ground water circulation (Breznik, 1973; 1998).

In an anisotropic karst aquifer, water flows through veins. The form of the veins is not defined: A vein can be a dissolution channel, a permeable fissured zone, a system of small connected cavities, etc. In seeking its course, water erodes paths through the least resistant rocks, so that the veins meander and ramify in many ways. Branching of vein or vein-branching is a place where the primary vein of karst massif branches off into the upper vein leading to the coastal spring and into the lower vein leading to the submarine estavelle. This is the conduit type of ground water circulation in karst. The mechanism of contamination with sea water, therefore, cannot be the same as in karst of isotropic permeability or in grained sediments of uniform porosity and with a semi-laminar diffused type circulation of ground water. In karst of anisotropic permeability, contamination occurs in the vein-branchings. This contamination was first explained by Gjurašin (1943), and in detail by Kuščer (1950), and Breznik (1973, 1978, 1990 and 1998), and Breznik & Steinman (2008).
In 1938, Prof. Gjurašin of Zagreb University developed a theory on the basis of the flow of sea water into the Gurdić spring on the Adriatic coast, that the various specific weights of sea and fresh water are the cause of the sea intruding into springs along the coast and coastal karst aquifers. The conduction channel splits into a larger upper vein and smaller lower vein, the mouth of which must be below sea level. Springs above sea level are only contaminated in a case in which the following equation is fulfilled:

$$\frac{\gamma_m - \gamma}{\gamma} \cdot h_S > h_V$$

where γ is the specific weight of fresh water, γ_m is the specific weight of sea water, h_S is the depth of the vein-branching below sea level, h_V is the height of the spring above sea level (Breznik, 1998). He also illustrated his theory pictorially for three hydrological conditions (Fig. 1; Gjurašin, 1943).

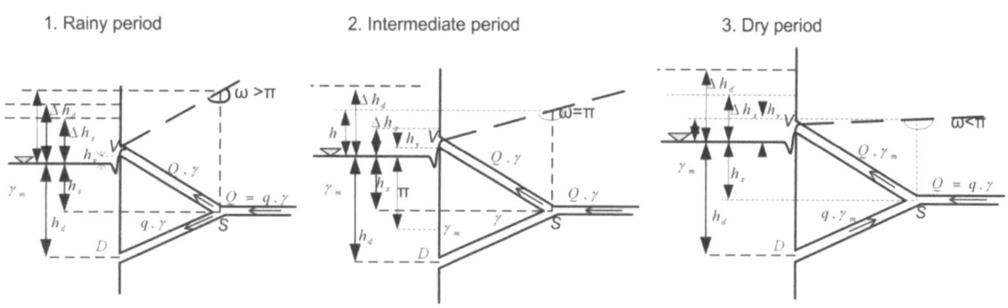

Fig. 1. Outflows of the Gurdić spring in the southern Adriatic (Gjurašin, 1943).

Field observations were performed by I. Kuščer and colleagues in 1938-1940 and in 1947. Seventy coastal and submarine springs, as well as thirty submarine estavelles, were registered near a sawmill at Jurjevo in the Northern Adriatic (Fig. 2). During rainy periods, all the springs deliver fresh water. The discharge of estavelle KEa is 1 m^3/s at a depth of 9 m below sea level. With the discharge decreasing in springtime, the outflow of estavelle KF stops and sea water intrudes into the vein. Estavelle KE and the springs KC and KD are contaminated by 700 mg/1 of Cl^-. In July, the springs KA and KB start delivering brackish water. In dry summers, the estavelles KEb and Kola swallow about 0.1 m^3/s of sea water, and the salinity of springs KB increases to 9000 mg/1 of Cl^-. The estavelle Kola changes

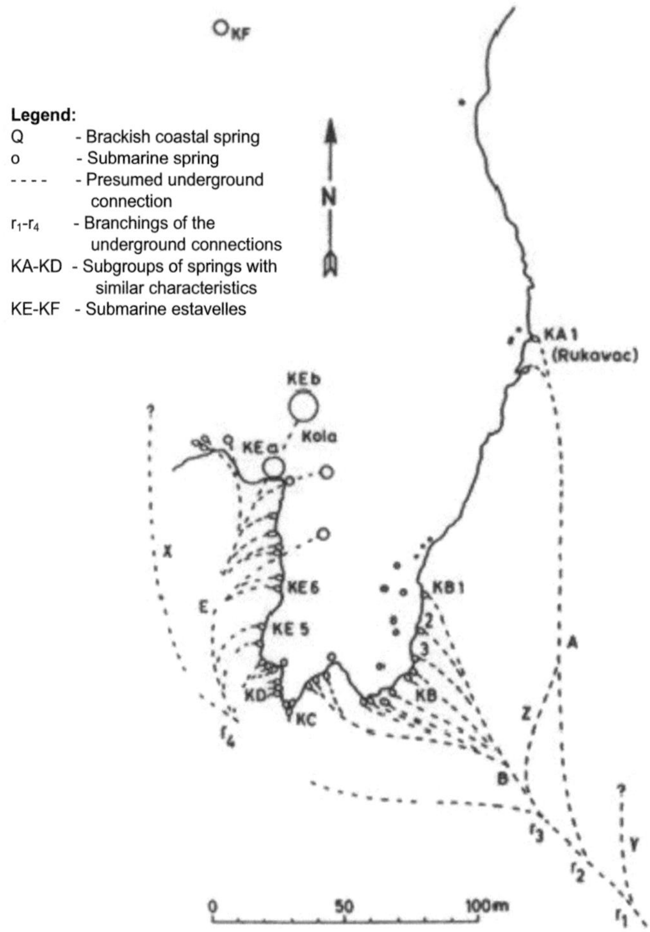

Fig. 2. Jurjevo bay in Northern Adriatic (Kuščer, 1950).

from a spring to a swallow hole very quickly, in 1 to 2 days, and the salinity of the KB springs also increases quickly. After the autumn rains and throughout the winter, all the springs and estavelles discharge fresh water (Kuščer, 1950).

A tracer test with 300 g of fluoresceine was performed on July 30th, 1947. The tracer was poured into the strongest submarine swallowhole, KEa. Colored water appeared after 5 hours in the springs KB, reached the highest concentration after 1 hour, and thereupon slowly decreased. After 6.5 hours, spring KA was also colored by a 2 to 3 times weaker concentration. Kuščer (Kuščer, 1950; Kuščer et al., 1962) indicates in his figure the estimated position of the veins and their important branchings. These field observations confirmed the type of contamination in vein-branchings. This scheme of sea water intrusion into a system of karst conduits is explained on a simplified section with the smallest number of necessary veins (Fig. 3).

Breznik examined the coastal springs and the estavelles in 40 karst places in the former Yugoslavia, Greece and Turkey since 1956 (Fig. 4).

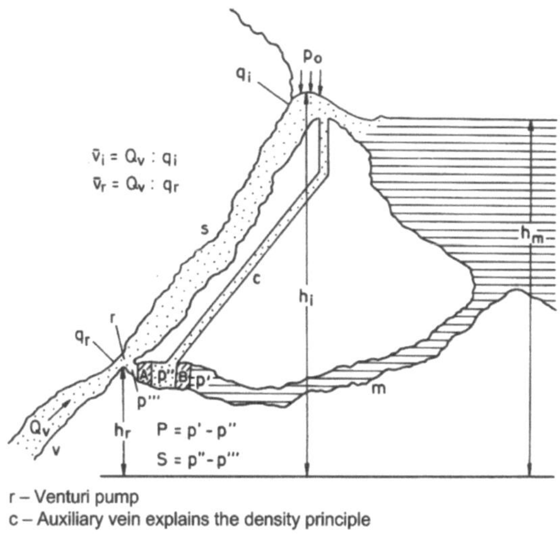

Fig. 3. Coastal spring of conduit type flow in karst aquifer (Kuščer, 1950).

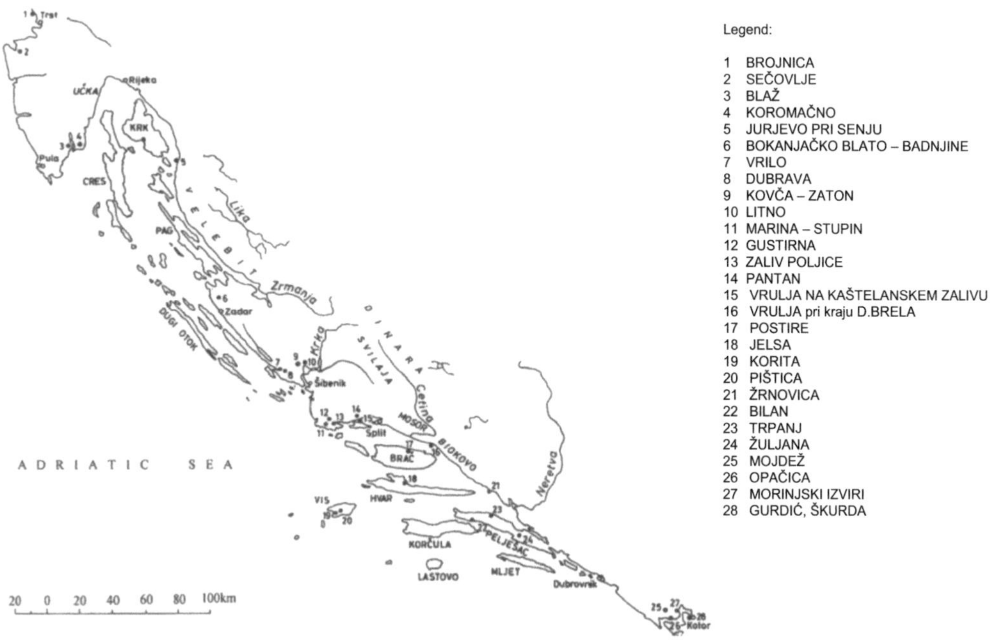

Fig. 4. Investigated karstic coastal springs along the eastern Adriatic coast (Breznik, 1973).

Fresh water from the karst massif is drained through the primary vein. This vein branches off into the lower vein, connected with the sea, and into the upper vein, leading to the spring. The present lower veins were formed in past geological periods by fresh water flowing towards the sea at a lower level, and were primary veins in these periods. Sea water later drowned them, either due to tectonic subsidence of the massifs in the Tertiary to Holocene periods, or due to the melting of Pleistocene ice. The level of the Mediterranean Sea in the Pleistocene period was initially 23 m higher, and then 120 m lower than at present

(Fig. 13). Karst water formed new channels to the actual sea surface, and these are the present upper veins and springs (Breznik, 1998).

All the changes in the direction of flow and salinities are shown in Figs. 7 and 8, whether there is fresh or brackish water in the same coastal spring or submarine estavelle, with either a fresh water outflow or sea water intrusion, are determined in the vein-branchings, and depend on the pressure of water in the veins forming the vein-branching. The piezometric head and density of water in each vein determine the pressure. In rainy periods, the head of water in the primary vein is high, and fresh water flows out of the lower vein as a submarine spring forming characteristic 'wheels' on the sea surface, and out of the upper vein as a fresh water coastal spring. In a dry period, the karst massif is drained and the piezometric head in the primary vein subsides. An equal or slightly higher pressure of sea water in the lower vein enables intrusion of sea water into the vein-branching. Brackish water flows through the upper vein to the spring. The energy for this flow pattern is derived from the fresh water head in the karst massif. Some submarine or coastal springs stop flowing in dry periods, since the fresh water head cannot counterbalance the sea water pressure. In such cases, the vein-branching and the lower part of the primary vein are flooded with sea water. This happens first in deeper vein-branchings (Figs. 5 and 7; Breznik, 1973; 1989; and 1998). Notations used in following figures and equations are shown in Fig. 5.

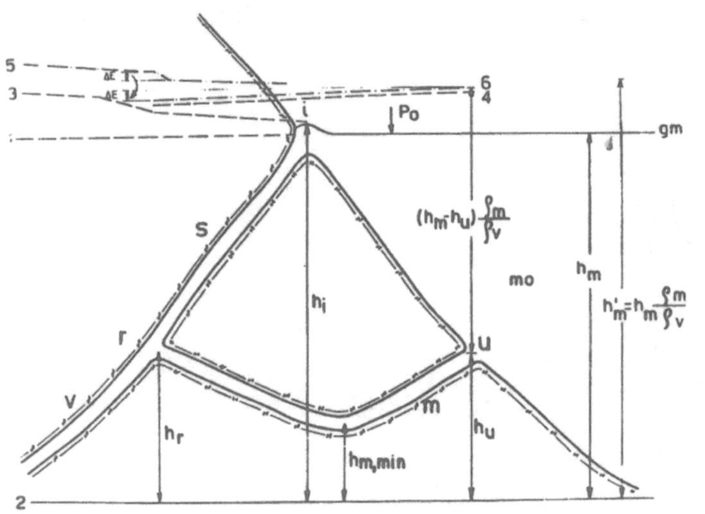

Legend:

m_0 - Sea
gm - Sea level
i - Coastal spring – brackish water
v - Primary vein with fresh water
r - Branching of veins
s - Upper vein – brackish water
m - Lower vein – sea water
m_{min} - Lowest point of lower vein
u - Mouth of the lower vein
h - Height above some reference level
ρ_v - Density of fresh water
ρ_m - Density of sea water
2 - Reference level
3 - Piezometric head line of the primary and upper veins
4 - Piezometric head line of the lower vein
5 - Energy head line of the primary and upper veins
6 - Energy head line of the lower vein
3-6 - All the heads are expressed through the head of fresh water

Fig. 5. Scheme of veins in the conduit type karst aquifer of a coastal spring (Breznik, 1973).

The pressure at the right side of the vein-branching is expressed by equation (1):

$$p' = p_0 + \left(h_m - h_r - \frac{v_m^{\;2}}{2g} - T_m \right) g\rho_m = f\left(Q_m\right) \tag{1}$$

and the pressure at the left side by equation (2):

$$p'' = p_0 + \left(h_i - h_r - \frac{v_r^{\;2}}{2g} + T_s \right) g\rho_s = f\left(Q_s\right) \tag{2}$$

Sea water can penetrate into a vein-branching if the pressure in the lower vein exceeds that in the upper one. In inequation (3) Breznik (1973) states this requirement:

$$h_i - h_r > \frac{\rho_m}{\rho_m - \rho_s} \cdot (h_i - h_m) + \frac{\rho_m T_m + \rho_s T_s}{\rho_m - \rho_s} - \frac{v_s^{\,2} \rho_s - v_m^{\,2} \rho_m}{2g(\rho_m - \rho_s)} \tag{3}$$

All the denominators in the right part of the inequation are differences in densities. The first numerator is the height of the spring above sea level, the second the sum of the head losses in the upper and lower veins, and the third the difference of the velocity heads in the two veins in the vein-branching.

There are certainly very few springs with only three veins, as in Fig. 5 which explains the mechanism of contamination. Many pairs of primary and lower veins, with branchings at different depths, must be expected for a single spring. This might explain the progressive contamination observed in the Almyros Irakliou spring in Greece (Fig. 6; Ré, 1968 in Breznik, 1973).

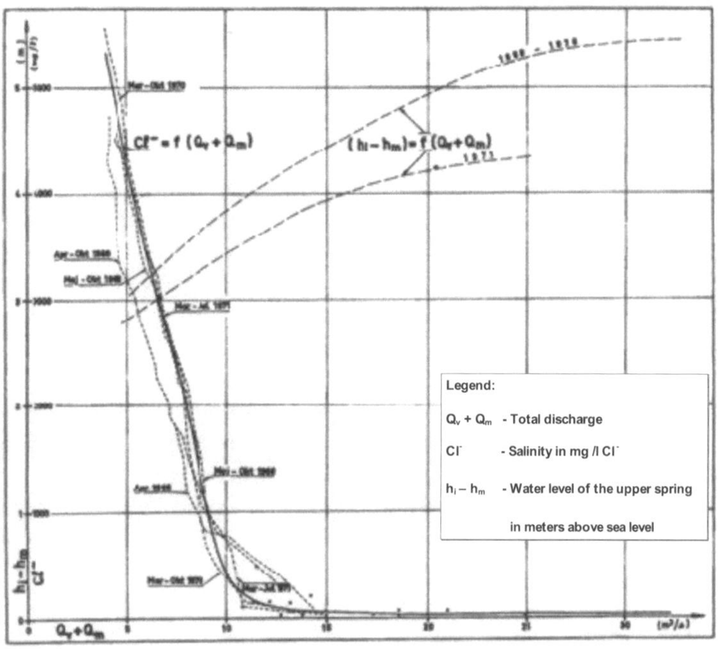

Fig. 6. Almyros Irakliou spring in Greece. Relation between discharge, water level and salinity (Ré, 1968, published in Breznik, 1973).

In rainy periods and during the melting of snow in 2500 high Psiloritis massif, spring water is fresh, with about 50 mg/l of Cl^- at discharges above 12,5 m^3/s. The relation between the discharge and progressive salinity is a curve for discharges from 12.5 to 9 m^3/s, and a straight line for discharges below 9 m^3/s. We can assume that at a discharge of 12,5 m^3/s, the deepest vein-branching starts to swallow some sea water. Shallower vein-branchings start swallowing sea water when the discharge decreases to 9 m^3/s. With smaller discharges, all the vein-branchings swallow sea water. The Almyros spring must have one upper vein, a long lower vein that divides into several channels at its end, some very deep vein-branchings at different depths, and one or several primary veins connected with different vein-branchings. The above system requires a conduit type flow pattern. Many drowned karst channels connected in many directions are a characteristic of a diffused flow pattern that cannot explain the very high level of the Almyros spring in 1977 and 1987 during tests with a spring level by a 1976 dam raised to 10 m ASL.

2.3.2 Equilibrium plane

Many karst springs are fresh during high discharges. When the discharge decreases, contamination begins. Let us suppose the discharge just before the beginning of the contamination is an equilibrium discharge Qeq. The lower vein is already filled with sea water which has not yet penetrated into the vein-branching. There are no losses of fresh water through the lower vein either. Hence

$$Q_m = 0, \quad T_m = 0, \quad \frac{v_m^2 \rho_m}{2g} = 0, \quad \rho_s = \rho_v = 1{,}0, \quad \rho_m = 1{,}028$$

$$h_i - h_r = \frac{\rho_m}{\rho_m - \rho_s} \cdot (h_i - h_m) + \frac{\rho_s}{\rho_m - \rho_s} \cdot \left(T_s - \frac{v_s^2}{2g} \right) \tag{4}$$

$$T_s = f(Q_{eq}), \quad \frac{v_s^2}{2g} = f(Q_{eq})$$

An equilibrium point in a karst vein filled with fresh water at one side, and with sea water at the other, is the point at which the pressure of sea water is equal to that of fresh water. In a karst aquifer of anisotropic permeability, an equilibrium plane is an interrupted plane that connects all the equilibrium points in the veins. It can be detected only in the veins in which it exists, and is found in very few boreholes (Breznik, 1973).

In a karst aquifer of isotropic permeability, or in an aquifer in granular soil, the sea water zone is separated from the fresh water zone by an interface, or a zone-of-mixing. The interface and the zone-of-mixing are continuous planes and can be detected in all boreholes in the area. The difference between an interface and an equilibrium plane is similar to the difference between the ground water table of a phreatic aquifer and the piezometric surface of a confined one. The first can be detected in any borehole in the area, while the second only in boreholes which have penetrated into the confined aquifer.

The elevation of the equilibrium plane changes in accordance with the elevation of the piezometric surface of fresh water. In rainy periods, the piezometric surface of fresh water is in a high position and the equilibrium plane in a low one. Fresh water flows out of the lower vein as a submarine spring and out of the upper vein as a coastal spring. In this period, the equilibrium plane is below the vein-branching and below the lower vein (Fig. 7, Phase A).

During the decline of the discharge, the piezometric surface of fresh water subsides and the equilibrium plane consequently rises (Phase B). When the equilibrium plane crosses the vein-branching, sea water from the lower vein intrudes into the vein-branching (Phase C). Brackish water fills the upper vein and flows out of the coastal spring. In the dry period, the piezometric surface in the karst massif continues to subside and the equilibrium plane in the coastal zone rises. When the equilibrium plane has risen above the vein-branching and crossed its primary vein, the outflow of fresh water through this vein-branching is blocked (Phase D). Fresh water drained from the karst massif flows through higher vein-branchings and is there contaminated. On the surface, we observe these phenomena in dry periods as a decrease in discharge of all springs, some coastal springs dry out, all submarine springs stop flowing, all submarine estavelles start swallowing sea water, and all springs deliver brackish water (Breznik, 1989; 1998).

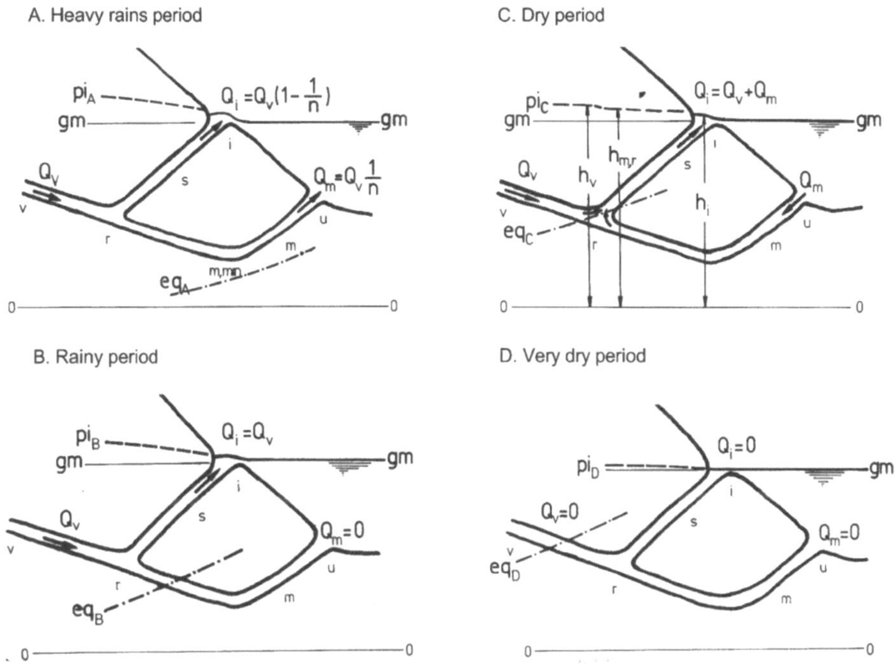

Fig. 7. Coastal spring with a siphon-like lower vein in a karst conduit flow aquifer, pi - piezometric surface, eq - equilibrium plane. Four Flow: Salinity regimes (Breznik, 1989; 1998).

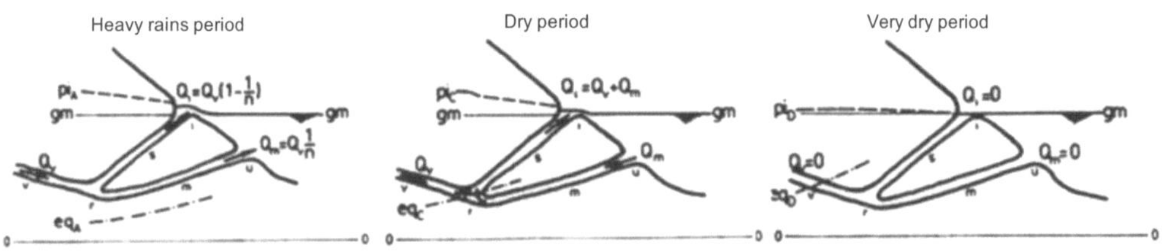

Fig. 8. Coastal spring with a rising lower vein in a karst conduit flow aquifer. Three Flow: Salinity regimes (Breznik, 1989; 1998).

There are certainly very few springs with only 3 veins. Many pairs of primary and lower veins, with the branchings of different depth, must be expected for a single spring. This explains the progressive contamination observed in the Almyros spring (Fig. 6).

2.3.3 The case of the 'Sea mills' on Kefalonia Island

Sea water flowing into swallow holes of the 'Sea mills' on Kefalonia Island in Greece was marked with 100 kg of uranine in 1963. The total inflow of sea water was about 1.7 m^3/s and

the brackish outflow of the Sami springs about 20 m^3/s. The brackish water of these springs contained from 10 to 12% sea water. The tracer reappeared after 16 to 23 days.
The distance between the mills and the Sami springs being 15 km, the mean velocity of tracer movement was 1 cm/s (Fig. 9). Glanz (1965) was of the opinion that this sea water intrusion of 1.7 m^3/s could not be ascribed to a Venturi-tube suction effect, because such an arrangement, together with the simultaneous dissolution of calceros Venturi pumps, would be too complex to resist under natural conditions. He explained the inflow of sea water by a natural injector effect of fresh karst water submersed in sea water, working on the principle of a water jet pump. A physical model supported that explanation (Fig. 10, Glanz, 1965; Fig. 11, Maurin, 1982). The phenomenon of the 'Sea mills' could be more easily explained by mixing in a deep vein-branching on the different densities principle (Breznik, 1998; Fig. 12, Breznik & Steinman, 2008).

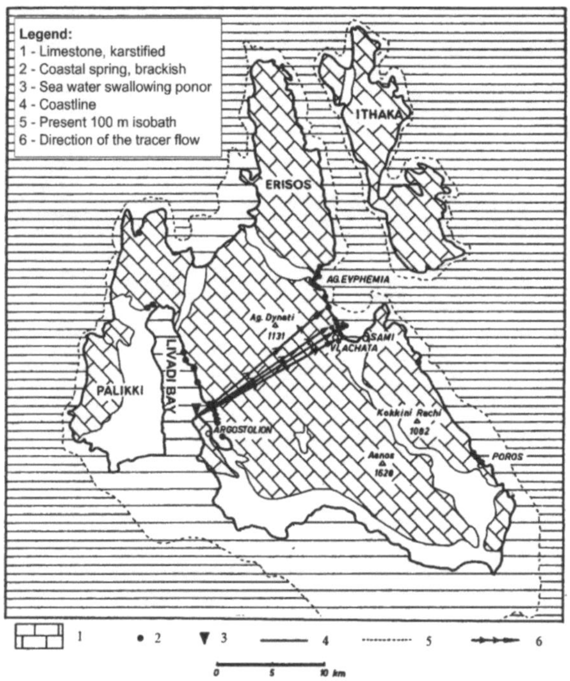

Fig. 9. Kefalonia Island in Greece. Ponor of sea water near Argostolion (Glanz, 1965).

Many terrain observations talk against the hydro-dynamic method of infusion. A flow of sea water has been noted in summer in the lower part of the source channel of the Port Miou springs in France. Even the construction of underground desalination barriers in the spring was not entirely successful, since source water still contained around 4000 mg/l CI. Further development of the springs was then abandoned.
A similar phenomenon, i.e. a powerful inflow of sea water into the estavelle on the floor of the gulf of Bali, and thus into the coastal karst aquifer, was observed in the summer of 1991 and the outflow in October 1970 and May 1983 on the island of Crete (Fig. 18).
All the above springs have fresh water in rainy periods and cannot have been contaminated by a hydrodynamic effect, which should be greatest at high discharges. These springs are contaminated in the vein-branchings in conduit flow aquifers because of different densities

of sea and fresh water. The vein-branchings can be deep in coastal karst aquifers, which is partially a result of the 120 m lower sea level in the Pleistocene period (Fig. 13) and partially the effect of inflow of sea water into deep syphons (Figs. 22 and 23).

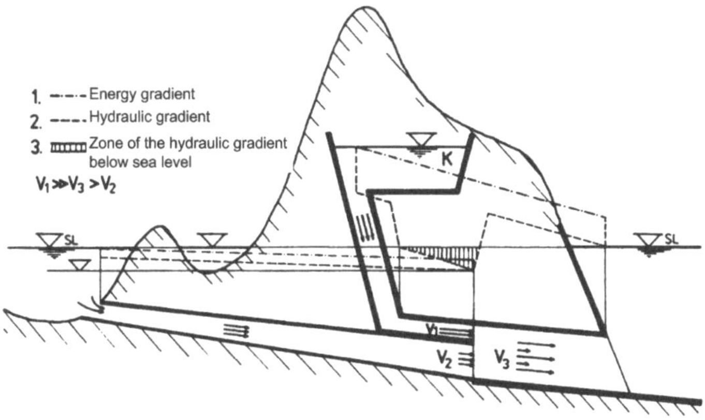

Fig. 10. "Sea water mill" on Kefalonia Island in Greece (Glanz, 1965).

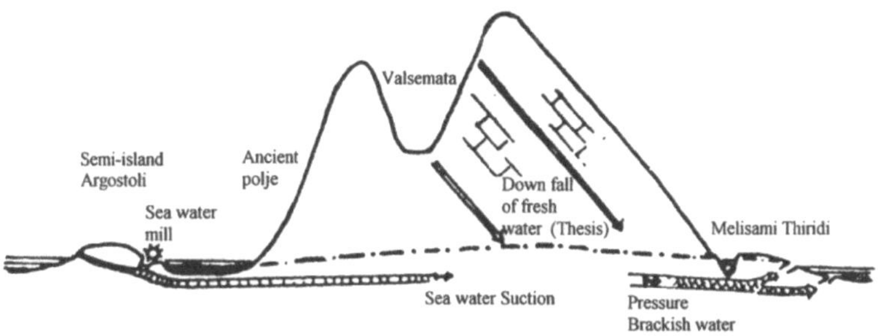

Fig. 11. Scheme of hydro-geological water circulation in the karstic rock mass of Kefalonia Island (Maurin, 1982).

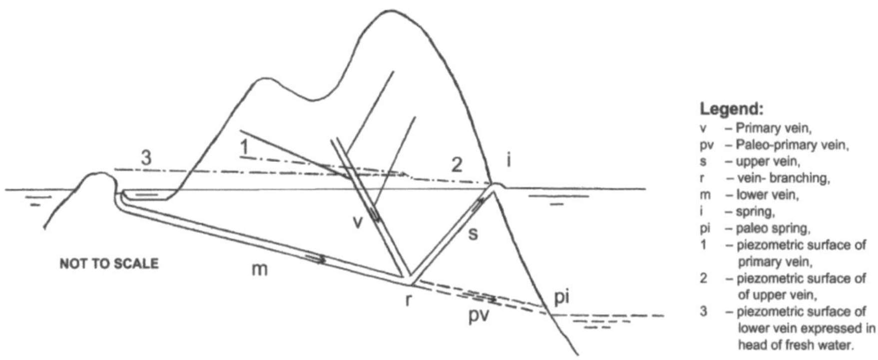

Fig. 12. "Sea water mill" on Kefalonia Island in Greece, explained by the different densities principle (Breznik & Steinman, 2008).

The consequence of deep vein-branching is high salt water contamination in dry periods, of springs such as Almyros Irakliou 10 m above sea level, Kournas lake 17 m and

Annavaloussa 12 m, on the island of Crete and Pantan 4 m in Croatia. The hydrodynamic method of salination cannot explain such high level of contaminated springs in dry periods, and this is the major indirect evidence of the method of contamination by the different densities of fresh and sea water, in deep vein-branchings in coastal conduit flow karst aquifers of anisotropic permeability (Breznik, 1973; 1998).

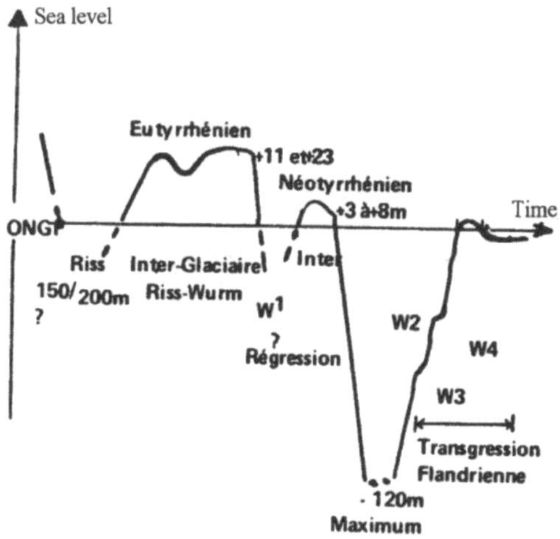

Fig. 13. Fluctuations of sea level in the Pleistocene period (Bonifay et al., 1974).

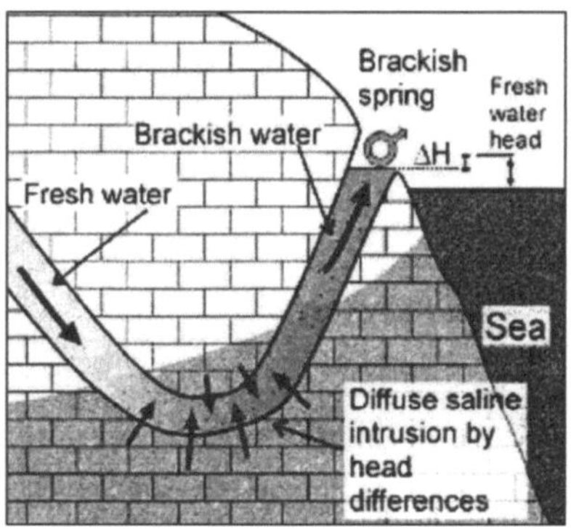

Fig. 14. Brackish spring contaminated by a diffuse saline intrusion in the main conduit (Arfib, 2005).

In the Pleistocene period was the sea level till 120 m deeper and the Livadi bay on Kefalonia Island, with a 30 m deep sea now, a karstic polje with a ponor and an outflow to the Sami coast. This paleo-primary vein is a lower vein now, and with a vein-branching at 100 m BSL. We can easily explain the inflow of sea water into the "mill ponor" by the different densities of fresh and sea water. The 100 m deep column of sea water in the vein-branching provides a 2,5 m rise of the piezometric head, expressed by the head of fresh water. Out of this are 0,8 m losses of the head, used for the flow of sea water through the 15 km long lower vein and 1,7 m is the denivelation of the sea level below the mill (Breznik & Steinman, 2008).

In the Arfib's one vein karst system, it would be difficult to explain a change from the outflow to the inflow of the estavelles during 1 to 2 days in the Jurjevo bay (Kuščer, 1950).

Also a 20 to 30 m high rise of the Almyros spring level would be an utopian proposal (Breznik, 1984 b; 1989).

2.3.4 Investigations

In such natural conditions, explorations are difficult and have to be scheduled by stages, with the less costly in the beginning and expensive ones undertaken only for important and promising springs. The first investigations include observations and recording of all natural phenomena over the course of a year. Discharge, salinity and duration of flow of coastal springs, intensity and duration of flow of submarine springs, and meteorological processes in the recharge area, must be observed. The second stage of explorations comprises measurements of spring level, discharge and salinity. Relations among discharge: salinity: level must be known for different periods of the year (Arandjelović, 1976).

The third stage of explorations must already be oriented towards the most promising development technique. For the 'isolation of fresh water from sea water method', geophysical measurements, drilling and salinity detection in boreholes have to ascertain the position and depth of the lower vein, which must be sealed. Tracer tests are also sometimes useful. Reliable values of salinity in boreholes are obtained only by 'bottom point' measurements, which means in boreholes with their casing open at the bottom only. It is not possible to distinguish between fresh and sea water zones in boreholes with a casing perforated throughout its length, because of the flow and mixing of water inside the casing.

For the 'interception of fresh water inside a karst massif method', trace lines, or lineaments, of karst and tectonic phenomena on the surface indicate the direction of underground water circulation. Fractured zones identified from aerial photos or satellite pictures may be chosen as favorable locations for drainage galleries or wells.

For the 'rise spring-level method', several discharge: salinity: level curves are needed, specifically for the original natural conditions, and for the conditions created by an artificial rise of 3 to several meters in spring level. Observations of the piezometric level in the recharge area of the spring during different periods of the year could give clues for ascertaining the depth of the vein-branching. A detailed program of exploration must be prepared and later modified according to the results obtained. For each proposed method of exploration, the program must be determined in advance: what is the aim and what are the possible results. The idea of 'let us explore and later see what that brings' consumes time and funds (Breznik, 1998).

The fourth stage of investigation starts with a partial and provisional construction of the proposed development structures. Their adaptability has to be foreseen in the program because final success can never be assured in advance by the first three stages only.

3. Methods of the desalination

3.1 Hydro-geological interception method

The idea of this method is to capture fresh water inside the karst massif, outside the present sea water influence. Successful are the deep drilled wells Klariči in Slovenia with 0,25 m^3/s, the system Zvir II with the dug and the drilled deep wells with 0,6 m^3/s in Croatia, Gonies and Krousonas deep wells in Greece and elsewhere. Unsuccessful are due to overexploitation and salination the deep wells in Tyllisos and Keri areas in Greece, many drilled deep wells in the Murgia, Salento and Taranto coastal aquifers in the southern Italy and in other places (Pavlin, 1990).

3.2 Hydrotechnical isolation method

The idea is to prevent the inflow of sea water into the karst massif by a diaphragm wall or a grout curtain. Positive examples are the grout curtains for the Žrnovica and Bačvice springs in Croatia and negative the grout curtain for the Tabačina spring in Montenegro, the diaphragm wall for the Malavra spring in Greece and the others (Breznik, 1998; Nonveiller, 1989; Vlahović, 1983).

3.3 Hydrotechnical rise-spring-level method

The idea of this method is to prevent the inflow of sea water by raising the spring level by a dam, a grout curtain or a diaphragm wall. This method could succeed only in an aquifer with a siphon shaped lower vein. But also there a too high rise could induce the losses of fresh water through the lower vein into sea, shown on Fig. 7 from the period C by a human rise of the spring level to the period B, and not by a to high rise, to the period A (Breznik, 1971, 1973; Breznik & Steinman, 2008).

3.4 Adaptable method of the reduced pumping discharge in dry periods

In many coastal aquifers a reduced pumping discharge could maintain a low salinity of water in the dry periods. A reduction of pumping prevents the salination of the drainage galleries Kovča - Zaton, Roman wells (Bakar, Trogir), Šipan in Croatia (Biondič, 2005), and Klariči in Slovenia (Breznik, 1998; Breznik & Steinman, 2008; and elsewhere) during the dry periods.

4. Cases

4.1 Investigation of the underground storage of the Rižana spring

The municipal water supply of the town of Koper in Slovenia, from the Rižana karst spring, was constructed 50 years ago. Observations over a period of about 100 years are available. The mean discharge of the spring is about 4 m^3/s, the maximum is about 80 m^3/s, the annual minimum 0.27 m^3/s and the absolute minimum 0.11 m^3/s, measured in 1921 after two very dry years. There is now a suggestion to increase the exploitation over the annual minimum, by over pumping the underground storage in dry periods. The recession curves of the spring discharge have indicated an underground storage of 7 million m^3 which has regulated the outflow at the present spring level. Several analyses using the stable isotopes deuterium and ^{18}O have ascertained a retention time of 4 months between winter precipitation and the outflow of this water the following summer. This pointed to a large underground storage of over 30 million m^3.

The Vaclusian type spring on a small limestone plateau is separated by a 1 km wide layer of marls from the main recharge area, an imbricate structure of about 180 km2 in limestone. The spring is at an elevation of 70 m above MSL and the recharge area is at 500 to 1100 m above MSL. Exploitation of this additional underground storage is possible only by lowering the present water table to a lower level in dry periods. In spite of this very clear principle, it is not clear where and how to capture water from this storage that is, at present, dead storage of the spring. In the event of this storage being a constantly rising water table, of it being uniform over a large area and connected with the spring by a deep siphon, it would be possible, during the dry period, to pump water out of the existing deep wells in the spring area.

A more probable supposition is that there are various dead storages at different levels, subdivided by either less permeable rock steps or contracted conduits. In such a case, each storage has to be captured by separate drainage tube wells, or a drainage gallery with a water gate and an access tunnel. We have noted that the locations of the underground storage are not known in advance (Fig. 15). They must be searched for individually, and individual deep wells drilled in each case for test pumping in dry periods. Eight new deep wells were built, with a total capacity of 0.92 m^3/s, at distances of 0.5 to 1.5 km upstream from the spring. In the dry summer of 1993, pumping began from the underground storage at a rate of 0.39 m^3/s from the wells, and a further 0.16 m^3/s flowed from the spring. Because of the rapid lowering of the water surface in the wells, by about 10 m, pumping had to be reduced after a few days to 0.15 m^3/s. The volume of the underground storage in the Podračje imbricate structure was assessed at about a quarter of a million m^3 of water. About 10 million m^3 of accumulated underground water are needed to increase the minimum flows of Rižana for some months. Since speedy solutions were required, we proposed that further exploration of this source be abandoned for 25 years (Breznik, 1993; 1998).

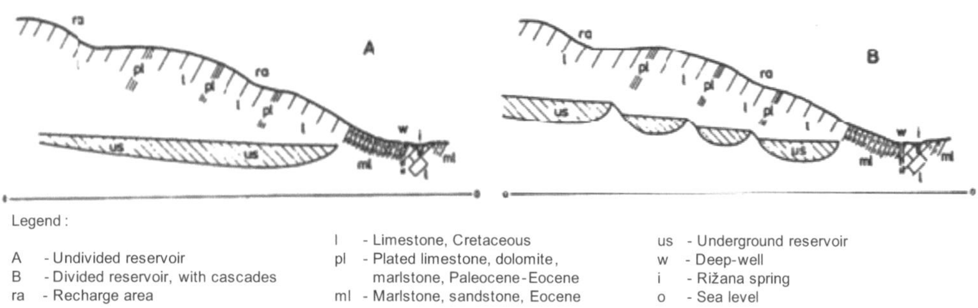

Fig. 15. Rižana springs in Slovenia. Possible type and situation of karstic underground reservoirs (Breznik, 1990; 1998).

4.2 Kras coastal aquifer in northern Adriatic

The surface of the Kras aquifer is 730 km^2, with a larger part in Slovenia and a coastal outflow area in Italy. Recharged is by the infiltration of the precipitations of 1100 mm/year, by the Notranjska Reka with 5 m^3/s (lowest measured: 0,18 m^3/s only) inflow in the ponors near Škocjan, by 1 m^3/s inflow in the ponors of the Vipava River and by the Isonzo (Soča) river ponors buried beneath gravel along the Doberdo karst. The mean outflow is 35 m^3/s in rainy periods and 10 m^3/s in dry periods - the majority of it in the Timavo springs area, a small part of 0,2 m^3/s out of small coastal springs and an important out of the estavelles in the Duino sea, which swallow sea water in the dry periods (Petrič, 2005; Steinman, 2007; Breznik, 2006).

In the Klariči pumping station 3 deep wells, VB-4 from 16 m ASL to 54 m BSL, were drilled near the B4 borehole and fresh water captured in a karst conduit of 1,3 m at a depth about 25 m below sea level (Krivic, 1982). The Klariči station with a limited pumping discharge of 250 l/s supplies drinking water to the Kras region with 25.000 people since 1986 and till 130 l/s of water is flown in dry periods to the Slovene coastal region with 40.000 people since 1994. Coastal region has also the Rižana karst spring with only 200 l/s m the dry periods (Petrič, 2005). Fresh water demand of the coastal region is 500 l/s in summer (Bidovec, 1965; Krivic and Drobne, 1980; Steinman et al., 2004; 2006; 2007). The level of the main Timavo spring

was risen about 1,5 m ASL with a small weir and was the main water source for the Trieste town until 30 years ago.
The lowest static water level in Klariči was 2,5 m ASL. Karst ground water pumped in Klariči is recharged in the eastern part from the Kras during rainy periods and from Isonzo river ground water mainly during dry periods. Isonzo river ground water infiltrates in paleo-ponors buried beneath gravel in the NW part of the Doberdo karst plateau. Three large estavelles about 1 km from the Duino (Devin) coast, are the main outflow of this water in rainy periods, and swallow sea water in dry periods. They were important springs of a part of the Isonzo river in a dry upper Adriatic land during the Lower Pleistocene (Šegota, 1968). The estavelles indicate a geologic border on sea bottom between the karstified carbonate rocks of the Cretaceous against impervious Flysh sediments of the Eocene (Breznik & Steinman, 2008).

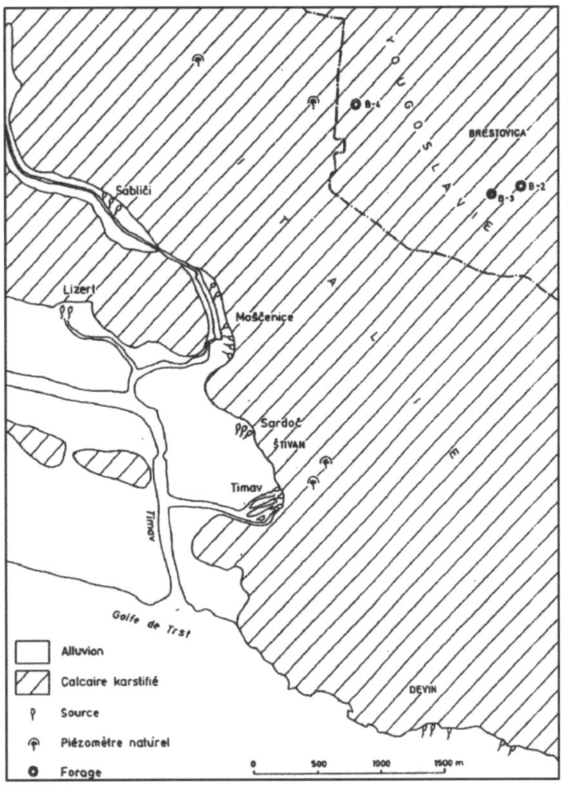

Fig. 16. Outflow area of Kras aquifer (Krivic, 1982).

How sustainable are good quantity and quality of this water? We explain these in Klariči station at 4 km from the coast and water pumped out a karst conduit at 25 m below sea level, by a human raised level of the Timavo spring, by a large outflow of Timavo springs of 10 m^3/s in dry periods, by a shallow karstifications due to an impermeable Flysh barrier on sea bottom of 20 m BSL at the outflow of the estavelles, by a low permeability of the karst rock mass of a conduit and by a chance of an absence of the human pollution until now.
Land reclamations in the Timavo springs area, f. e. for new storage places of the shipyard, with drainage ditches and a destruction of the small weir would lower water level there and also in the Klariči station and could induce a salination of water.

Mercury ore was excavated in the Idrija mine, which is closed now, for 500 years. Idrijca and Soča rivers transport 1500 kg of Hg to sea every year, washed out of old mine's deposits. In water pumped out of VB4 well in Klariči 1,2 ng/l of Mercury was measured. This very small quantity of Hg is not harmful for the health, but could accumulate in the cave deposits (Doctor et al., 2000). A third threat could be human pollution as there are no protection areas.

We propose to pump 2-3 m^3/s of the Rižana river to a 100 m higher large Dragonja storage reservoir in the rainy period and to flow it out in the dry periods to Rižana with an existing drinking water treatment plant and towards the pipeline from the Gradole spring in Croatia, where the contract of an obligatory water supply has expired in 2005. The elevation of water in Dragonja reservoir at about 170 m ASL enables always a gravitational outflow to the supply system without electricity. In the dry periods, could be the lower layer of euthropic water of Dragonja reservoir released, in cascades enriched with oxygen and used for the irrigation of Dragonja and Sečovlje plains, what is a 25 years old proposal. The mixing of reservoir water, with the aim to prevent eutrophication, will not be necessary. Two inflows of water, from the northern Rižana and southern Dragonja directions, into the coastal water supply system, will increase its safe operation.

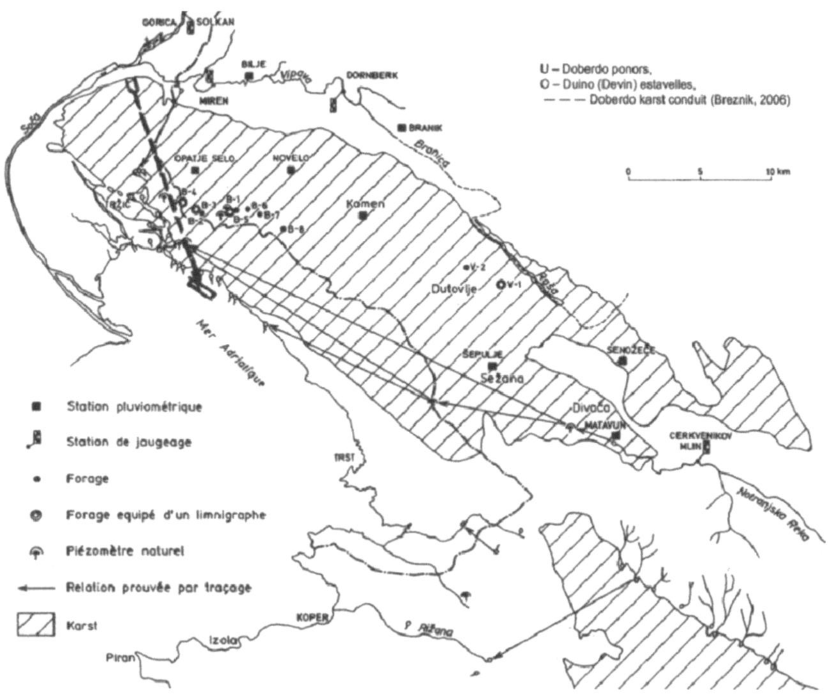

Fig. 17. Doberdo karst conduit (Breznik & Steinman, 2008), Kras aquifer (Krivic, 1982).

The proposed, and designed new Padež or Suhadolca storage reservoirs with a 35 km long new pipe-line, will be "on the other side of the hill" and will need a 250 m high pumping, all around the year. The electrical energy is however not always at disposal as few-days-lasting black outs in 2006 and 2007 in California, New York and Western Europe have demonstrated. These reservoirs would need also an artificial mixing of water with the aim to prevent eutrophication (Breznik & Steinman, 2008).

4.3 Bali bay coastal aquifer in Greece

The catchment area of the Bali bay aquifer is the Talea Ori karstic massif with 50 km2. In the wet period fresh water flows out of coastal springs and estavelles in the Bali bay (Economopoulos, 1983). We explored Syphona spring No 3 at 12 m BSL with an outflow of some m^3/s of brackish water with around 10.000 mg/l CI in the late summer of 1970. Divers led by P. Economopoulos were hampered by poor visibility in the outflow funnel of the spring, what indicates the mixing of fresh and sea water there (Breznik, 1998). French divers studied estavelles-ponors in the Bali bay in late summer 1991. At a depth of 12 m there were a number of smaller and 3 larger estavelles which swallow more than 1 m^3/s of sea water (Barbier et al., 1992). Position of estavelles in the wet period and the outflow of the Syphona spring in the dry period indicate the direction of the Talea Ori underground flow. Geophysical methods: Map of electrical potentials "mise à la masse" with one electrode in the Syphona spring and the Very low frequency (VLF) of radio waves (Mueller et al., 1986) should determine the position of the main Talea Ori water conduit to the Syphona spring (Breznik, 1998; Breznik & Steinman, 2008).

The next exploration works are: capture ground water of the conduit with drilled interception wells; excavate a ditch with a regulation valve from the interception wells to sea at 2 m ASL; plug with a concrete and a grouting the conduit between the interception wells and the Syphona spring; construct a one row grout curtain, distance of the boreholes 2-3 m, till 60 m and 120 m BSL; rise the water level by the regulation valve to 5 m and 10 m ASL; find out possible water losses along the coast by registration of new springs and by the temperature logging of coastal water from a helicopter; construct additional grout curtains if necessary.

We evaluate there is a 70% probability to desalinate 0,5 to 1,0 m^3/s of the Talea Ori ground water that could be used in the Rhetimnon city area.

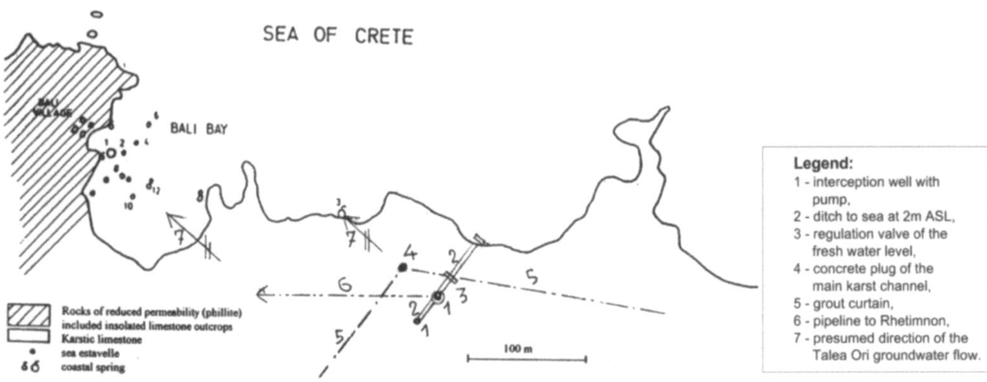

Fig. 18. Structures for the desalination of the Bali bay springs and estavelles (Breznik & Steinman, 2008). Bali bay springs and estavelles (Economopoulos, 1983).

4.4 Anavalos-Kiveri coastal springs in Greece

Tripoli and other poljes are drained by coastal and submarine springs along the NE coast of the Peloponnesus peninsula in Greece (Gospodarič et al., 1986). A 180 m long semicircular dam was founded on calcareous breccia at 10 m BSL and with a top at 4 m ASL in 1968. We visited the place in the spring 1969. The sluices of the dam were open and a river of greenish

color flowed out, that clearly differed from the blue sea. We observed a typical circle of ground water flowing out of an estavelle at a distance about 0,5 km.
Prof. Ständer from Germany, who proposed the isolation of springs, answered in a letter that a major development was achieved by the isolation of the springs area with the dam, thereupon the salinity decreased to 200-300 mg/l CI. A second phase of the development was completed with a rise of the pool level to 3 m ASL at a discharge of 12 m^3/s and the inflow of sea water stopped (Ständer, 1971). A photo shows a present outflow of ground water outside the Kiveri dam (Lambrakis, 2005). The average springs discharge is 6 m^3/s. During the irrigation periods 1955-1990 the ground water quality worsened due to the over pumping and the sea water intrusion (Monopolis et al., 1997; Tiniakos et al., 2005).
A short analysis of the available data indicates that the isolation of the Kiveri springs against sea water inflow is not completed. A dam founded on much karstified breccia without a consolidation of the limestone mass and without a grout curtain, is not a completed structure. Prof. Ständer estimated the depth of the karstification at 90 m BSL. We suppose this depth to be either 30 m deeper of the sea bottom at the estavelle observed in 1969, or 30 m deeper than a 120 m BSL deep sea level in the Pleistocene if the Argos bay is deep enough (Breznik, 1998; Tiniakos et al., 2005).

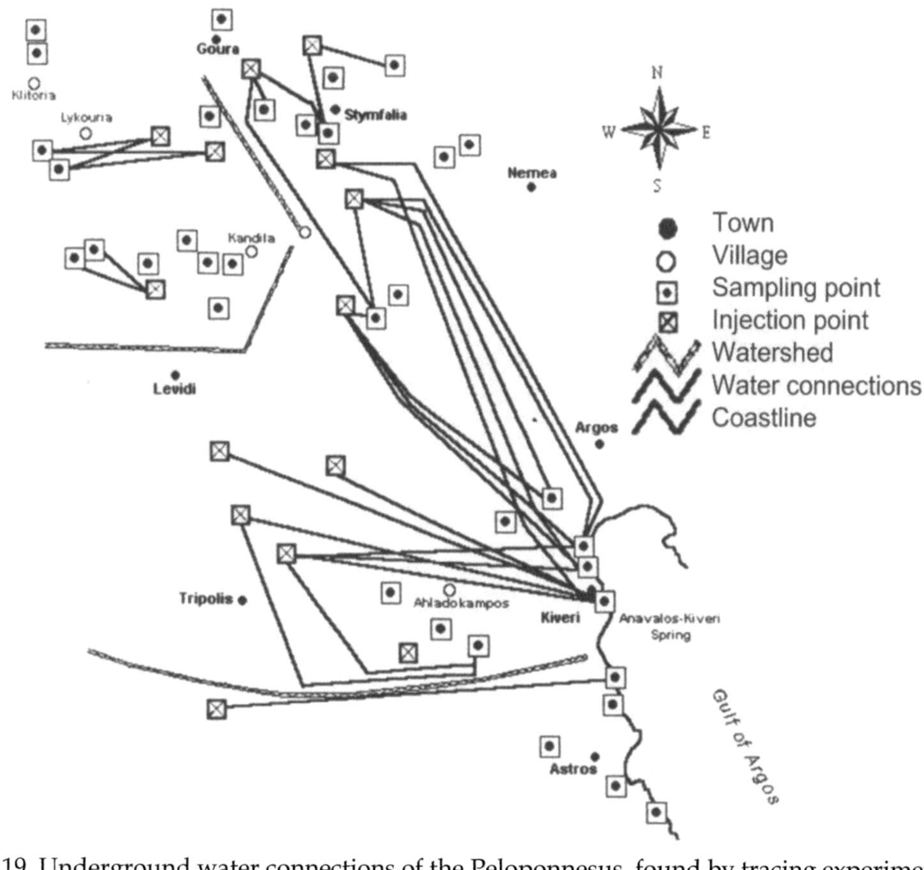

Fig. 19. Underground water connections of the Peloponnesus, found by tracing experiments (Gospodarič & Leibungut, 1986).

We propose to prevent the sea water inflow by a grout curtain. The exploratory works should be done in phases:

- First phase: boreholes drilled at a distance of 4 m along the crest of the dam and grouted to a depth of 65 m BSL, then consolidation grouting of the karstified breccia below the dam from 10 m to 35 m BSL.
- Second phase: boreholes, in between boreholes of the first phase, drilled and grouted till 130 m BSL.
- Third phase: grout curtain below the road extended for 100 and later 200 m on both sides of the dam.
- Forth phase: additional grout curtains behind the smaller springs to the north if needed and a higher rise of the pool's level. In all this exploratory phases a testing with a rise-spring-level to be made, the results analyzed and the next phases adjusted. A 4 m rise enables the existing dam (Breznik, 1998).

Fig. 20. Kivery dam. Desalination structures proposed (photo Breznik, 1969; Breznik, 1998; Breznik & Steinman, 2008).

This is a general proposal for exploration activities and they should be adapted to the partial results obtained. A final success with a 70% probability is to desalinate spring's water to 50 mg/l Cl^- in dry periods, and a 90% probability in wet periods.

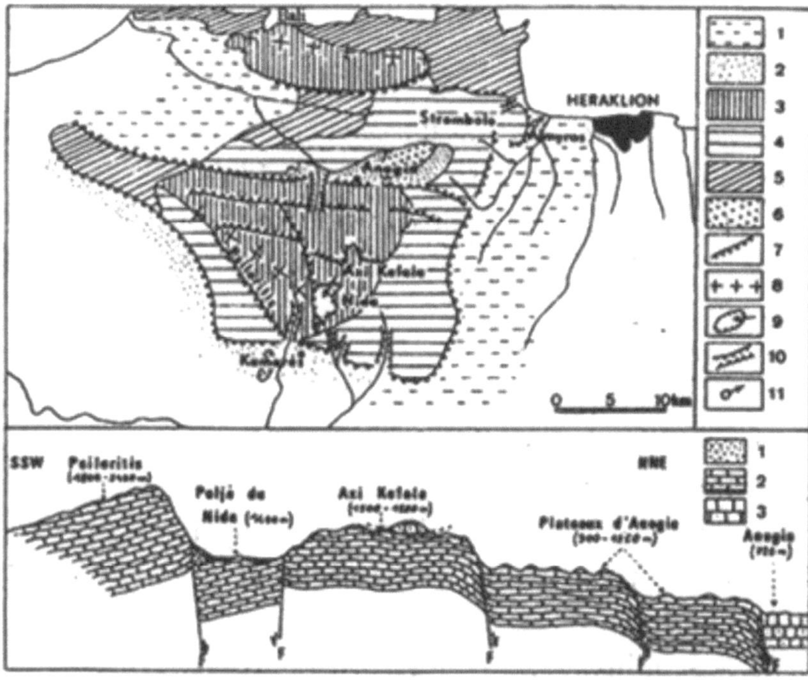

Fig. 21. Morpho-structural sketch of the Psiloritis massif and the Psiloritis-Anogia geological section with Almyros Irakliou and Bali springs (Bonnefont, 1972).

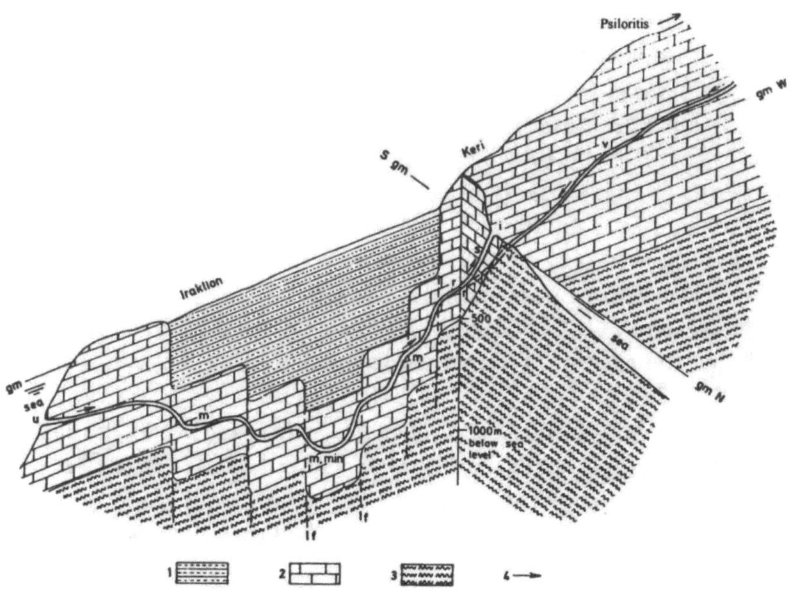

Fig. 22. Almyros Irakliou spring in Greece. Schematic geological block diagram with the supposed disposition of the spring veins in the conduit-flow karstic aquifer (Breznik, 1978).

4.5 Almyros Irakliou brackish spring in Greece

The characteristic of this spring, at 1 km from the sea coast, with many primary veins, of a 300 km[2] karstic recharge area and with very deep vein-branchings at differed depths, is a very slow increase of the salinity during a decrease of the discharge (Ré, 1968; Fig. 6; Breznik, 1971; 1973; 1998; Breznik & Steinman, 2008; Monopolis et al., 2005; Panagopoulos, 2005; Soulios, 1989).

All the veins are in Mesozoic limestone and the lower veins below the Festos-Irakliou graben filled with Neogene deposits. This spring was investigated by the United Nations - UNDP-FAO and Greek Government in the years 1967-1972. Between the spring and sea coast 15 deep boreholes, with a mean depth of 240 m, were drilled, with the aim to find, and to seal with a grout curtain, a conduit with sea water inflow. The result of investigation was that this conduit is not between the nearest sea and the spring, but is below Neogene deposits at about 800 m BSL and about 14 km long.

Almyros spring has a mean discharge of 8 m^3/s, a temperature of water 16° C and had a tritium content of 45 T. U. of samples taken in August 1969, analyzed at IAEA in Vienna.

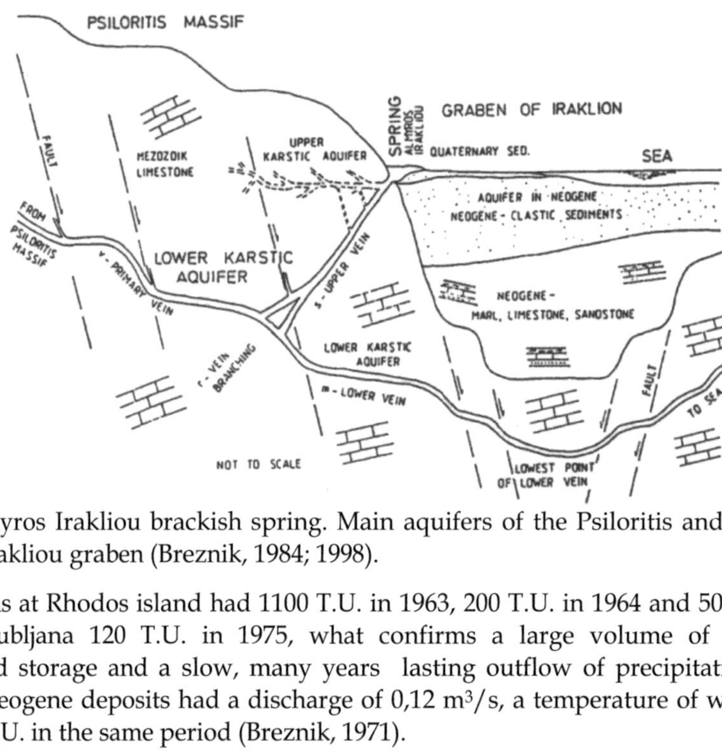

Fig. 23. Almyros Irakliou brackish spring. Main aquifers of the Psiloritis and Keri massifs and of the Irakliou graben (Breznik, 1984; 1998).

Precipitations at Rhodos island had 1100 T.U. in 1963, 200 T.U. in 1964 and 50 T.U. in 1969, while in Ljubljana 120 T.U. in 1975, what confirms a large volume of the Psiloritis underground storage and a slow, many years lasting outflow of precipitations. A week aquifer in Neogene deposits had a discharge of 0,12 m^3/s, a temperature of water 19-20° C and 19-13 T. U. in the same period (Breznik, 1971).

We proposed to explore the desalination of the Almyros spring by the isolation, rise-spring-level and interception methods. A 10 m rise of spring level was proposed (Breznik, 1971). A new dam was constructed (1976) and spring level was raised at 10 m ASL for some month in 1977 and 1987. Spring water remained brackish (negative result) but the discharge diminished only slightly and no estavelles appeared in the Irakliou Sea (positive results). We concluded that a higher elevation of the level should be determined by a winter test with a larger discharge of water (Breznik, 1978), proposed a 20 to 30 m rise (Breznik, 1984) and calculated a 28,76 m, however with uncertain data (Breznik, 1989).

4.6 Rise-spring-level method of the development

This method requires a siphon shaped lower vein. Almyros has indeed a very deep lower vein, formed by a gradual subsidence of the Festos-Irakliou graben. We propose a 25 to 35 m ASL spring level with a construction of an underground dam.

The exploration phases with testing are: First phase: excavate a shaft, of 8 m diameter, with reinforced concrete lining, from surface to 5 m ASL with 2 table valves; drill interception wells into the main karst conduit till 30 m BSL; excavate 2 bottom outlets, of 5 m^2 with reinforced concrete lining, with valves at the outlets; seal the conduit with a concrete plug and a consolidation grouting. Raise the spring level, register the salinity and locate water losses. Second phase: construct a grout curtain of one row boreholes at a 4 m distance, till a depth of 80 m BSL. Raise the spring level and register the results. Third and other phases: condense and extain the grout curtain, with boreholes at 2 m distance, till a depth of 120 m BSL, construct a small dam around an expected overflow karst spring in the Keri ravine. Raise the spring level to 25-35 m ASL. When the salinity is below 50 mg/l CI and losses of water are small the exploration phases are completed. We expect, with an 80% probability, a safe yield of fresh water of about 2 m^3/s in dry periods and a 90% probability of fresh water in wet periods.

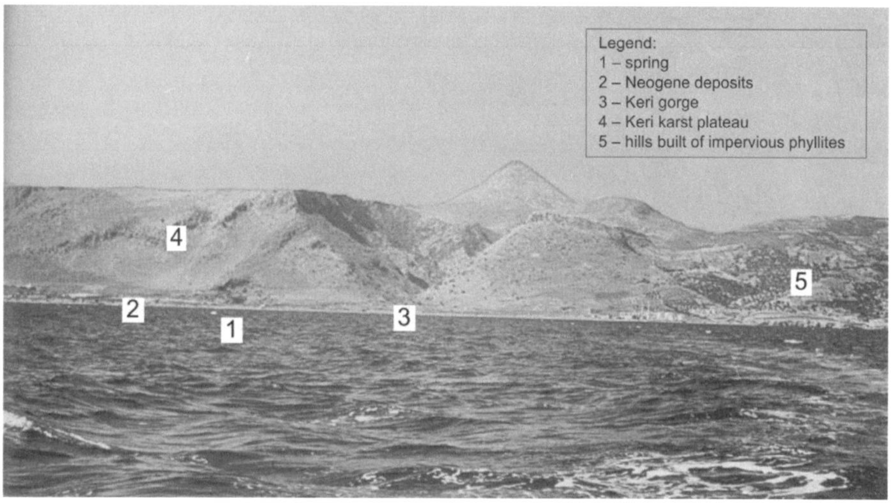

Fig. 24. View of the Almyros Irakliou spring area (photo Breznik, 1970).

The main object of the underground dam is the concrete plug with grouting (No. 5 in Fig. 25). The ground water flow through the place of the proposed plug could be blocked, by diverting the flow through the extended bottom outlet (No. 3), bellow the 1976 dam. This is achieved by raising the water level of the spring pool to about 6 m ASL, by regulating the water valves of the 1976 dam. Down flow of the fresh concrete into the steep main karst conduit (No. 1), could be prevented by a downside planking of the plug.

The first possibility is a planking of closely drilled boreholes of 60 m depth, of 30 cm diameter with casing filled with concrete. Two to three additional boreholes with pipes will enable pumping of concrete from the surface.

The second possibility is a new access shaft of 3 m diameter and 45 m depth at a 10 m distance upstream from the plug. Divers have to construct steel planking and install pipes for pumping concrete from a surface to a depth of about 25 m below water level. Constructors could propose other solutions.

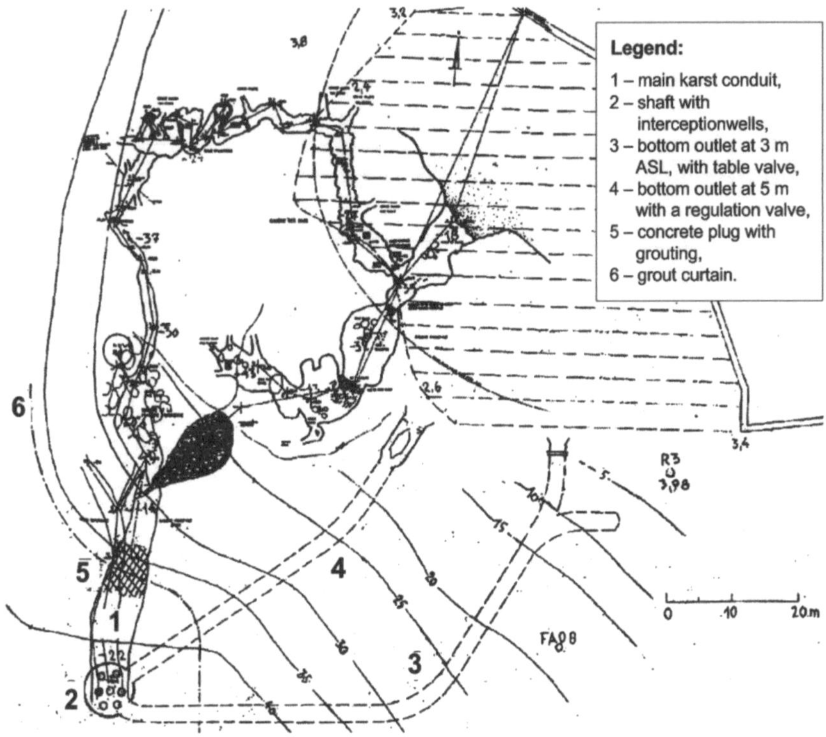

Fig. 25. Structures for the exploration phases of the desalination of the Almyros brackish spring by the rise of the spring level with an underground dam (Breznik & Steinman, 2008). Map of karst conduit (Barbier et al., 1992).

The structures for the final exploitation of fresh water are: Spillway for the high water outflow, small hydropower station for the regulation of the required water level for the desalination and for the production of the electricity, fresh water pipeline to Iraklion. Hydropower stations regulate the level of water in the irrigation canal along the Durance River in France. Rise-spring-level method could desalinate also ground water of the Keri, Tylissos and other low altitude areas. The existing Iraklion power station could be cooled by hyperbolic cooling towers used in Europe, or by sea water pumped out of deeper layers, used for cooling nuclear power stations in Japan. This unique desalination plant will be very attractive for tourists in Crete and should be economically exploited by the presentation of the underground hydrogeology, of the desalination structures and the restoration of the old scenery with mills (Breznik & Steinman, 2008).

4.7 Interception method of development

In the years 1968-1971 were the piezometric levels of fresh water in the Gonies area in boreholes at about 44 m ASL at the distance of 8 km from the Almyros spring and in the Koubedes-Tylissos area in the boreholes at about 29 m ASL, at the distances 3-4 km (Breznik, 1971; 1973; 1990; 1998; Breznik & Steinman, 2008). The municipal DAYAH Company had drilled 40 deep wells in the areas Keri, Tylissos, Gonies and Krousonas at 13 km since 1987. In the year 2000 fresh water was pumped out of 17 deep wells (Arfib, 2000). A normal consequence of a pumping many years out of coastal aquifers is a decline of the piezometric surface and the inflow of sea water. In Tylissos area the piezometric surface declined from about 29 m in seventies to about 15 m in 1997 and induced a salination of wells (Monopolis et al., 1997; 2005).

The important question is now; could water of wells in the Gonies and Krousonas areas remain fresh? Ground water of these areas flows to Almyros spring through a very deep vein-branching at 800-1000 m BSL, where is a fresh water outflow and a sea water inflow which depends upon the piezometric surface of fresh water. An expected overpumping of the Gonies-Krousonas wells, due to the loss of the Keri-Tylissos salinated wells, will lower the fresh water piezometric surface and induce a sea water inflow. Only moderate pumping yields could prevent the salination of this water. An over pumping of Malia wells will have similar consequences (Breznik & Steinman, 2008).

5. Conclusions and recommendations

Many desalination methods were proposed and many scientific papers published but, the important Greek springs: Bali, Kiveri and Almyros Irakliou, are still brackish after 30 years of attempts. In a karst underground are so many unknown data, needed for a mathematical ground water model, that the results are not reliable. We propose to achieve the desalination with physical-field tests: by the isolation method for the Bali and Kiveri springs with grout curtains and by the rise-spring-level method for the Almyros Irakliou spring with an underground dam. We estimate there are 70-80 % probabilities of the success in dry periods and 95% for Bali and Kiveri and 90% for Almyros Irakliou springs in wet periods.

The Dragonja river storage reservoir with 20 - 30 millions m^3 of fresh water pumped out of Rižana river, could solve water shortage of SW Slovenia. The Intergovernmental Panel on Climate Change (IPCC) warns about still smaller precipitations and higher temperatures in the Southern Europe in the future. So, the supply of fresh water will become increasingly important.

The proposed methods are intended to intercept fresh water before it is mixed with salt water, allowing the accumulation of water in wet seasons. No doubt, proposed solutions require greater initial investment, but have low operating costs. Besides, water supply is not exposed to the imported high-technology and is not high energy demanding.

We reserve author's rights for the proposed desalination methods and structures (Breznik, 1998; Breznik & Steinman, 2008).

6. Glossary

Admissible salinity: The quantity of salts in drinking or irrigation water which is harmless to people, animals or vegetation. Slovene and other countries' standards for drinking water is 250 mg/l of Cl^-. In dry areas drinking water with 500 mg/l of Cl^- is considered as harmless. Many villages in the Mediterranean area use water with more than 500 mg/l of Cl^-, the Bedouins of the Sahara up to 2000 mg/l of Cl^-.

Aerated zone: Zone above ground water surface in which karstic pores are filled partially with air and partially with water.

Aquifer: A formation, group of formations or part of a formation that bears water which is not bound chemically or physically to the rock.

Brackish spring: General term which means a spring with brackish water but also the vein and a place of such a spring.

Brackish water zone (also called zone-of-mixing or transition zone): Part of aquifer saturated with brackish water.

Doline: A depression that has a funnel-shaped hollow with a diameter of 10 to 100 m, formed by the dissolving of limestone or dolomite. It is an international term. The local term is vrtača, the English term being sinkhole.

Drowned zone: Zone below ground water surface in which karstic pores are saturated with water.

Equilibrium plane: Nominal plane in a karst of anisotropic permeability connecting those points of veins and branchings where the water pressures from fresh water and sea water sides are equal.

Fresh water zone: Part of aquifer saturated with fresh water.

Interface: The surface bordering the fresh water and sea water in an aquifer of isotropic permeability. This border could be sharply defined but is usually a transition zone.

Karst aquifer of anisotropic permeability: Karst region with isolated karstified zones with unkarstified blocks between them. Ground water moves along veins or conduits, which means along well-karstified zones. The aquifer is highly permeable in the direction of veins, but poorly permeable or impermeable in the transverse direction. Ground water movement is similar to the movement of water in a system of pipes which are not densely disposed, known as 'conduit type circulation'.

Karst aquifer of isotropic permeability: Karst region with many solution fissures, small channels which are all well connected in all directions. Movement of water is possible in all directions and is analogous to the ground water movement in granular sediments, known as 'diffused type circulation'.

Karstic ground water, karst aquifer: Water which fills karstic pores and veins in the drowned zone and is not bound physically or chemically to the rock.

Polje: An international term that refers to the largest karst hollow with a flat floored linear depression. In its typical form it has a steep side and steep circumference.

Ponor: This is the largest entry in the base or in the side of the polje in which water flows, an international term. Schwinde (Ger), swallow hole (Eng) and perte (Fr).

Salinity: Quantity of salts in water. In this paper expressed as content of chlorine ions (Cl^-) in mg/l. The salinity of the Mediterranean Sea is about 21000 mg/l of Cl^-.

Sea estavelle: A submarine spring with fresh water which ceases to flow in each dry season and starts to swallow sea water.

Sea ponor: Hole in the sea bottom or seashore which swallows sea water.

Sea water zone: Part of aquifer saturated with sea water.

Storage coefficient of the karst is the volume of water which a karstic aquifer releases from storage or takes into storage.

Submarine spring: A spring with either fresh or brackish water rising from the sea bottom.

Uvala: A coalescence of two or three dolines, an international term.

Vein or conduit: General term for a zone which is highly permeable in the flow direction and poorly permeable or impervious in the transverse direction. Ground water moves through veins in a karst of anisotropic permeability. The form of the vein is undefined; it could be a solution channel, a permeable fissured zone, a system of small connected cavities, etc.

Vein-branching or branching: The place where the primary vein branches off into a lower vein, connected with the sea, and an upper vein, leading to the spring.

7. Acknowledgments

We thank the Governments of Slovenia, Croatia, Montenegro, Greece and Turkey for the presentation of their unpublished investigation results.

8. References

Arandjelović, D. (1976). *Geofizika na karstu,* Geozavod Beograd, Beograd.

Arfib, B.; de Marsily, G. & Ganoulis, J. (2000). Pollution by seawater intrusion into a karst system: New research in the case of the Almyros source (Heraklio, Crete, Greece). *Acta carsologica,* Vol. 29, No. 1, pp. 15-31, ISSN 0583-6050.

Arfib, B. & Bonacci, O. (2005). Particular aspects of discharge in coastal karstic aquifers, In: *Groundwater management of coastal karstic aquifers (EUR 21366 En, COST Action 621, Final Report, Part II),* Tulipano, L.; Fidelibus, M.D. & Panagopoulos, A. (Ed.), pp. 87–104, Office for Off. Publ. of the EC, ISBN 92-898-0002-X, Luxembourg.

Barbier, J. L.; Therond, R. & Paloc, H. (1992). Source d'Almyros d'Heraklion, Synthese des etudes depuis 1988, Rapport general, *Unpublished report of GERSAR to the Greek Ministry of Agriculture,* pp. 65, Paris.

Bidovec, F. (1965). The hydrosystem of karstic springs in the Timavo basin, *UNESCO - IHD Symposium on Hydrology of Fractured Rocks,* pp. 263-274, Dubrovnik, October 1965, UNESCO, Paris.

Biondić, B. (1988). Tapping and protection of underground water in the Adriatic region related to the new conception of the structure of Dinarides. In: *Proceedings of the IAH 21st Congress, Guilin, China,* Daoxian, Y., (Ed.), pp. 187-193, Guilin, China.

Biondić, B.; Gunay, G.; Marinos, P.; Panagopoulus, A.; Potié, L.; Sappa, G. & Stefanon, A. (2005). Protection and remediation practices, In: *Groundwater management of coastal karstic aquifers (EUR 21366 En, COST Action 621, Final Report, Part II),* Tulipano, L.; Fidelibus, M.D. & Panagopoulos, A. (Ed.), pp. 231-241, Office for Official Publications of the European Community, ISBN 92-898-0002-X, Luxembourg.

Boegan, E. (1906). Le sorgenti d'Aurisina. Rassegna bimestrale della Societa Alpine delle Giulie. Trieste.

Bonacci, O. (1987). *Karst Hydrology with Special Reference to the Dinaric Karst,* Springer Verlag, Berlin.

Bonnefont, J.C. (1972). La Crete, etude morphologique, Ph. D. Thesis, Paris, University of Paris IV.

Bonifay, E. (1974) in Potie, L. & Ricour, J. 1974. Etude et capitage de résurgences d éau douce sous - marines. - Ressources en Eau, pp. 5-27, Paris.

Borelli, M. & Pavlin, B. (1965). On the underground water leakage from the storages in Karst region. Karst storages Buško blato, Peruća and Kruščica, *UNESCO - IHD Symposium on Hydrology of Fractured Rocks,* pp. 32-63, Dubrovnik, October 1965, UNESCO, Paris.

Bosi et al. (1996). Eustatic curve related to Quaternary and curve of max regressions. - EUR 21366 En, COST Action 621, p. 165, Brussels.

Božičević, S. (1976). Losses of Grout in Caves, *1st Yugoslavian Symposium of Soil Conservation,* pp. 17-23, Zagreb, 1976.

Breznik, M. (1961). Akumulacija na Cerkniškem in Planinskem polju (Water accumulation in the Cerknica and Planina polje.), *Geologija,* Vol. 7, pp. 119-149, ISSN 0016-7789.

Breznik, M. (1971). Geology and Hydrogeology of the Almyros spring area, *Unpublished note No. 103 of UNDP – FAO, presented to Greek Gov.*, pp. 1-94, Iraklion.

Breznik, M. (1973). Nastanek zaslanjenih kraških izvirov in njihova sanacija (The Origin of Brackish Karstic Springs and their Development; Summary of the Doctor of Geology Thesis, 1972), *Geologija*, Vol. 16, pp. 83-186, ISSN 0016-7789.

Breznik, M. (1976). Mogućnost saniranja zaslajenih kraških izvora sa injektiranjem (Possibility of Development of Brackish Karstic Springs by Grouting), *1st Yugoslavian Symposium of Soil Conservation*, pp. 293-296, Zagreb, 1976.

Breznik, M. (1977a). Exploration and Development of Coastal and Submarine Brackish Springs in Turkey, *Unpublished report of UN-OTC presented to Turkish Government*, pp. 1-35, Ankara.

Breznik, M. (1977b). Test to Raise the Water-level of the Almyros Irakliou Spring - Evaluation of Results of the 1977 Summer Test, *Unpublished report presented to Greek Gov.*, pp. 1-11, Iraklion.

Breznik, M. (1978). Mechanism and Development of the Brackish Spring Almyros Irakliou. *Ann. Geol. Des Pays Hell.*, pp. 29-46, Athens.

Breznik, M. (1979). The Reliability of and Damage to Underground Dams and Other Cut off Structures in Karstic Regions, *13th International Congress on Large Dams*, pp. 57-79, ISBN, New Delhi, Paris.

Breznik, M. (1981). Groundwater hydrology and hydraulics – Flow in aquifiers. *International course: Water resources engineering*, pp. 1-54, Beograd.

Breznik, M. (1984a). Bansagar project - Protection of Kuteshwar limestone deposit, *Unpublished report presented to the Central Water Commission of Government of India*, pp. 1-15, New Delhi.

Breznik, M. (1984b). Development of the Almyros Irakliou brackish spring, *Unpublished report presented to the Greek Government*, pp. 1-34, Ljubljana.

Breznik, M. (1984c). Exploration of the Bali brackish spring, *Unpublished report presented to the Greek Government*, pp. 1-5, Malia.

Breznik, M. (1985a). Exploration, design and construction of cut offs in karstic regions, *15th International Congress on Large Dams*, pp. 1111-1129, Lausanne, 1985, Paris.

Breznik, M. (1985b). Neka iskustva o bušenju bunara u krasu (Some experiences on drilling wells in the karst, in Serbo-Croatian), *Conference 'Voda i krš'*, pp. 159-164, Mostar.

Breznik, M. (1985c). Perspektiva in problematika izkoriščanja podzemnih voda. (Perspective and problematics of ground water exploitation, in Slovene), *Acta hydrotechnica*, Vol. 3, Special edition [1], pp. 1-45, ISSN 0352-3551.

Breznik, M. (1989). Explorations, mechanism and development of brackish karst spring Almyros toy Irakleioy, *Unpublished report presented to the Greek Ministries of Agriculture and Research and Technology and Universities of Athens and Crete*, pp. 1- 59, Ljubljana.

Breznik, M. (1990a). Development of Brackish Karstic Spring Almyros in Greece. *Geologija*, Vol. 31/32, pp. 555-576, ISSN 0016-7789.

Breznik, M. (1993). Evaluation of exploration results and development possibilities by underground dam and other methods of karst spring Almyros Irakliou, *Unpublished report presented to the Greek Gov.*, Eastern Crete Development Organisation, pp. 1-90, Ljubljana.

Breznik, M. (1996). *Vodni viri za Obalo in Kras Slovenije (Water sources of Coastal Region and Slovenian Karst, in Slovene)*, UL-FGG, Ljubljana.

Breznik, M. (1998). *Storage reservoirs and deep wells in karst regions*, A.A. Balkema, ISBN 9789054106883, Rotterdam/Brookfield.

Breznik, M. (2005). *Vodni viri Slovenske Istre (Water sources of Slovene Istria, in Slovene)*, Unpublished report to the Ministry for Environment and Physical Planning, pp. 1-35, Ljubljana.

Breznik, M. & Steinman, F. (2008). Hydromechanism and desalination of coastal karst aquifers: Theory and cases. *Acta carsologica*, Vol. 37, No. 2-3, pp. 197-209, ISSN 0583-6050.

Doctor, D.; Lojen, S. & Horvat, M. (2000). A stable isotope investigations of the classical Karst aquifer: Evaluating karst groundwater components for water quality preservation. *Acta carsologica* Vol. 29, No. 1, pp. 79-92.

Economopoulos, P. (1983 and 1989). Situation of estavelles and coastal spring in Bali bay, scale 1:25000, *Unpublished map, photo, written communications to Breznik.*

Ghyben-Badon, W. (1888-1889). Nota in verband met de voorgenomen putboring nabij Amsterdam. - Tijdschrift van het Koninklijk Institut van Ingenieurs, Hague. 27 pp.

Gjurašin, K. (1943). Prilog hidrografiji primorskog krša. *Tehnički vjesnik*, Vol. 60, No. 1-2, pp. 1-17, Zagreb.

Glanz, T. (1965). Das Phaenomen der Meermuhlen von Argostolion. *Steirische Beitrage zur Hydrogeologie*, Vol. 17, pp. 113-127, Graz.

Gospodaric, R. & Leibungut, C. (1986). Evaluation and Interpretation of the Tracing data, In: *Hydrogeology of the Eastern Pele-ponissos, Greece*, Morphia, A. & Zojer, A. (Ed.), pp. 100-110, 5th Int. Symposium on Underground Water Tracing, New York.

Herzberg, A. (1901). Die Wasserversorgung einiger Nordseebader. *Zeitung für Gasbeleuchtung und Wasserversorgung*, Vol. 44, pp. 815-819.

Kajfež-Bogataj, L. (2006). Glede spreminjanja podnebja Slovenija ni izjema (in Slovene). *Proteus*, Vol. 69, No. 2, pp. 54-61.

Kajfež-Bogataj, L. (2007). Poročili, ki spreminjata poglede na podnebne spremembe. Pentagon and Stern reports (in Slovene). *Proteus*, Vol. 69, No. 7, pp. 294-302.

Kohout, F. A. (1966). Submarine Springs. A neglected Phenomenon of Coastal Hydrology, *Symposium on Hydrology and Water Resources*, ISBN, Ankara, 1966.

Komatina, M. (1975). Hidrogeološke oblike slivova centralnodinarskog karsta, *Rasprave Zavoda za geološka i geofizička istraživanja*, Vol. 16, pp. 1-105, Beograd.

Krivic, P. (1982). Naravna nihanja gladine podtalnice kraskega vodonosnika (Variations naturelles de niveau piezometrique d' un aquifføre karstique.), *Geologija*, Vol. 25, No. 1, pp. 129-150, ISSN 0016-7789.

Krivic, P. & F. Drobne. (1980). Hidrogeološke raziskave Tržasko - Komenskega krasa, (Hydrogeologische Untersuchungen des Karstes von Trieste und Komen), *6th Yugoslavian Symposium Of Hydrogeology and Engineering Geology*, pp. 233-239, Portorož, 1980, Yugoslavia.

Kuščer, I. (1950). Kraški izviri ob morski obali (Karst Sources at the Sea Coast, in Slovene), *Dissertationes Academia Scientarium et Artium Slovenica*, Vol. 3, No. 1, pp. 97-147.

Kuščer, I. & Kuščer, D. (1962). Observations of Brackish Karst Sources and Swallowholes in the Yougoslav Coast. - Mem. de l' Ass. Intern. Des Hydrogeologues. V. Reunion d'Athens.

Lambrakis, N. (2005). Kiveri coastal springs, Greece. - EUR 21366 En, COST Action 621, p. 262, Brussels.

Maurin, V. & Zoetl, J. (1965). Salt Water Encroachment in the Low Altitude Karst Water Horizons of the Islands of Kephallinia. In: *Dubrovnik Symposium, AISH-UNESCO*, pp. 423-438.

Monopolis, D.; Lambrakis, N. & Perleros, B. (2005). The brackish karstic spring of Almiros of Heraklion, In: *Groundwater management of coastal karstic aquifers (EUR 21366 En,*

COST Action 621, Final Report, Part II), Tulipano, L.; Fidelibus, M.D. & Panagopoulos, A. (Ed.), pp. 321-328, Office for Off. Publ. of the Europ. Commun., ISBN 92-898-0002-X, Luxembourg.

Mueller, I. & Schotterer, U. (1986). Electromagnetic VLF Resistivity Prospection in the Region of Tripolis and the Coastal area of Argos - Astros, *5th International Symposium of Underground Water Tracing*, pp. 425–440, Athens, 1986.

Nonveiller, E. (1989). *Grouting Theory and Practice*, Elsevier, ISBN 0444874003, Amsterdam.

Panagopoulos, A. (2005). Almiros spring, Heraklion, Greece. - EUR 21366 En, COST Action 621, p. 258, Brussels.

Pavlin, B. (1990). Yugoslav littoral belt karst springs used for water supply, *6th International Congress of International Association for Engineering Geology*, pp. 1387-1394, ISBN 90 5410 130 3, Amsterdam, 1990, Balkema, Rotterdam.

Petrič, M. (2005). Klariči, Brestovica, Slovenia. - EUR 21366 En, COST Action 621, p. 264, Brussels.

Potie, L. & Ricour, J. (1974). Etudes et captage de résurgences d'eau douce sous-marines (Studies and Capture of Resurgences of Submarine Freshwater), *Ressources en eau*, pp. 5-26.

Ré, R. & Breznik, M. (1968). The problems of the Almyros spring of Iraklion, *Unpublished note No. 2 of UNDP – FAO, presented to Greek Government*, pp. 1-114, Iraklion.

Rošker, J. (2007). Živimo v najtoplejši dobi v zadnjih 150 tisoč letih (in Slovene), *Polet, supplement of Delo*, 05.04.2007, pp. 10-13, ISSN 1580-8041.

Soulios, G. (1987). System karstique aquifére d'Almyros, Iraklion, un case interessant de functionement de systéme littorale, *Bull. Centre d'Hydrogeol.*, Vol. 7, pp. 169-191.

Steinman, F.; Gosar, L. & Banovec, P. (2004). Preparation of Sensitivity Maps of the Slovenian Coast, Final report, *CAMP Slovenia*, pp. 1-24 & maps.

Steinman, F.; Kozelj, D. & Banovec, P. (2006). Regional Water Supply in State Development Plan (in Slovene). - Slovenian Water Protection Association - Proceedings I, 24-32, Portorož.

Steinman, F. et al., (2007). Expert group assessment of the project: Drinking Water Supply of the Slovenian Istra and Karst hinterland, *Unpublished Report to the Ministry of Environment and Physical Planning (in Slovene)*, pp. 1-64, Ljubljana.

Ständer, W. (1971): written answer to Breznik.

Šegota, A. (1986). Surface of the Adriatic Sea throughout the last glacial period with part of the present Adriatic Sea emerged as dry land 25.000 years ago. - EUR 21366 En, COST Action 621, p. 116, Brussels.

Thérond, R. (1973). *Recherche sur l'étancheité des lacs des barrages en pays karstiques*, Eyrolles, ISBN 978-2-212-01550-8, Paris.

Tiniakos, L.; Tavitian, J. & Livaniou-Tianikou, A. (2005). The Anavalos-Kiveri coastal spring (Argolis, E. Peloponnesus, Greece): Hydrogeology and drought-water quality relation, In: *Groundwater management of coastal karstic aquifers (EUR 21366 En, COST Action 621, Final Report, Part II)*, Tulipano, L.; Fidelibus, M.D. & Panagopoulos, A. (Ed.), pp. 312-320, Office for Off. Publ. of the Europ. Commun., ISBN 92-898-0002-X, Luxembourg.

Vlahović, V. (1981). *Kraška akumulacija Slano (Karstic accumulation in Slano)*, Črnogorska akademija nauka i umjetnosti (The Montenegrian Academy of Sciences and Arts, in Serbian), Podgorica.

4

Corrosion Control in the Desalination Industry

Michael Schorr[1], Benjamín Valdez[1], Juan Ocampo[2] and Amir Eliezer[3]
[1]*Instituto de Ingeniería, Departamento de Materiales, Minerales y Corrosión, Universidad Autónoma de Baja California, México, Blvd. Benito Juárez S/N, CP. 21900, Mexicali, Baja California,*
[2]*Facultad de Ingeniería Mexicali, Universidad Autónoma de Baja California, México, Blvd. Benito Juárez S/N, CP. 21900, Mexicali, Baja California,*
[3]*Sami Shamoon College of Engineering Corrosion Research Center, Ber Sheva ,*
[1,2]*México*
[3]*Israel*

1. Introduction

The environment quality, worldwide water scarcity and clean energy have been established today as central disciplines in modern science, engineering and technology. They are already being linked to the crucial, actual problems of climate change, global warming and greenhouse-gas emissions, all interrelated phenomena (Valdez & Schorr, 2010). Innovative desalination technology of saline water (SW) contributes to alleviate these problems by producing abundant fresh water, from SW, mainly seawater and brackish water (I.D.E Technologies, 2004; Charash et al, 1991).

Desalination plants (DPs) have a high level of corrosion risk since they handle and process aggressive SW under severe operating conditions which include filtration, heat exchange, distillation, evaporation, agitation and circulation and high flow velocities, often turbulent. These SW: sea, brackish and brines cause localized corrosion such as pitting, crevice, galvanic and stress corrosion. In addition, biological fouling and mineral scaling are frequent nuisances that alter the equipment surface performance and induce corrosion (Malik, 2000).

Desalination has been practiced since ancient times for providing drinking water on seafaring ships, using solar or fuel heat for distillation. Aristotle, the Greek philosophic scientist (384-322 B.C.) mentions desalting seawater with solar energy. Natural gas was used as fuel in ancient China to evaporate water from salt brine.

Moses, the prophet, wandering through the Sinai Desert found water that the people could not drink because it was bitter. Then, Moses threw a piece of wood into the water and the water become sweet (Exodus 15:22-25). Perhaps, these are some antique practices on water treatment, their details lost through the eons…

Public water supplies are recorded in the Bible: Genesis 26, II Kings 20:20, John 4; community wells and water works where built by the Hebrews, Egyptians, Mesopotamians, Phoenicians, Persians, Greeks and Romans, including canals, aqueducts, reservoirs, distribution pipes and flood-control facilities.

Desalination is not a new technology; in 1790 the US Government received a proposal to install a distillation method to convert salt water to fresh water. In 1952 the US Congress passed "The Saline Water Act" to provide federal support for desalination, as a new mean for supplementing long deficient supplies of fresh water.
Use of suitable corrosion resistance alloys (CRAs): titanium, stainless steels (SS), Ni-base alloys, Cu -Ni alloys and Al-Mg alloys is the most direct means of preventing corrosion. Corrosion resistance is the main property to be considered in the choice of materials for plant equipment. Today about 15,000 DPs operate worldwide with an estimated total production capacity of 32 million m^3/day, in the Mediterranean Sea coast countries, the Middle East, South America deserts, the Canary and Caribbean islands; all places with limited water supplies. In the USA there are 1,500 desalination facilities constituting a 30b USD business which is expected to double in capacity by 2016. A limited number of DPs have been built on the California coast, primarily because of desalination cost is generally higher than the cost of other water supply alternatives, however, as drought conditions occurs desalination large projects are being planned, e.g., the Carlsbad project.
The world largest plant in Saudi Arabia produces 1 Mm^3/day. An advanced seawater DP was installed in 2005 in Ashkelon, Israel with a capacity of 100 Mm^3/year. It is operated by IDE Technologies; uses Seawater Reverse Osmosis (SWRO) technology and employs state-of-the-art means for recovery of energy from independent, combined cycle electricity station, with a capacity of 80 MW (Kronenberg, 2004).

1.1 Water resources

There is an almost unfathomable amount of water on earth: about 1.4 billion km^3 (330 million cubic miles), (Barlow and Clark, 2002). Of this total, less than 3% is fresh water (about 35,000,000 km^3), much of which (about 24,000,000 km^3) is inaccessible due to the fact that it is frozen in ice caps and glaciers (Figure 1). It is estimated that just 0.77% (about 11,000,000 km^3) of all the earth's water is held as groundwater, surface water (in lakes, swamps, rivers, etc.) and in plants and the atmosphere (Shiklomanov, 1993).

2. The desalination industry

Due to an increased population growth and the expectation of high living standards, the demand for water and electricity in the desertic and arid regions of the world is soaring. Placing DPs combined with power generating units allows the heat extracted from the process to evaporate seawater. Desalination is the most viable solution to the 21th century´s shortage of fresh water for human consumption and irrigation obtained from sources of SW (Kowitt, 2009).
The desalination industry is in the middle of an expansion and modernization program designed to construct more efficient and larger DP's, that will reduce production costs. The maintenance of its infrastructure assets requires a robust understanding of the integration between global climate change and the materials engineering-structure-climate-interaction, induced by variations in humidity, temperature, solar radiation, drought and pluvial precipitation mainly during extreme events (Valdez & Schorr, 2010). Recently the Institute of Materials, Minerals and Mining (IOM3), London has published a special issue of its journal which brings together papers examining climate change induced corrosion (Valdez & Schorr, editors 2010; Roberge, 2010).

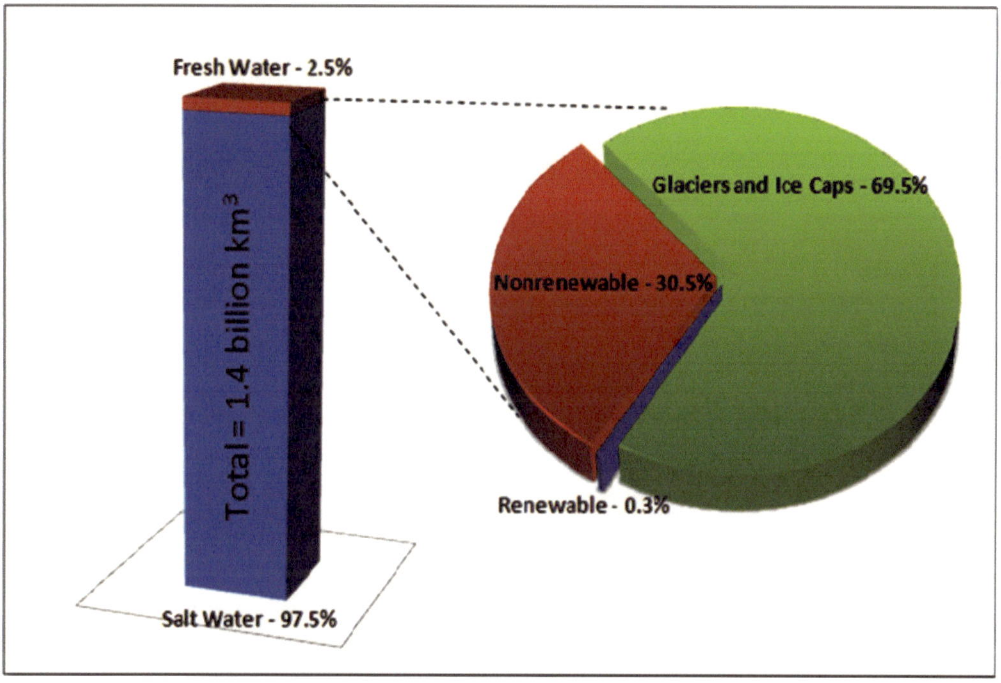

Fig. 1. Distribution of world water

This industry is based on the principles and practices of water chemistry, chemical engineering and efficient energy management. The most widely utilized technologies are thermal and membrane but solar "green" energy is applied, without relying on fossil fuels: oil, gas and coal.

Actual innovation desalination technology is less energy-consuming and more environmentally friendly. DP's require varied engineering materials, structures, installations, equipment and machinery that should function with industrial efficiency and labor safety to assure its economic performance.

The economic and social relevance of the desalination industry is evident by the activities of the diverse international and national professional associations, R&D institutions and industrial enterprises involved in all the aspects of desalination science, engineering and technology (Table 1). It includes authorities from government, industry, and academia that address progress of vital importance for the national and global prosperity.

Lately, the threat of bioterrorism, have pushed desalination to the forefront of efforts to preserve the available supply of water.

3. Desalination processes and plants

DP's have a high level of corrosion risk since they handle and process aggressive SW under severe operating conditions which include filtration, heat exchange, distillation, evaporation, agitation, and high flow velocities, often turbulent(Dillon, 1994). There is no universal desalination process; every type of SW requires a process adapted to its physicochemical characteristic and performance. The DPs are feed with seawater, containing 35 g/l of total dissolved solids (TDS) or brackish water with TDS in the range 2 to 5 g/l,

Association, Organization, Enterprise		Website
International Desalination Association	IDA	www.idadesal.org
European Desalination Society	EDS	www.edsoc.com
American Desalting Association	ADA	www.webrom.com/ada
Asociación Española de Desalación y Reutilización	AEDyR	www.aedyr.com
Middle East Desalination Research Center	MEDRC	www.medrc.org
Office of Water Research and Technology, USA	OWRT	www.ntis.org
US Bureau of Reclamation, DOI	USBR	www.usbr.gov/water/desaltin
International Atomic Energy Agency	IAEA	
Kuwait Institute of Scientific Research	KISR	
UNESCO-IHE Institute for Water Education, The Netherlands		
Encyclopedia of Desalination and Water Resources		www.desware.net
Israeli Desalination Society	IDS	www.ids.org.il
Bureau for Use of Saline Water, SCT, Mexico		
Commissariat a l'Energie Atomic	CEA	www.cea.fr
Desalination and Water Treatment Lab.		www.bgm.ai.il
Veolia Water Solutions and Technologies		www.veoliawater.com
GE Water and Process Technologies		www.gewater.com
Siemens Water Technologies		
Japan desalination Association JWWA	JWWA	www.k4.dion.ne.jp/~jda-hp21/
IDE Technologies		www.ide-tech.com
Belsa Agua, Spain		
Doosen, South Korea		
Fisia Italimpianti		
Dow Water Solutions		www.dowwater.com
California Coastal Commission		www.coastal.ca.gov
MeKorot National Water Co.		Mekorot.com.il
Hydranautics , CA, USA		www.membranes.com
Desalination Directory Online		www.desline.com
Caribbean Desalination Association		www.caribda.com
Indian Desalination Association InDA	InDA	Magnum.base.esnet

Table 1. Associations, organizations and enterprises dealing with desalination science, engineering and technology

taken from briny wells or wells infiltrated by seawater. Many DP's are located in desertic/arid regions with a harsh climate and limited rainfall. About one-fifth of the DP's operate in the Middle East with Saudi Arabia, producing half of the world desalted water.
Desalination is used to produce potable water from water sources containing dissolved salts, such as sea water or brackish water. Natural waters are classified according to their total dissolved solids (TDS) values (Table 2):

Type of water	TDS value (mg/l)
Sweet water	0 - 1000
Brackish waters	1000 - 5000
Moderately saline water	5000 - 10 000
Severely saline water	10 000 - 30 000
Seawater	More than 30 000

Table 2. Natural water classification

The main application of desalination techniques is the production of fresh water on ships, islands, and in the coastal regions of arid Middle East countries. The water obtained is so pure that consumers do not like the lack of taste; therefore small quantities of salt water are then added to improve the flavour. Two main desalination technologies (membrane and thermal) are implemented worldwide (Table 3).

Process	Characteristics
a. Membrane Reverse Osmosis RO	Pressure is applied to the SW forcing it trough a semipermeable plastic membrane that separates brine from water.
b. Thermal Multistage Flash Distillation MFD	SW is heated and the pressure is lowered in several stages so the water flashes into steam, to be cooled.
Multieffect Distillation MED	Low pressure steam, 60 °C is handled in a train of evaporative-condensers (effects) with heat rejection condensers.
Mechanical Vapour Compression MVC	Distillation is effected by an electrically driven centrifugal compressor mounted on the evaporator

Table 3. Membrane and thermal desalination processes

- Membrane separation process e.g. Reverse Osmosis (RO). Under high pressure the water molecules contained in seawater pass through a selective membrane while the dissolved salt ions do not pass through the membrane. (Figure 2), Some RO membranes are made from high-grade polymeric PVDF material to form a hollow fiber membrane that is very durable and less prone to breakage. Special membrane incorporate a brush layer of hydrophilic polymer chain anchored to the membrane surface which blocks foulants such as bacteria, mineral crystal and protein from adhering to the membrane. It also resists mineral scaling by preventing its nucleating on the surface. Others membranes are made from polymers specially developed and manufactured to serve in DP's. RO desalting devices are used also to upgrade the quality of industrial water.

- The thermal processes are based on improved distillation, evaporation and condensation technologies with the aim to save energy and to obtain fresh water with a low level of TDS and at a low cost operation. In general, the thermal processes are more expensive than RO but distillation produces pure water independent of the quality and salinity of the feed water. The cost of desalting brackish ground water is generally less than the cost of desalting seawater due to a lower TDS content. The average expense for desalting brackish water is 0.50 USD/m^3 and for seawater 1.5 USD/m^3.

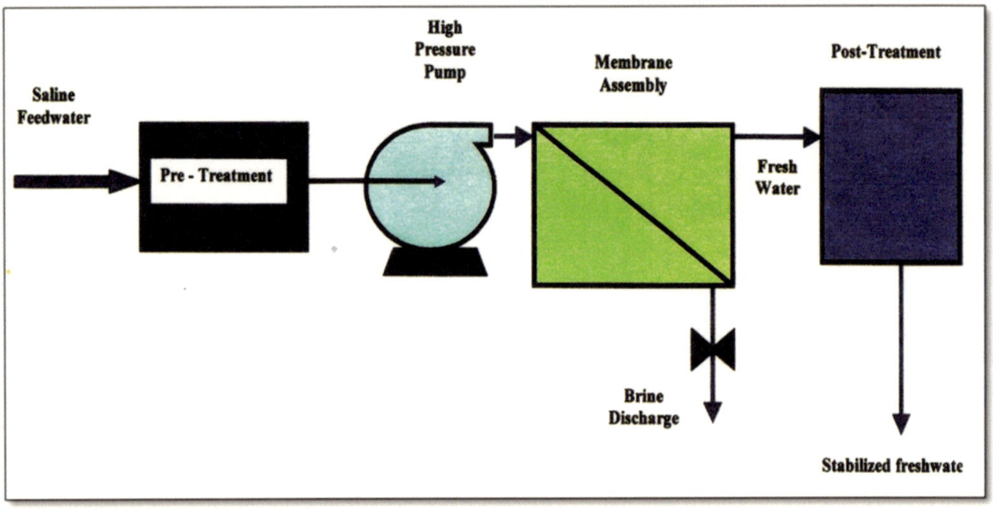

Fig. 2. Flow diagram of RO desalination process

4. Desalination equipment

To maintain this continuous and effective operation, a diversified assembly of equipment is employed in DPs including significant quantities of storage tanks, pumps, heat exchangers, pipes of varying sizes for both operational processes and transmission lines. Table 4 presents the varied equipment utilized in different DPs, both thermal distillation and membrane processes. Electric steam generators, made of SS UNS S30400 or S31600, are ideal for clean steam, reducing steam contamination introduced by steam piping. To improve boiler efficiency, an electrical conductivity sensor is utilized to determine steam conductivity in the range of 0 to 10 mS/cm.

Reactive piping materials such as plain carbon steel, galvanized or cadmium plated steel and cast iron should not be used in DP's due to risk or corrosion and consequent contamination by corrosion products (Malik, 2001).

Flash evaporators, low pressure turbines, evaporators and condensers, operated in contact with steam, are particularly prone to corrosion by salts. Corrosion inhibitors are applied to avoid or minimize corrosion of steel, and other alloys provoked by corrosive impurities from steam or water such as chlorides, caustic, inorganic and organic acids, carbonates, sulphates, hydrogen sulphide and their mixtures (Table 4).

Pipes, tubes and ducts	Evaporators
Saline water pumps, vertical and centrifugal	Vapor condensers, Diesel engines
Valves, diverse types	Flash chambers
Gasketed plate-and-frame heat exchanger	Demisters
Filters, diverse types	Steam generators
Fittings and flanges	Condensers
Steam and gas turbines	Dearators
Compressors	Chlorinators
Control and flow instrumentation	Storage tanks

Table 4. Equipment for desalination plants.

5. Corrosion resistant alloys

The varied equipment of DP is fabricated from a wide spectrum of engineering materials, metallic and non metallic, which display a reasonable endurance to the fluids (liquids, vapor and gases) handled and processed in the plant installations and environments.
The prime consideration during the selection of materials of construction is their corrosion characteristics. CRAs used in the desalination industry are classified into two large groups:

- Ni- containing alloys e.g. Ni-base alloys, Cu-base alloys and SS.
- Titanium and aluminum alloys. e.g. UNS A95052

Their UNS (Unified Numbering System) designation and chemical composition are displayed in Table 5. In addition, nonmetallic materials such as plastics: polyethylene (PE), polypropilene (PP), polyvinylchloride (PVC) and composites, in particular fiber reinforced plastic (FRP) based on polyester and/or epoxy resins are employed mainly for piping and storage vessels.
Those CRAs have an outstanding corrosion resistance, mechanical strength and weldability. This corrosion resistance is due to the tenacious, durable and self-healing, protective film of metal oxides formed in the presence of air and moisture. According to ASTM standard D 4194, SS UNS S31600 should be used for all wetted parts of RO devices. Furthermore, it warns about the use of piping made of carbon steel, cast iron and galvanized or cadmium plated carbon steel to avoid contamination by corrosion products. Two most popular austenitic SS S30400 and S31600, several superaustenitics and duplex are in service in DPs, in seawater and brackish water applications, as material of construction for centrifugal pumps. In recent years, grades of SS with high resistance to pitting and crevice corrosion in seawater have been developed, e.g. the 6 Moly families of alloys with increased nitrogen content (Table 4). The CRAs industry employs advanced equipment and facilities, highly skilled engineering staffs and sophisticated quality control procedures, to develop and to produce suitable alloys for industrial applications. CRAs producers provide engineering and development services to Original Equipment Manufacturers (OEMs) to ensure their suitability to the DPs processes and fluids.

5.1 Steel

Carbon steel is the main material for the erection of plant structures, water storage, tanks and pipelines. Because of its useful mechanical properties, easy of machining and welding

as low cost, it is preferred. But due to its limited corrosion resistance it should be protected by paints and coatings and in many structures in contact with water by cathodic protection.
SS are iron-based alloys containing Cr as the main alloying element at a concentration of at least 12%. They have outstanding corrosion resistance, mechanical strength and weldability; their corrosion resistance is due to the tenacious, durable and self-healing protective film of Chroming oxide (of 5 nanometer thickness) formed in the presence of air and moisture Unfortunately, this passive film can break down in SW containing chlorides, but it is enhanced when the SS contains Mo. Table 4 presents a great variety of SS applied in DPs around the world.

5.2 Ni-base alloys

Ni- alloys are among the most important because they resist corrosion in a wide variety of environments, including SW rich in chlorides (Table 5). They are divided in two groups: those constituted mainly by Ni and those which employ Cr as a major alloying element. Their corrosion resistance depends upon the presence of Nickel and Chromium oxides imparting a passive state. They are used in DPs for the fabrication of high-pressure pumps and brine concentrators in thermal DPs.

5.3 Copper-base alloys

The main alloys used in DPs are: Cu-Ni for tube and shell-heat exchangers and condensers (Table 5); bronze (Cu-Sn) for ship propellers and parts of pumps for seawater transportation and brass (Cu-Zn) for hot and cold water circulation. The blue, green layer of corrosion products which form on the Cu-alloys surface does not provide effective protection.

5.4 Titanium and aluminum alloys

Ti has good corrosion resistance in strongly oxidizing environments, e.g. nitric acid and wet chlorine but not with reducing acids, e.g. hydrochloric acid. It can be readly shaped and formed; is available in conventional forms. Ti shows excellent resistance to seawater and SW but it does not tolerate even trace amounts of fluorides which cause severe corrosion. Its corrosion resistance is due to a stable, protective, strongly adherent film of titanium oxide (TiO_2). Alloying of Ti with palladium (Pd) and other noble metals yields corrosion resistant alloys (Table 5).
Aluminum (Al) corrodes under both acidic and alkaline conditions yielding Al^{+3} and AlO_2^- aluminates ions, respectively. When Cl^- penetrates the passive film of Al_2O_3, it initiates pitting and crevice corrosion at localized sites with breakdown of passivity. The Al potential, about -1.65 volts (SHE) indicates its natural tendency to corrode but the oxide film imparts corrosion resistance to Al equipment in contact with SW, including seawater. In thermal desalination equipment Al tubes are utilized in huge heat-exchangers for condensation of water vapor with seawater.

6. Corrosion, scaling and fouling

Frequently, corrosion, scaling and fouling phenomena appear simultaneously in DPs; they interact and influence each other. Scaling and fouling have marked effect on corrosion, often associated with SW velocities. They originate in the SW, depending on their chemical and biological composition, their interaction with the equipment surface and plant operational conditions such as pH, flow regime, temperature and pressure (The Newsletter, 2003).

UNS* Number	Chemical composition%							
	Cr	Ni	Mo	Al	Cu	Ti	C max.	Other
Aluminium alloys								
A95052				Bal.				2.5 Mg, 0.25 Cr
A95054				Bal.				2.7 Mg, 0.8 Mn
Copper alloys								
C70600		10			90			1.0 Zn
C71900		30			70			2.6 Cr
Stainless steels								
Austenitic chromium-nickel steels								
S30400	18-20	8-12					0.08	
S30403	18-20	8-12					0.03	
S30908	22-24	12-15					0.08	
S31600	16-18	10-14	2-3				0.10	
S31603	16-18	10-14	2-3				0.03	
S31700	18-20	11-14	3-4				0.08	
High-alloyed austenitic								
S31254	20	18	6.1				0.02	0.2N
S32654	24	22	7.3				0.015	0.5N
N08367	20.5	24	6.3				0.03	0.22N
N08904	20	25	4.5				0.02	1.5Cu
N08926	20	25	6.8				0.02	0.2N, 1.0 Cu
N08020	21	25	4.5				0.03	0.03Cb
N08028	27	31	3.1				0.02	1..0 Cu
N08031	27	31	6.5					0.2 N
N08932	20	25	4.8				0.01	0.2N, 1.5Cu
Austenitic, castings								
J92500	19	10					0.03	
J92800	19	11	2.5				0.03	0.2N
J95150	20	29	2.5				0.07	3.5Cu
Duplex: ferritic-austenitic								
S32250	25	6.5	3.0				0.02	0.1 7N, 1.5Cu
S31803	22	5.0	3.0				0.03	0.15N
Ni-Base alloys								
N06600	16	Bal.					0.08	8.0 Fe
N08825	21	Bal.						29Fe, 2.0Cu
N06030	30	Bal.	5.5					15Fe, 2.0Cu
Titanium								
R50250						99.8	0.10	
R50450						99.8	0.10	
R52400						Bal.	0.10	0.15 Pd

*UNS: Unified Numbering System

Table 5. CRAs used for manufacture of equipment in desalination plants

Unless prevented, corrosion and the buildup of scale and biological fouling impact on the DP operation and in extreme cases even lead to equipment shutdown. Given the huge number of heat exchangers, condensers, pumps that handle SW; corrosion, and scale inhibitors and biocides should be applied as needed.
Pollution and corrosion are interrelated processes since many atmospheric pollutants, e.g. SOx, NOx, COx, H_2S accelerate corrosion, and corrosion products such as rust also pollute water bodies. Both: pollution and corrosion are pernicious processes which impact the environment quality and the structures durability, particularity in coastal areas, near large cities and commercial-industrial ports contaminated with municipal, industrial and agricultural effluents (Schorr & Valdez, 2005; Wiener et. al., 2006). As a result of these effluents, coastal pollution has reached crisis levels in many areas due to population growth, poor planning of land used in exaggerated development of tourism and its coastal facilities (Rasoanandrasana, 2010).
The pH values of SW are in the range of 5 to 8 and the concentration of dissolved oxygen (DO), the main corrodent, varies from 4 to 6 mg/l as a function of temperature and flow regime. SW contaminated with H_2S, a reductant, are slightly acidic and corrosive towards some CRAs, therefore H_2S should be eliminated by mechanical or chemical methods.

6.1 Corrosion

Corrosion is an electrochemical process that takes place upon the metallic surface by reaction with the components of the SW. The dominant factors are DO concentration, temperatures and salinity. The overall corrosion process is the combination of two reactions: anodic and cathodic, that takes place at the metal-SW interface. For Fe the reaction is:

$$2\ Fe \rightarrow 2\ Fe^{+2} + 4\ e^- \quad (1)$$

The cathodic reaction involves the reduction of DO:

$$O_2 + 2\ H_2O + 4\ e^- \rightarrow 4\ OH^- \quad (2)$$

The overall corrosion reaction is:

$$2\ Fe + O_2 + 2\ H_2O \rightarrow 2\ Fe(OH)_2 \quad (3)$$

Ferrous hydroxide $Fe(OH)_2$ will further oxidize to ferric hydroxide $Fe(OH)_3$ and eventually turn to rust ($Fe_2O_3.xH_2O$).
Corrosion inhibitors that slow down either the anodic or cathodic reactions or both, by interaction with corrosive ions, or form a protective film on the metallic surface, will reduce the extent of corrosion.
On the basis of the appearance of corrosion phenomena on the metal surface or in the geometry of the equipment, corrosion is classified as uniform when the metal corrodes at the same rate resulting in a uniform decrease of thickness or localized when corrosion appears at a specific area in a part of the equipment. Corrosion results from differences in aeration, concentration, pH, water velocity. Types of localized corrosion which are often encountered in DPs, are illustrated in Figure 3.
Atmospheric corrosion in DPs installed in coastal zones is driven by the effect of marine aerosols containing humid particles of NaCl. It is promoted by the formation of moisture

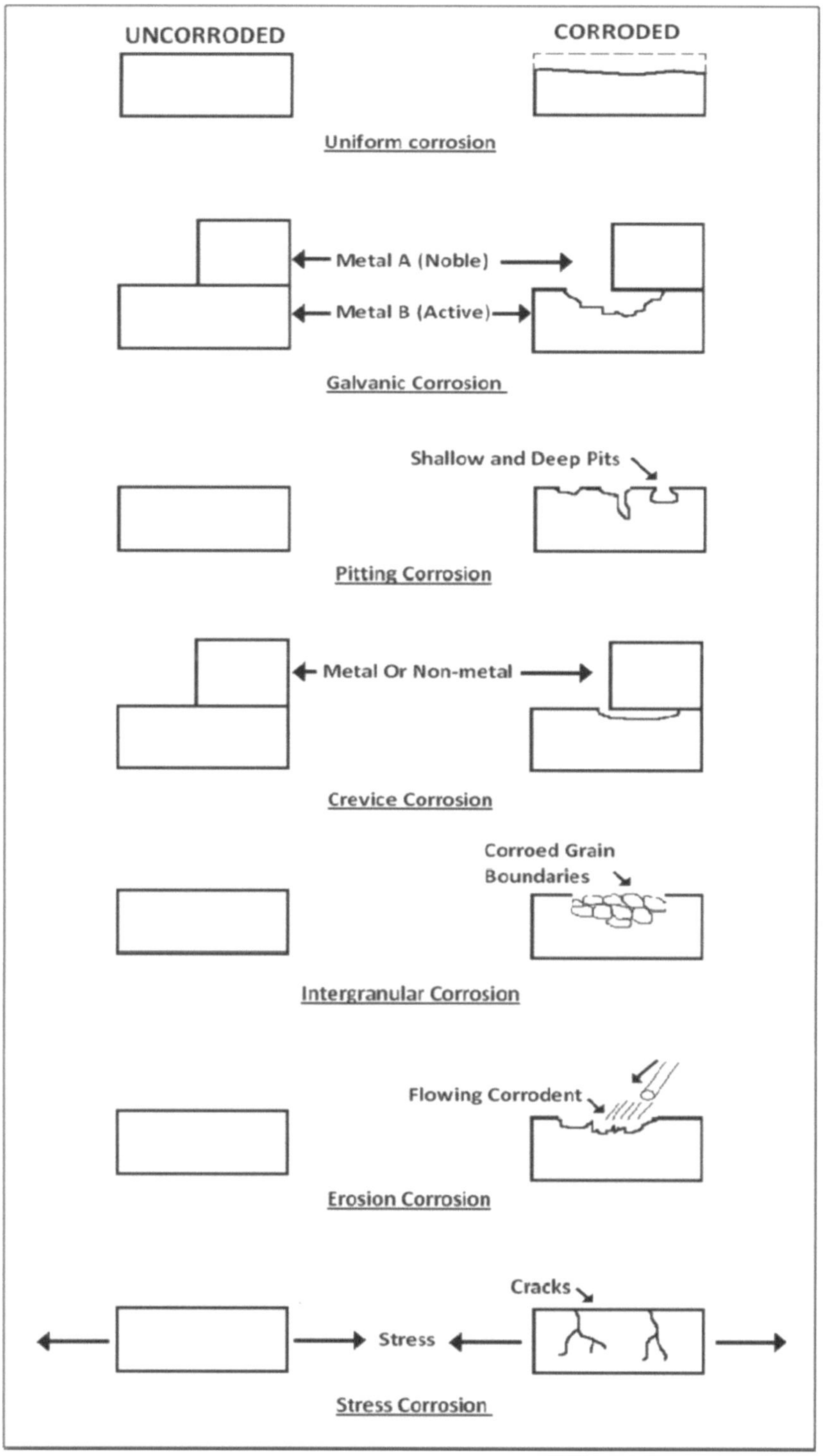

Fig. 3. Types of corrosion in desalination plants.

layers on metal surfaces and by the chemistry of these layers .Climate change could influence the macro- and micro-surface effects controlling either the chemistry or the duration of these wet surfaces (Cole, 2010). Furthermore, hygroscopic salts absorb moisture and form deliquescent salt particles which can greatly accelerate the corrosion of metals (Li, 2010). Corrosion damage increases maintenance expenses and generate problems in DP operation. In a recent report, (Koch, 2002) it was estimated that cost of corrosion in the USA is 276 billion USD on an annual basis, which represents 3.1% of the USA Gross Domestic Product (GDP), including 36 billion USD corrosion in drinking water and sewage systems.

6.2 Scaling

Highly concentrated SW tends to form thick scales by deposition of dissolved and suspended solid such as carbonates, silicates and hydroxides. As SW is circulated the last compound that tends to come out of the water is calcium carbonate. To prevent $CaCO_3$ scaling the feed water is treated with sulfuric acid converting the carbonate to carbon dioxide CO_2 that escapes from the water. When phosphates and phosphonates are applied to provide corrosion protection, careful control is implemented to avoid scaling by calcium phosphate [$Ca_3(PO4)_2$] adding organic polymers containing carboxilic acids for control of calcium phosphate deposition.

Chemical and physical pre-treatment of feed water is required to remove substances that would interfere with the desalting operation and will damage the equipment, in particular, the plastic membranes of the RO process. Pretreatment with ozone, a powerful oxidant and biocide will remove sulphur, iron, manganese and other water-soluble heavy metals compounds, bacteria, odor and color.

Some alkaline chemicals e.g. soda ash neutralize the acidity found in some brackish waters, helps reduce corrosion and extends the life of equipment. Citric acid removes iron and polyphosphates reduce iron staining but these pretreatments are rather expensive.

Scaling is controlled by introducing additives to inhibit crystal growth, reducing temperatures and salt concentration. Inorganic, colloidal particles, e.g. silica or silicic acid, hydrous iron oxide, aluminum oxide and organic substances in the feed water by special pretreatments. Furthermore, a particular type of corrosion, known as microbiologically influenced corrosion (MIC) develops under these complex organic and inorganic deposits. Chelating agents and acid are injected into the feed water to prevent precipitation and scaling on the RO membrane surface. Some DPs combine distillation with RO to produce both power and water.

6.3 Fouling

By their very nature SWs provide and ideal environment for macro- and micro- organisms to thrive. Unless properly controlled through the use of biocides; bacteria, algae, fungi, mollusk will grow on metallic surfaces. These organisms secrete polymeric substances, forming a film that generates acids and other harmful compounds that induce corrosion. Furthermore, they deplete and establish oxygen concentration cells causing localized and pitting corrosion. Many biocides are applied to prevent the formation of biological fouling; such as gaseous chlorine (Cl_2) sodium hypochlorite (NaOCl), chlorine dioxide (ClO_2) and

bromide salt (NaBr). These chemicals should be used with appropriate regulations approval to avoid the proliferation of toxic agents.
The CRAs desalination equipment should be maintained clean and smooth to avoid calcareous scaling on heat transfer surfaces and to diminish the propensity to biological fouling on rough or polished surfaces. Acidic and alkaline cleaning is a mechano-chemical operation easily implemented in CRA equipment, to remove biological fouling and mineral scale since they alter the equipment surface performance and induce corrosion.

7. Corrosion protection, monitoring and control

Corrosion engineering and technology develop and apply methods and techniques of prevention and protection to avoid the interaction of the equipment and its construction materials with the corrosive factors of the DPs environment. Practical methods that minimize or eliminate corrosion include selection of suitable CRAs (Encyclopedia of Desalination, 2010; Habib, 2004; Malik, 2004) application of coatings, paints and linings to carbon steel and galvanized steel equipment and cathodic protection.
The technical process of selection is usually divided in to three main stages:

- Analysis of the requirements and collection of the relevant information about the conditions imposed by the desalination process and the corrosion resistance required by the equipment.
- Selection and evaluation of candidate materials by screening of the information collected in the first stage. Laboratory and pilot plant corrosion tests are performed by exposing suitable materials in the desalination process fluids and environments (ISO 845 and ASTM G4, G31).
- Selection of the most appropriate material based on its costs, availability, easy fabrication and repair, maintenance and safety.

Corrosion resistance is the main property to be considered in the choice of materials for DPs equipment but the final selection must be a compromise between technological and economic factors. It is sometimes more economical to use a high-priced CRA that will provide long and trouble-free service than to use a lower priced material that may require frequent maintenance or replacement. The selected CRA should be able to perform its function safely for a reasonable period of time and at a reasonable cost.
Corrosion monitoring (CM) is the practice of measuring the corrosion events and rate by continuously exposing materials probes in a body of water or a operating DP. Modern electrochemical, electronic, mechanical, non-destructive and computational devices are applied in the field of CM such as potenciometry, multielectrode probes, electrical resistance, communication networks, remote CM, expert programs and artificial neural networks. CM techniques provide daily warning of costly corrosion damage and critical information, where the damaging event is occurring and about the rate of deterioration. This information is essential to take decisions about the type, urgency and cost of preventive and curative measures to be applied on site without delay. (Robenge, 2000).
Chloride ion (Cl^-), a main component of SW, can breakdown the passive film of CRAs. Table 6 shows the upper limit of Cl^- to assure corrosion resistant performance and how their increment in Cr, Ni and Mo content enhances resistance to pitting and crevice corrosion. This graphic display might serve as a guideline for selection of SS and Ni-base

alloys for desalination equipment handling brackish water with varied Cl^- content (Valdez & Schorr, 2000).
Today, the main and fastest source of information on corrosion control of industrial equipment, plants and facilities is the Internet. Data bases and computers based expert systems dealing with selection of materials, their properties and corrosion control for many environments and industries are listed in Roberge´s Handbook (Roberge, 1999).
Important professional institutions dealing with corrosion research and control are the European Federation of Corrosion (EFC), NACE (National Association of Corrosion Engineers, USA), National Institute of Standards and Technology (NTIS), Materials Technology Institute (MTI), and other materials entities surveying the different computer-assisted programs for corrosion control in industrial facilities. Other software is based on Artificial Neuronal Network (ANN) for the prediction, analysis and solving corrosion problems (Roberge, 1999). All these data-base expert systems and software are applicable for the plants, equipment installations and environments of the desalination industry.
Additional sources of corrosion information appear in data collections, handbooks and standards, in particular those published by the International Organization of Standardization (ISO); the American Society for Testing and Materials (ASTM), NACE International, USA, etc, (Mattsson,1996).

Cl, pm	CRAs
2000	N06625 S31803
1500	N06007 S31254
1000	N08904
500	S31700 S31600
0	S30400

Table 6. Cl^- content in brackish water for corrosion resistance of CRAs.

8. Closure

In recent years, desalination experts have been playing an increasingly active role by refining desalination process, saving energy resources, selecting long-lasting CRAs and improving the efficiency of DPs. The worldwide increase in population, rising standards of living and the extension of water pollution evolves into a critical demand of potable water. Desalination technology is the obvious response to this challenge, to provide the water supply needed for the future generations.

9. References

Barlow, M and Clark, T. (2002). in Blue Gold, New Press, New York, 2002.

Charach, J.; Schorr, M. & et al. (1991). Corrosion and Scaling Behavior in Dead Sea Basin Saline Waters. *Corrosion Reviews*, Vol. 9, no 3-4, 1991.

Dillon, C.P.(1994). Corrosion Control in the Chemical Process Industries. *Materials Technology Institute*, ISBN: A-8779914-58-4.

Encyclopedia of Desalination and Water Reuse. (2010). Materials Selection and Corrosion, www.desware.net

Habib, K. et al (2001). Risk Assessment and Evaluation of Materials Commonly Used in Desalination Plants, Desalination 139, pp. 249-253.

I.D.E. Technologies. (2004), Fresh Water from the Sea, Israel, www.ide-tech.com

I.S. Paterson, D.A. (2010). Possible effects of climate change on atmospheric corrosion in Australia. *Corr. Eng. Sci. and Tech*, Vol.45(1) pp 19-26.

ISO 11845. (1995). Corrosion of metals and alloys, general materials for corrosion testing.

Koch G.H. et (2002). Corrosion Costs and Preventive Strategies in the USA. *Supplement to Materials Performance*. July, 2002. Also: www.corrosioncosts.com

Kowitt, B. (2009). The future of water. Fortune, pp 111-116. October 12, 2009.

Kronenberg, G. (2004). Ashkelon 100 MC/year, V.I.D Desalination Company Ltd, International Conference on Desalination Costing, Cyprus, December 2004.

Li, S.X. Hihara, L.H. (2010). Atmospheric corrosion initiation on steel from predeposition of NaCl particles in high humidity atmospheres. *Corr. Eng. Sci. and Tech.* Vol.45(1) pp 49-56.

Matteson, E. (1996), Basic Corrosion Technology for recientest and engineers, 2nd. Ed. The Institute of Materials, ISBN 86125 0118, London.

Malik, A.U. (2001) Case histories on the failure of pipelines in desalination plants, Saline Water Conversion Corporation, Saudi Arabia.

Miller, J. (2003). Review of Water Resources and Desalination Technologies. Materials Chemistry, Department Sandia National Laboratories SAND 2003-0800. http://www.sandia.gov/water/docs/MillerSAND2003_0800.pdf

Albuquerque, NM 87185-1349 P.A Roberge (2000), Handbook of Corrosion Engineering, Mc Graw Hill.

Roberge, P.R. (2010) Impact of climate change on corrosion risks, ditto.

Shiklomanov, I.A. (1993). Water in Crisis: A Guide to the World's Fresh Water Resources, P.H. Gleick, ed. Oxford University Press, New York, 1993

Schorr, M; Valdez, B. (2005), Corrosion of the marine infrastructure in polluted seaports, *Corr. Eng. Sci and Tech.* Vol 40.

The Newsletter,(2003), Critical Assessment of Fouling and Scaling Indices, Middle East Desalination Research Center, Oman, September 2003.

Valdez, B.; Schorr, M; et al. (2010). Effect of climate change on the durability of engineering materials in hydraulic infrastructure; an overview. *Corrosion Engineering Source and Technology*, Vol 45(1), pp.34-41/155 and 1478.

Valdez, B.; Schorr, M. (2010), Editors, Special Issue: Relationship of corrosion with climate change. *Corr. Eng. Sci. Tech.* Vol 45(1).

Valdez B.; Schorr, M. (2000). Stainless steel for corrosion control in desalination plants. *Stainless Steel World*, May, pp 40-44.

Wiener, M.S.; Valdez, B.; et al. (2006). Effect of H_2S in corrosion in polluted water. *Corr. Eng. Sci. Technical*, Vol. 41(3) pp 221-227.

Zhw, X.R., Huang, L.Y. Lim, L.Y., Lim, D.Y. (2008). *Corr. Eng. Sci. Tech.* Vol 43(4), pp 328-334.

Part 2

Novel Trends and Technologies

5

Application of Renewable Energies for Water Desalination

Mattheus Goosen[1], Hacene Mahmoudi[2], Noreddine Ghaffour[3] and Shyam S. Sablani[4]
[1]Office of Research and Graduate Studies, Alfaisal University, Riyadh,
[2]Renewable Energy Development Centre (CDER), Bouzareah, Algiers, Hassiba Ben Bouali University, Chlef,
[3]Water Desalination & Reuse Centre, King Abdullah University of Science and Technology (KAUST),
[4]Dept. of Biological Systems Engineering, Washington State University, Pullman, Washington,
[1,3]Saudi Arabia
[2]Algeria
[4]USA

1. Introduction

Renewable energy technologies such as wind, solar, and geothermal and even alternatives such as nuclear show great promise for water desalination (Serpen et al., 2010; Goosen et al, 2010; Stock Trading, 2010; Khamis, 2009; Misra, 2010). These energy driven desalination systems fall into two categories. The first includes distillation processes driven by heat produced directly by the renewable energy system (RES), while the second includes membrane and distillation processes driven by electricity or mechanical energy produced by RES. With the world's fresh water demands increasing, much research has been directed at addressing the challenges in using renewable (and environmentally friendly) energy to meet the power needs for desalination plants. Lack of water, for instance, has caused great distress among the population in large parts of the MENA countries (Middle East and North Africa). The economic and industrial potentials of renewable energies, such as geothermal, solar and wind, as well as the environmental advantages have been pointed out in several recent studies (Serpen et al., 2010; Mahmoudi et al., 2010; Huang, 2010; Lund, 2007; and Cataldi, et al., 1999). Lund (2007) noted that recorded accounts show uses of, for example, geothermal water by Romans, Japanese, Turks, Icelanders, Central Europeans and the Maori of New Zealand for bathing, cooking and space heating. The first use of geothermal energy for electric power production occurred in Italy a century ago with the commissioning of a commercial power plant (250 kWe). Small decentralised water treatment plants can also be connected to a wind energy convertor system (WECs). The wind turbines as well as the desalination system can be connected to a grid system (Eltawil et al., 2009). The Kwinana Desalination Plant, for example, located south of Perth in Western Australia, produces nearly 140 megalitres of drinking water per day, supplying the Perth metropolitan area (BlurbWire 2010). Electricity for the plant is generated by the 80 MW Emu Downs Wind Farm located in the state's Midwest region.

Solar energy can also be converted to thermal or electrical (i.e. photovoltaic) energy and then used in water desalination directly or indirectly, respectively (Mahmoudi et al., 2008, 2010; Goosen and Shayya, 1999). Thermal energy, for instance, can be employed in solar stills, collectors, or solar ponds. Electrical energy can be produced from solar energy directly by PV conversion or via solar thermal power plant. The coupling of renewable energies such as wind, solar and geothermal with desalination systems holds great promise for water scarce regions (Mahmoudi et al., 2008, 2010; Goosen and Shayya, 1999; Tester et. al., 2007). We can argue that an effective integration of these technologies will allow countries to address water shortage problems with a domestic energy source that does not produce air pollution or contribute to the global problem of climate change. Furthermore this approach will help to bypass the problems of rising fuel prices and decreasing fossil fuel supplies. Desalination plants, for example, may be run with geothermal of energy being employed directly to heat the saline or brackish water in multiple effect distillation units and/or it could be used indirectly to generate electricity for operating reverse osmosis units (Kalogirou, 2005). In addition, alternative energy sources such as nuclear also need to be considered (Khamis, 2009; Misra, 2010). The Shevchenko BN350 nuclear fast reactor and desalination plant, for instance, situated on the shore of the Caspian Sea, in Kazakhstan, during its lifetime of some 27 years could generate 135 MWe of electric power and provide steam for an associated desalination plant which produced 80,000 m^3/day of potable water (Kadyrzhanov et al., 2007). About 60% of the plants power was used for heat and desalination.

Bourouni et al. (1999a, 1999b, 2001) reported on installations using humidification dehumidification processes in the form of evaporators and condensers made of polypropylene and operated at a temperature between 60 and 90 ^{0}C. Bouchekima (2003) reported on the use of brackish underground geothermal water to feed a solar still installed in the South of Algeria. Furthermore, with the recent progress in membrane distillation technology, the utilization of direct geothermal brine with temperature up to 60 ^{0}C has shown promise (Houcine, et al., 1999). Iceland is widely considered as the most successful state in the geothermal community. The country of just over 300,000 people is fully (i.e. 100%) powered by renewable forms of energy, ranking the highest in the 15 top countries that generate electricity from geothermal resources. Wright (1998) has estimated that given that the worldwide energy utilization is equal to about 100 million barrels of oil per day, the Earth's thermal energy to a depth of 10 kilometers could theoretically supply all of mankind's power needs for several million years.

The aim of this chapter is to provide a critical review of recent trends in water desalination using renewable as well as alternative energy resources. After providing an overview of desalination using renewable energies, specific case studies will be presented as well as an assessment of environmental risks and sustainability. The chapter will conclude with a section on market potential and risk management.

2. Water desalination using renewable and alternative energies

The combination of renewable energy with desalination systems holds immense promise for improving potable water supplies in arid regions (Mahmoudi et al., 2008, 2009a, 2009b, 2010). We can argue that an efficient amalgamation of these technologies will allow nations to deal with water shortage problems with a domestic energy source that does not produce air pollution or contribute to the global crisis of climate change. Furthermore, while fuel prices are rising and fossil fuel supplies are decreasing, the fiscal outlay for desalination and renewable energy systems are steadily decreasing. The latter is due in part to a variety of possible arrangements that can be envisaged between renewable power supplies and desalination technologies (Rodriquez et al, 1996).

2.1 Applications of solar energy for water desalination

Desalination by means of solar energy is a suitable alternative to conventional methods (e.g. fossil powered thermal distillation) to providing fresh water, especially for remote and rural areas where small quantities of water for human consumption are needed (Al-Hallaj et al., 1998). Attention has been directed towards improving the efficiencies of the solar energy conversions, desalination technologies and their optimal coupling to make them economically viable for small and medium scale applications. Solar energy can be used directly as thermal or it can be converted to electrical energy to drive reverse osmosis units. The thermal energy can be achieved in solar stills, collectors, or solar ponds. Electrical energy can be produced from solar energy directly by photo-voltaic (PV) conversion or via a solar thermal power plant.

Solar stills, for example, which have been in use for several decades, come in a variety of options (Figure 1) (Goosen et al., 2000). The simple solar still (Figure 1A) is a small production system yielding on average 2 - 5 L/day. It can be used wherever fresh water demand is low and land is inexpensive. Many modifications to improve the performance of the solar stills have been made. These include linking the desalination process with the solar energy collectors (Figure 1E), incorporating a number of effects to recover the latent heat of condensation (Figures 1D & 1F), improving the configurations and flow patterns to increase the heat transfer rates (Figures 1B, 1C, 1E, and 1F), and using low-cost materials in construction to reduce the cost. Nevertheless these systems are not economically viable for large-scale applications. One of the more successful solar desalination devices is the multiple-effect still (Figure 1F) (Al-Hallaj et al., 1998). Latent heat of condensation is recovered, in two or more stages (generally referred to as multi-effects), so as to increase production of distillate water and improve system efficiency. A key feature in improving overall thermal efficiency is the need to gain a better understanding of the thermodynamics behind the multiple use of the latent heat of condensation within a multi-effect humidification-dehumidification solar still (Al-Hallaj et al., 1998). In addition, while a system may be technically very efficient it may not be economic (i.e., the cost of water production may be too high) (Fath, 1998). Therefore, both efficiency and economics need to be considered when choosing a desalination system. We can further argue that desalination units powered by renewable energy systems are uniquely suited to provide water and electricity in remote areas where water and electricity infrastructures are currently lacking.

Solar collectors are usually classified according to the temperature level reached by the thermal fluid in the collectors (Table 1) (Kalogirou, 2005). Low temperature collectors provide low-grade heat, only a few degrees above ambient air temperature and use unglazed flat plate collectors. This low-grade heat is not useful to serve as a heat source for conventional desalination distillation processes (Fahrenbruch and Bube, 1983; Kalogirou, 2005). Medium temperature collectors provide heat of more than 430C and include glazed flat plate collectors as well as vacuum tube collectors using air or liquid as the heat transfer medium. They can be used to provide heat for thermal desalination processes by indirect heating with a heat exchanger. High temperature collectors include parabolic troughs or dishes or central receiver systems. They typically concentrate the incoming solar radiation onto a focal point, from which a receiver collects the energy using a heat transfer fluid. The high temperature energy can be used as a thermal energy source in thermal desalination processes or can be used to generate electricity using a steam turbine. As the position of the sun varies over the course of the day and the year, sun tracking is required to ensure that the collector is always kept in the focus of the reflector for improving the efficiency. For large-scale desalination applications, these systems need large collector areas.

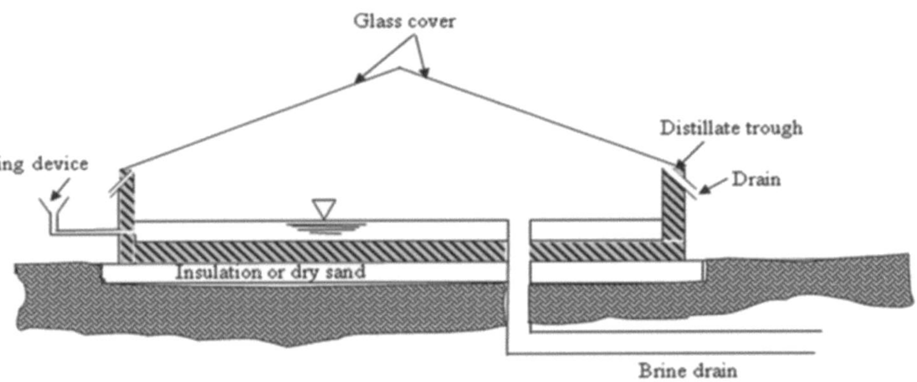

A

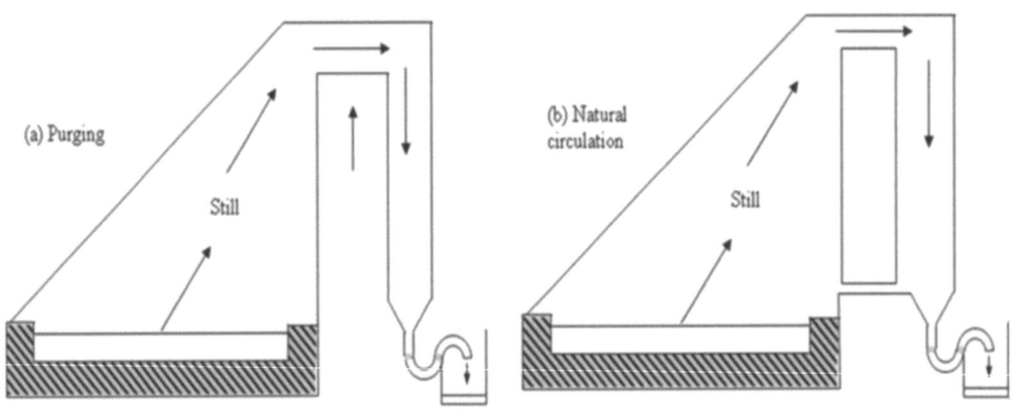

B

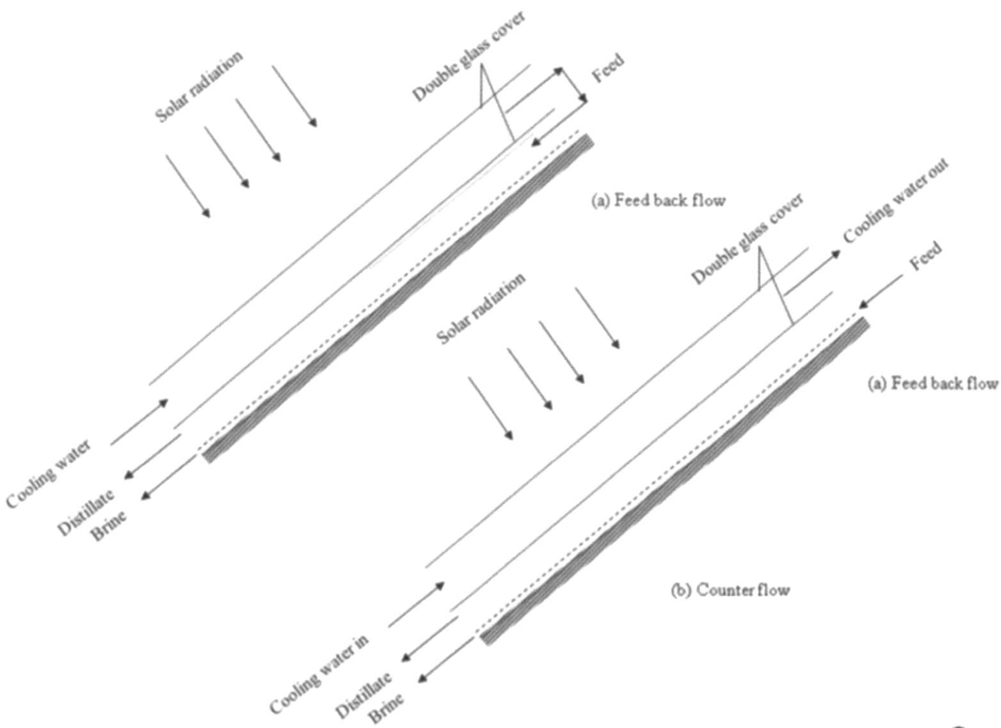
Solar radiation
Double glass cover
Feed
(a) Feed back flow
Cooling water
Distillate
Brine
Solar radiation
Double glass cover
Cooling water out
Feed
(a) Feed back flow
(b) Counter flow
Cooling water in
Distillate
Brine

C

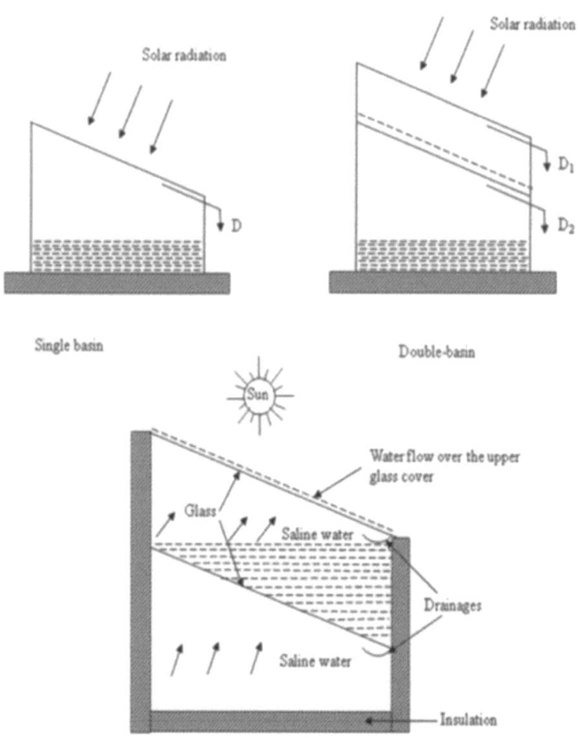
Solar radiation
D
Single basin
Solar radiation
D1
D2
Double-basin
Sun
Water flow over the upper glass cover
Glass
Saline water
Drainages
Saline water
Insulation

D

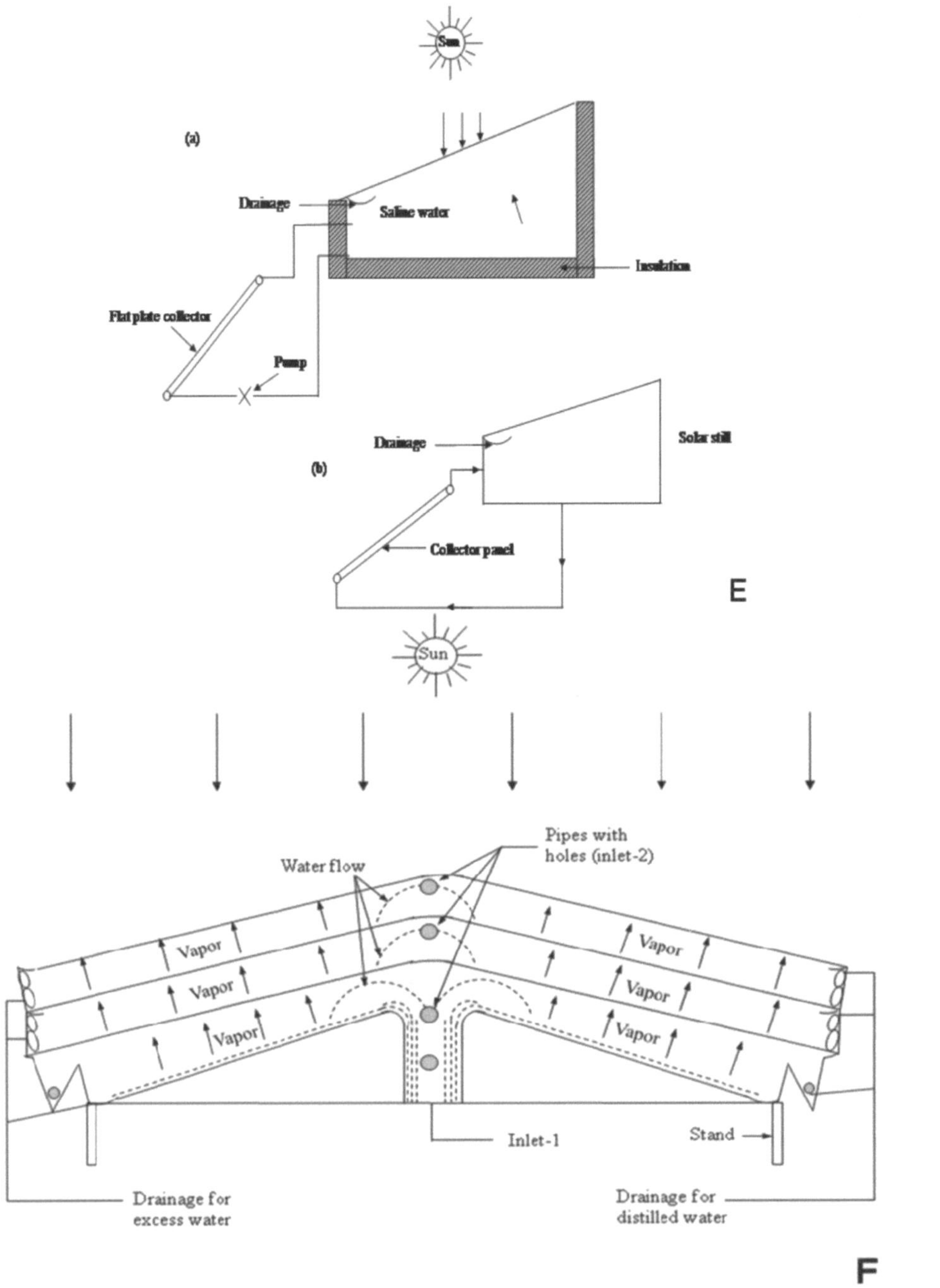

Fig. 1. Solar desalination systems (Goosen et al., 2000; adapted from Fath, 1998). A. Single-effect basin still. B. Single-sloped still with passive condenser. C. Cooling of glass cover by (a) feedback flow, and (b) counter flow. D. Double-basin solar stills: (a) schematic of single and double-basin stills and (b) stationary double-basin still with flowing water over upper basin. E. Directly heated still coupled with flat plate collector: (a) forced circulation and (b) natural circulation. F. Typical multi-effect multi-wick solar still.

Indicative temperature range (8C)	Concentration ratio	Absorber type	Collector type	Motion
30–80	1	Flat	Flat plate collector (FPC)	Stationary
50–200	1	Flat	Evacuated tube collector (ETC)	
60–240	1–5	Tubular	Compound parabolic collector (CPC)	
60–300	5–15	Tubular	Compound parabolic collector (CPC)	Single-axis tracking
60–250	10–40	Tubular	Linear Fresnel reflector (LFR)	
60–300	15–45	Tubular	Parabolic trough collector (PTC)	
60–300	10–50	Tubular	Cylindrical trough collector (CTC)	
100–500	100–1000	Point	Parabolic dish reflector (PDR)	Two-axes tracking
150–2000	100–1500	Point	Heliostat field collector (HFC)	

Table 1. Solar Energy Collectors (Kalogirou, 2005) Note: Concentration ratio is defined as the aperture area divided by the receiver/absorber area of the collector.

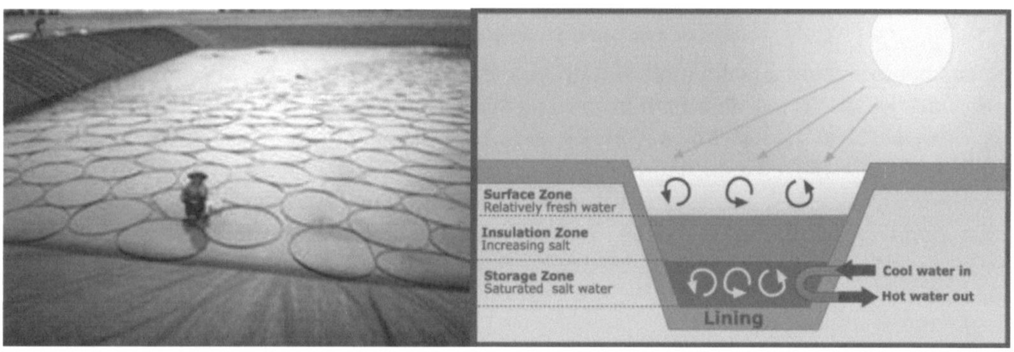

Fig. 2a. (*Left*) Solar pond for heating purpose demonstration in Australia (http://www.aph.gov.au/library/pubs/bn/sci/RenewableEnergy_4.jpg). 2b. (*Right*) Solar Ponds Schematic The salt content of the pond increases from top to bottom. Water in the storage zone is extremely salty. As solar radiation is absorbed the water in the gradient zone cannot rise, because the surface-zone water above it contains less salt and therefore is less dense. Similarly, cooler water cannot sink, because the water below it has a higher salt content and is denser. Hot water in the storage zone is piped to, for example, a boiler where it is heated further to produce steam, which drives a turbine. (Wright, 1982; and www.energyeducation.tx.gov/.../index.html)

Solar ponds (Figure 2) combine solar energy collection with long-term storage. Solar ponds can be used to provide energy for many different types of applications. The smaller ponds have been used mainly for space and water heating, while the larger ponds are proposed for industrial process heat, electric power generation, and desalination. A salt concentration gradient in the pond helps in storing the energy. Whereas the top temperature is close to ambient, a temperature of 90 °C can be reached at the bottom of the pond where the salt concentration is highest (Figure 2b). The temperature difference between the top and bottom layer of the pond is large enough to run a desalination unit, or to drive the vapour generator of an organic Rankine cycle engine (Wright, 1982). The Rankine cycle converts heat into work. The heat is supplied externally to a closed loop, which usually uses water. This cycle generates about 80% of all electric power used throughout the world including virtually all solar thermal, biomass, coal and nuclear power plants (Wright, 1982). An organic Rankine cycle (ORC) uses an organic fluid such as n-pentane or toluene in place of water and steam. This allows use of lower-temperature heat sources, such as solar ponds, which typically operate at around 70–90 °C. The efficiency of the cycle is much lower as a result of the lower temperature range, but this can be worthwhile because of the lower cost involved in gathering heat at this lower temperature.

Solar ponds have a rather large storage capacity. This allows seasonal as well as diurnal thermal energy storage. The annual collection efficiency for useful heat for desalination is in the order of 10 to 15% with sizes suitable for villages and small towns. The large storage capacity of solar ponds can be useful for continuous operation of desalination plants. It has been reported that, compared with other solar desalination technologies, solar ponds provide the most convenient and least expensive option for heat storage for daily and seasonal cycles (Kalogirou, 2005). This is very important, both from operational and economic aspects, if steady and constant water production is required. The heat storage allows solar ponds to power desalination during cloudy days and night-time. Another advantage of desalination by solar ponds is that they can utilize what is often considered a waste product, namely reject brine, as a basis to build the solar pond. This is an important advantage for inland desalination. If high temperature collectors or solar ponds are used for electricity generation, a desalination unit, such as a multistage flash system (MSF), can be attached to utilize the reject heat from the electricity production process. Since, the standard MSF process is not able to operate with a variable heat source, a company ATLANTIS developed an adapted MSF system that is called 'Autoflash' which can be connected to a solar pond (Szacsvay, et al., 1999). With regard to pilot desalination plants coupled to salinity gradient solar ponds the seawater or brine absorbs the thermal energy delivered by the heat storage zone of the solar pond. Examples of different plants coupling a solar pond to an MSF process include: Margarita de Savoya, Italy: Plant capacity 50–60 m^3/day; Islands of Cape Verde: Atlantis 'Autoflash', plant capacity 300 m^3/day; Tunisia: a small prototype at the laboratoire of thermique Industrielle; a solar pond of 1500 m^2 drives an MSF system with capacity of 0.2 m^3/day; and El Paso, Texas: plant capacity 19 m^3/day (Lu et al., 2000).

Solar photo-voltaic (PV) systems directly convert the sunlight into electricity by solar cells (Kalogirou, 2005). Solar cells are made from semiconductor materials such as silicon. Other semiconductors may also be used. A number of solar cells are usually interconnected and encapsulated together to form a PV module. Any number of PV modules can be combined to form an array, which will supply the power required by the load. In addition to the PV module, power conditioning equipment (e.g. charge controller, inverters) and energy storage equipment (e.g. batteries) may be required to supply energy to a desalination plant. Charge controllers are used for the protection of the battery from overcharging. Inverters are used to convert the direct current from the photovoltaic modules system to alternating current to the loads. PV is a mature technology with life expectancy of 20 to 30 years. The

main types of PV systems are the following: - Stand-alone systems (not connected to the utility grid): They provide either DC power or AC power by using an inverter. - Grid-connected systems: These consist of PV arrays that are connected to the electricity grid via an inverter. In small and medium-sized systems the grid is used as a back-up source of energy, (any excess power from the PV system is fed into the grid). In the case of large centralized plants, the entire output is fed directly into the grid.- Hybrid systems: These are autonomous systems consisting of PV arrays in combination with other energy sources, for example in combination with a diesel generator or another renewable energy source (e.g. wind).There are mainly two PV driven membrane processes, reverse osmosis (RO) and electrodialysis (ED). From a technical point of view, PV as well as RO and ED are mature and commercially available technologies at present time. The feasibility of PV-powered RO or ED systems, as valid options for desalination at remote sites, has also been proven (Childs et al., 1999). The main problem of these technologies is the high cost and, for the time being, the availability of PV cells. Many of the early PV-RO demonstration systems were essentially a standard RO system, which might have been designed for diesel or mains power, but powered from batteries that were charged by PV.
Burgess and Lovegrove (2005) compared the application of solar thermal power desalination coupled to membrane versus distillation technology. They reported that a number of experimental and prototype solar desalination systems have been constructed, where the desalination technology has been designed specifically for use in conjunction with solar thermal collectors, either static or tracking. To date such systems are either of very low capacity, and intended for applications such as small communities in remote regions, or else remain unproven on a larger scale. Several systems which are of some interest were discussed. Schwarzer et al (2001) described a simple system which has flat plate collectors (using oil as a heat transfer fluid) coupled to desalination "towers" in which water evaporates in successive stages at different heights (similar to the multi effect still shown in Figure 1F). The condensation of vapour in one stage occurs at the underside of the next stage, transferring heat and increasing the gain output ratio. A very similar system (not mentioned by Schwarzer), called a "stacked plate still", is described by Fernandez (1990). Furthermore, the Vari-Power Company, based in California, has developed an RO based desalination system which is specifically tailored to solar thermal input (Childs et al, 1999). A patented direct drive engine (DDE) converts heat to the hydraulic power required by RO. Desalinated water production using the DDE is projected to be more than 3 times greater (for an identical dish collector) than that which would be obtained by RO driven by a dish-Stirling electricity generation system or PV power. Burgess and Lovegrove (2005) noted that the project remains at the pilot stage with the DDE not commercially available: it has perhaps become less attractive due to the advances in conventional RO. The choice of the RO desalination plant capacity depends on the daily and seasonal variations in solar radiation levels, on the buying and selling prices for electricity, and on the weight given to fossil fuel displacement. A conceptual layout for a solar dish based system with power generation and RO desalination is shown in Figure 3.

2.2 Wind power and desalination

Kalogirou (2005) in a rigorous review on renewable energy sources for desalination argued that purely on a theoretical basis, and disregarding the mismatch between supply and demand, the world's wind energy could supply an amount of electrical energy equal to the

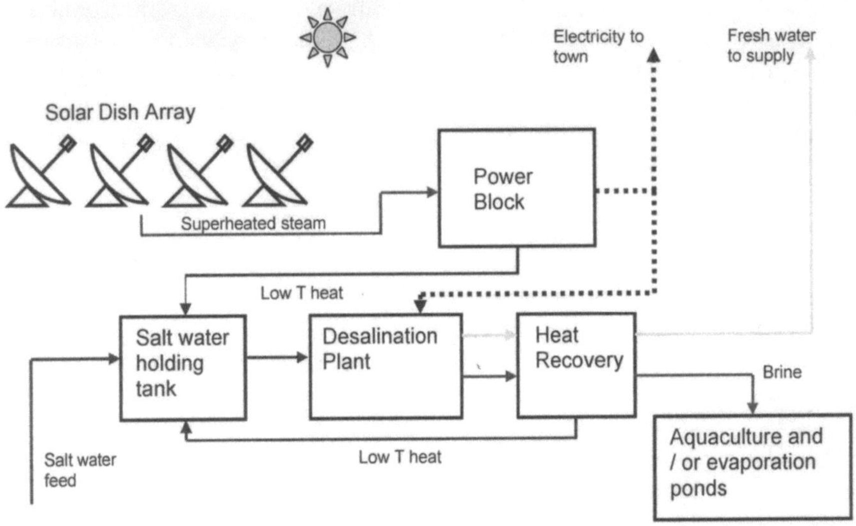

Fig. 3. Combined dish based solar thermal power generation and RO desalination (Burgess and Lovegrove, 2005)

present world electricity demand. Wind is generated by atmospheric pressure differences, driven by solar power. Of the total 173,000 TW of solar power reaching the earth, about 1200 TW (0.7%) is used to drive the atmospheric pressure system (Soerensen, 1979). This power generates a kinetic energy reservoir of 750 EJ with a turnover time of 7.4 days. This conversion process mainly takes place in the upper layers of the atmosphere, at around 12 km height (where the 'jet streams' occur). If it is assumed that about 1% of the kinetic power is available in the lowest strata of the atmosphere, the world wind potential is of the order of 10 TW, which is more than sufficient to supply the world's current electricity requirements.

Small decentralised water treatment plants combined with an autonomous wind energy convertor system (WECs) (Figure 4a) show great potential for transforming sea water or brackish water into pure drinking water (Koschikowski and Heijman, 2008) Also, remote areas with potential wind energy resources such as islands can employ wind energy systems to power seawater desalination for fresh water production. The advantage of such systems is a reduced water production cost compared to the costs of transporting the water to the islands or to using conventional fuels as power source. Different approaches for wind desalination systems are possible. First, both the wind turbines as well as the desalination system are connected to a grid system. In this case, the optimal sizes of the wind turbine system and the desalination system as well as avoided fuel costs are of interest. The second option is based on a more or less direct coupling of the wind turbine(s) and the desalination system. In this case, the desalination system is affected by power variations and interruptions caused by the power source (wind). These power variations, however, have an adverse effect on the performance and component life of certain desalination equipment. Hence, back-up systems, such as batteries, diesel generators, or flywheels might be integrated into the system.

Regarding desalinations, there are different technologies options, e.g. electro-dialysis or vapour compression. However, reverse osmosis is the preferred technology due to the low specific energy consumption. The only electrical energy required is for pumping the water

Fig. 4a (Upper Left) Wind Farm (Kalogirous, 2005); 4b (Right) Wind turbines and PV cells of Sureste SWRO plant (Sadhwani, 2008; IDA Conference , 2008) ; 4c (Lower Left) Wave Energy. The Aquabuoy 2.0 is a large 3 meter wide buoy tied to a 70-foot-long shaft. By bobbing up and down, the water is rushed into an acceleration tube, which in turn causes a piston to move. This moving of the piston causes a steel reinforced rubber hose to stretch, making it act as a pump. The water is then pumped into a turbine which in turns powers a generator. The electricity generated is brought to shore via a standard submarine cable. The system is modular, which means that it can be expanded as necessary (Chapa, 2007).

to a relatively high operating pressure. The use of special turbines may reclaim part of the energy. Operating pressures vary between 10 and 25 bars for brackish water and 50–80 bars for seawater (Eltawil et al., 2009). The Kwinana Desalination Plant, located south of Perth in Western Australia, is one example where wind power and reverse osmosis desalination have been successfully combined. The plant produces nearly 140 megalitres of drinking water per day (BlurbWire 2010). Electricity for the plant is generated by the 80 MW Emu Downs Wind Farm located in the state's Midwest region. The reverse osmosis plant was the first of its kind in Australia and covers several acres in an industrial park.

Recently, many medium and large scale water treatment and desalination plants are partially powered with renewable energy mainly wind turbines, PV cells or both. The energy demand of Sureste seawater reverse osmosis (SWRO) plant located in Gran Canaria, Canary Islands, of a capacity of 25,000 m^3/d is provided by a combination of PV cells (rooftop) with minor share of RO energy demand and the rest from the grid which consist of energy mix including wind energy (Figure 4b) (Sadhwani, 2008; IDA Conference, 2008).

2.3 Wave power desalination

Wave-powered desalination offers an environmentally sensitive solution for areas where there is a shortage of water and sufficiently energetic waves. Energy that can be harvested from oceans includes waves, tides and underwater oceanic currents (Figure 4c). Most of the work on wave energy conversion has focused on electricity production (Davies, 2005); any such converter could, in principle, be coupled to electrically-driven desalination plant, either with or without connection to the local electricity grid. Worldwide exploitable wave energy resource is estimated to be 2 TW, so it is a promising option for electricity generation. Thus, there is a potential option of coupling wave power with seawater reverse osmosis. A study by Davies (2005) focused on the potential of linking ocean-wave energy to desalination. They found that along arid, sunny coastlines, an efficient wave-powered desalination plant could provide water to irrigate a strip of land 0.8 km wide if the waves are 1 m high, increasing to 5 km with waves 2 m high. Wave energy availabilities were compared to water shortages for a number of arid nations for which statistics were available. The maximum potential to correct these shortages varied from 16% for Morocco to 100% for Somalia. However, the author noted that wave energy is mainly out-of-phase with evapotranspiration demand leading to capacity ratios of 3–9, representing the ratios of land areas that could be irrigated with and without seasonal storage. In a related study, Magagna and Muller (2009) described the development of a stand-alone, off-grid reverse osmosis desalination system powered by wave energy. The system consisted of two main parts; a high pressure pump (Wave Catcher) that allows generation of a high pressure head from low head differences, and a wave driven pump to supply the necessary head to the Wave Catcher. The high pressure pump could produce 6 MPa of pressure which is necessary to drive a RO membrane for desalination of water. We can argue that wave energy technology is still at prototype stage, there is no standard technology. Wave energy has also an intermittent and variable behaviour similar to wind energy.

2.4 Geothermal desalination

Geothermal energy is widely distributed along the world (White, 2002). This energy can be used for heat and electricity generation. Thus, there is a potential use for thermal (MED, MSF, MD, VC) and membrane (RO, EDR) desalination processes. Geothermal reservoirs can produce steam and hot water. Superheated dry steam resources are mostly easily converted into useful energy, generally producing electricity, which can be cheaper than that from conventional sources. Geothermal production of energy is 3rd highest among renewable energies. It is behind hydro and biomass, but before solar and wind. In the case of Iceland, 86% of space heating and 16% of electricity are supplied by geothermal energy.

Lack of water causes great distress among the population in large parts of the MENA countries (Middle East and North Africa). Small decentralised water treatment plants with an autonomous energy convertor system (WECs) can help solve this problem by transforming sea water or brackish water into pure drinking water (Koschikowski and Heijman, http://www.sciencedirect.com.scopeesprx.elsevier.com/science/journal /09582118 2008).

Considering that the energy requirements for desalination continues to be a highly influential factor in system costs, the integration of renewable energy systems with desalination seems to be a natural and strategic coupling of technologies (Tzen et al., 2004). As an example of the potential, the southern part of the country of Algeria consists almost entirely (i.e. 90%) of the great expanse of the Sahara Desert. This district has fresh water shortages but also has plenty of solar energy (Bouchekima, 2003), wind energy (Mahmoudi et al., 2009a, 2009b) and important geothermal reservoirs (Fekraoui and Kedaid, 2005; Mahmoudi et al., 2010). The amalgamation of renewable resources with desalination and water purification is thus very attractive for this district (Table 2). We will discuss this example in more detail in the Case Study section.

When using geothermal energy to power systems such as desalination plants we avoid the need for thermal storage. In addition, the energy output of this supply is generally stable compared to other renewable resources such as solar and wind power (Bourouni and Chaibi, 2005). Kalogirou (2005) has shown that the ground temperature below a certain depth remains relatively constant throughout the year. Popiel et al (2001) reported that one can distinguish three ground zones; surface, shallow and deep, with geothermal energy sources being classified in terms of their measured temperatures as low (<100 C), medium (100-150 C) and high temperature (>150 C), respectively.

Geothermal wells deeper than 100 m can reasonably be used to power desalination plants (Kalogirou, 2005). We can also envisage the utilization of geothermal power directly as a stream power in thermal desalination plants. Furthermore, with the recent progress on membranes distillation technology, the utilization of direct geothermal brine with temperature up to 60 ^{0}C has become a promising solution (Houcine, et al., 1999). Fridleifsson et al. (2008) has reported that electricity is produced by geothermal means in 24 countries, five of which obtained up to 22% of their needs from this source. Furthermore, direct application of geothermal energy for heating and bathing has been reported by 72 countries. By the end of 2004, the worldwide use of geothermal energy was 57 TWh/yr of electricity and 76 TWh/yr for direct use. Six developing countries are among the top fifteen states reporting direct use with China on the top of the list. Fridleifsson et al. (2008) goes on to argue that it is considered possible to increase the installed world geothermal electricity capacity from the current 10 GW to 70 GW with present technology, and to 140 GW with enhanced technology.

Desalting technologies						
Compression (MVC)	**Electrodialysis (ED)**	**Reverse osmosis (RO)**	**Multi-stage flash (MSF)**	**Multiple effect boiling (MEB)**	**Feed water salinity**	**Renewable energy sources technology**
			✓	✓	Seawater	Solar thermal
		✓			Seawater	Photovoltaic
	✓	✓			Brackish water	
✓		✓			Seawater	Wind
		✓			Brackish water	
				✓	Seawater	Geothermal

Table 2. Renewable energy sources (RES) desalination combinations (Mahmoudi et al., 2010)

2.5 Desalination using hydrostatic pressure

The potential exploitation of the hydrostatic pressure of seawater at a sufficient operative depth was considered by several investigators in the 1960s in view of increasing the energy efficiency of the then developing RO industrial desalination technology (Drude, 1967; Glueckstern, 1982). More recently, several configurations were proposed for fresh water production from seawater using RO and hydrostatic pressure: submarine, underground and ground-based (Reali et. al., (1997). In conventional surface-based industrial desalination plants applying RO technology, the freshwater flow at the membrane outlet is approximately 20–25% of the inlet seawater flow, depending on membrane type and characteristics. The resulting brine is disposed off in the sea. While RO installations generate the required pressure with high-pressure pumps, the submarine approach uses seawater hydrostatic pressure. The desalinated water, produced at about atmospheric pressure and collected in a submarine tank at the same working depth, is pumped to the sea surface. It was shown that this approach saves about 50% of the electricity consumption with respect to an efficient conventional RO plant (about 2–2.5 kWh/m3) since only the outlet desalinated water is pumped instead of the inlet seawater, thus reducing the pumping flow rate by 55–80% (Pacenti et al., 1999). The advantage of this configuration is also to avoid the pre-treatment of the inlet seawater, therefore saving costs for chemicals and equipment (Charcosset, 2009).

3. Case studies

3.1 Capacity building strategies and policies for desalination using renewable energies in Algeria

Among the major challenges facing the region are limited water and energy resources as well as risk management of the environment (Mahmoudi et al., 2009b; Laboy et al., 2009). Mahmoudi et al. (2010) has noted that due to the world economic crisis and the decreasing oil and gas reserves, decision makers in arid countries such as Algeria, need to review their policies regarding the promotion of renewable energies. Algeria is an oil and gas producer; hence decision makers believed that encouraging using renewable energies can affect the country's oil exports (Mahmoudi et al., 2009b). The country is also Africa's second-largest nation and the eleventh in the word in terms of land area, being bordered in the north by 1200 km of Mediterranean coastline.

In 1988, an ambitious program was established with the aim to expand the utilization of geothermal heated greenhouses in regions affected by frost; sites in eastern and southern region of the state. Unfortunately, this program was has been hampered by security concerns (Fekraoui and Kedaid, 2005). In the last few decades, much effort has also been expended to exploit the numerous thermal springs of the North and the hot water wells of the Saharian reservoir (Figure 5b). More than 900 MWt is expected to be produced in the future (Fekraoui and Kedaid, 2005). Geothermal energy represents one of the most significant sources of renewable energies in the case study area. This can be divided into two major structural units by the South Atlas Fault (Figure 5a); with Alpine Algeria in the north and the Saharan Platform in the south. The northern region formed by the Tellian Atlas, the High Plains and the Saharian Atlas. This part is characterized by an irregular distribution of its geothermal reservoirs (Figure 5b).

The Tellian nappes, constitute the main geothermal reservoirs. Hot ground water is generally at neutral pH, total dissolved salts (TDS) are up to 10 g/l and can reach a temperature in the

Fig. 5. a. Geological units of Northern Algeria (Mahmoudi et al., 2010; Kedaid, 2007)

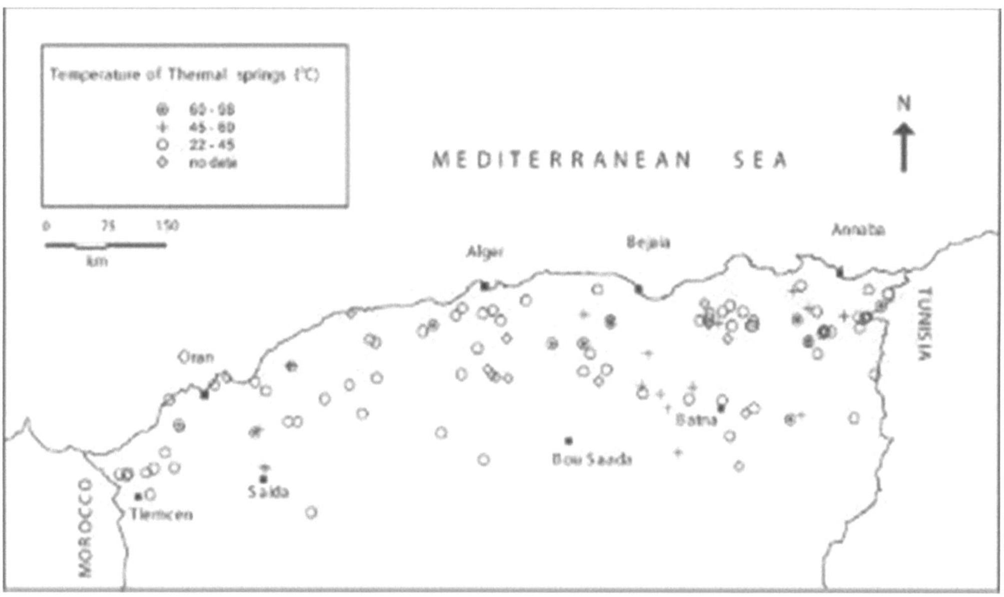

Fig. 5. b. Main thermal springs in northern Algeria (Mahmoudi et al, 2010; Kedaid, 2007)

range of 22°C to 98°C (Fekraoui and Kedaid, 2005). The southern region formed by the Algeria northern Sahara is characterized by a geothermal aquifer which is commonly named 'Albian reservoir'. The basin extends to Libya and Tunisia in the East and covers a total surface of 1 million km². This part of Algeria is estimated at 700 000 km² and contains approximately 40 thousand billion m³ of brackish groundwater water. The depth of the reservoir varies between 200 m in the west to more than 1000 m in the east. Deeper wells can provide water at 50 to 60 °C temperature, 100 to 400 L/s flow rate and average TDS (total dissolved solids) of 2g/L.

Mahmoudi et al., (2010) in a recent report proposed the application of geothermal sources to power a brackish water greenhouse desalination system for the development of arid and relatively cold regions, using Algeria as a case study (Figure 6a). He noted that countries which have abundant sea/brackish water resources and good geothermal conditions are ideal candidates for producing fresh water from sea/brackish water. The establishment of human habitats in these arid areas strongly depends on availability of fresh water. Geothermal resources can both be used to heat the greenhouses and to provide fresh water needed for irrigation of the crops cultivated inside the greenhouses.

University of Queensland's Geothermal Energy Center's director Hal Gurgenci was quoted as saying that geothermal-powered desalination systems could be a boon for small towns facing water shortage (Wash Technology, 2009). He went on to state that this is a clever combination where desalination is coupled with an agricultural function which is both cost-efficient and environmentally-friendly. Gurgenci said that while some of the geothermal resources may not be hot enough for power generation, they would be a perfect fit for thermal desalination of underground brackish aquifers. Studies indicate that for plants in the range of one to 100 megalitres (megalitre is one million liters) per day, thermal desalination technologies are more suitable than reverse osmosis especially if there is a cheap and abundant supply of heat. Geothermal heat can be used to heat and to humidify a greenhouse and produce fresh water at the same time.

The innovative idea of a seawater greenhouse was originally developed by Seawater Greenhouse Ltd in 1991 (Paton and Davies, 1996; Sablani et al., 2003). The first pilot was built and tested in the Canary Island of Tenerife in 1992, once known as the 'Garden of the Gods', but now arid and gravely damaged by excessive abstraction of ground water (Paton and Davies, 1996). The early results showed were promising and demonstrated the possibility to develop the technology in other arid regions. A modified and improved novel seawater greenhouse was constructed on Al-Aryam Island, Abu Dhabi, United Arab Emirates in 2000 (Davies and Paton, 2005). For both pilot studies the production of crops was excellent, and fresh water was successfully produced for the greenhouse irrigation proposes. In 2004 Seawater Greenhouse Ltd in collaboration with Sultan Qaboos University built a new pilot Seawater Greenhouse near Muscat, Oman (Figure. 6b) (Mahmoudi et al, 2008). The aim of the project was to demonstrate the technology to local farmers and organizations in the Arabian Gulf.

The brackish water is pumped and filtered from a well and sent into a ground heat exchanger where it absorbs heat from a geothermal fluid. This heat exchanger can be built of polyethylene to conserve costs. The heated brackish water is then fed in a cascade to the first evaporator then to the second evaporator. The brine can be circulated in the circuit several times until its concentration increases over an acceptable dissolved salt concentration. The concentrated brine is finally collected in a tank, where it is stored for later treatment or processing or reinjection. The evaporator is the entire front wall of the greenhouse structure. It consists of a cardboard honeycomb lattice and faces the prevailing wind. Hot brackish water trickles down over this lattice, heating and humidifying the ambient cooler air passing through into the planting area and contributing to the heating of the greenhouse. Fans draw the air through the greenhouse. Air passes through a second evaporator and is further humidified to saturation point. Air leaving the evaporator is nearly saturated and passes over the passive cooling system with a condenser (IC) immersed in a water basin. The fresh water condensing from the humid air is piped for irrigation or other purposes. This design can be scaled up to provide 10-20 kL/day while also helping greenhouse plant growing.

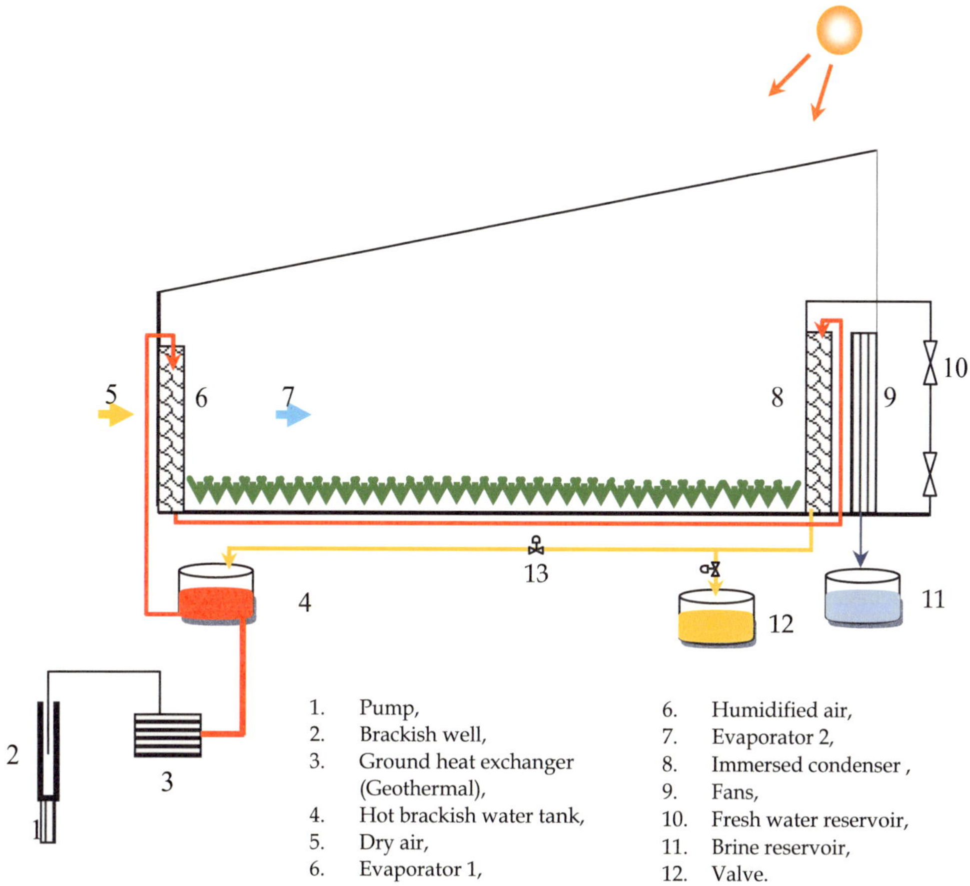

Fig. 6. Top (a): Process schematic for Brackish Water Greenhouse coupled to geothermal system (Mahmoudi et al., 2010). Greenhouse cover is normally plastic sheeting and the feed water to the pump is either brackish groundwater or sea water from a beach well; Bottom (b): The Seawater Greenhouse at Al-Hail, Muscat, Oman (Mahmoudi, 2008).

3.2 Seawater greenhouse development for the Arabian Gulf

A key feature in improving overall efficiency is the need to gain a better understanding of the thermodynamics of the processes and how the designs can be made more efficient. Goosen et al. (2003) determined the influence of greenhouse-related parameters on a desalination process that combines fresh water production with the growth of crops in a greenhouse. A thermodynamic model was used based on heat and mass balances. A software program developed by Light Works Limited, England was used to model thermodynamic analysis of the humidification/dehumidification Seawater Greenhouse system. The computer program consisted of several modules: Seapipe, Airflow, Evaporator 1, Roof, Planting Area, Evaporator 2, Condenser (air/water heat exchanger) (Figure 6a.).

Weather data for the year 1995 obtained from the Meteorological Office situated at Muscat, Sultanate of Oman, were used. The software needs a weather data file and a bathymetric (seawater temperature) file. These are specific to a location. The file contained transient data on solar radiation on a horizontal surface, dry bulb temperature and relative humidity of air, wind speed and wind direction. The bathymetric file contained temperature of the seawater at distance along the sea bed from the coast. The software program predicts the inside air conditions and water production for a given configuration/dimension of the greenhouse, and weather and bathymetric data. The program allows many parameters to be varied. These variables can be grouped into following categories: Greenhouse (i.e. dimension of the greenhouse, and its orientation, roof transparency of each layer, height of front and rear evaporative pads, height of the planting area, condenser); Seawater pipe; Air exchange Three parameters i.e. dimension of the greenhouse, roof transparency and height of the front evaporator were taken as variables. These parameters were varied as follows: Dimensions of Greenhouse (width x length): Area was kept constant at 10^4 m^2; 50 x 200, 80 x 125, 100 x 100, 125 x 80 and 200m x 50m; Roof Transparency 0.63 x 0.63 and 0.77 x 0.77; Height of the Front Evaporator 3 and 4m . The parameters kept constant were: Height of Planting Area = 4m; Height of the Rear Evaporator = 2m; Height of the Condenser = 2m, Orientation of Greenhouse = 40° N; Seawater pipe diameter = 0.9m, length = 5000m; Volumetric flow = 0.1m^3/s; Pit depth = -3m, height = 7.5m, wall thickness = 0.1m; Air change = 0.15 (fraction)/min; Fin spacing = 0.0025m and depth = 0.1506m.

Three climatic scenarios were considered. In the *temperate version* the temperature in the growing area is cool and the humidity high. This version is suited to lettuces, French beans, carrots, spinach, tomatoes, strawberries and tree saplings, for example. In the *tropical version* the temperature in the growing area is warm and the humidity very high. Examples of suitable crops include aubergines, cucumbers, melons, pineapples, avocados, peppers and pineapple. The design is similar to that of the temperate version, but the airflow is lower. The *oasis version* allows for a diversity of crops. This version is separated into temperate and tropical sections of equal area. The areas covered by these Greenhouses were 1080 m^2 (Temperate and Tropical) and 1530 m^2 (Oasis).

The overall water production rate increased from 65 to 100 m^3.d^{-1} when the width to length ratio increased from 0.25 to 4.00. Similarly the overall energy consumption rate decreased from 4.0 to 1.4 kW.h. m^{-3} when the width to length ratio increased from 0.25 to 4.00. Analyses showed that the dimensions of the greenhouse (i.e. width to length ratio) had the greatest overall effect on water production and energy consumption. The overall effects of roof transparency and evaporator height on water production were not significant. It was possible for a wide shallow greenhouse, 200 m wide by 50 m deep with an evaporator height of 2 m, to give 125 m^3.d^{-1} of fresh water. This was greater than a factor of two compared to the worst-case scenario with the same overall planting area (50 m wide by 200 m deep) and same evaporator height, which gave 58 $m^3.d^{-1}$. For the same specific cases, low power consumption went hand-in-hand with high efficiency. The wide shallow greenhouse consumed 1.16 kW.h. m^{-3}, while the narrow deep structure consumed 5.02 kW.h. m^{-3}.

While the Tropical version produced water most efficiently (i.e. lowest power consumption), it also produced the smallest surplus of water over that transpired by the crop inside. In this case the fan and pumps are run at their lowest rate to maintain high temperatures and humidity while still producing water in all conditions. Total fresh water production for the three climate scenarios was also calculated. One year's detailed meteorological data from Seeb Airport, Muscat was entered into the model to test the performance sensitivity for the various designs. The model results predicted that the Seawater Greenhouse would perform efficiently throughout the year, but with measurable variations in performance between the alternative versions. For example, the water production rate and energy efficiency results from the simulations using optimized and constant values for fan and pump speeds showed that the *temperate scenario* had almost double the water production rate per hectare compared to the *tropical scenario* (i.e. 20,370 m^3 /hectare compared to 11,574 m^3 /hectare) while the power consumption for the former was only slightly higher (i.e. 1.9 and 1.6 kWh/ m^3, respectively).

3.3 Water desalination with renewable energies in Baja Peninsula Mexico

In Mexico, specifically in the arid region of Baja California, there is not only an abundance of traditional renewable resources like sun and wind but also hot springs, tidal currents and tidal amplitudes of over six meters in the upper part of the Gulf of California (Alcocer et al., 2008). The National University of Mexico (UNAM) assessed the extent of these renewable resources and looked at ways to use them for desalinating sea water. It was established that at only 50 m depth, very high temperatures could be obtained, sufficient for use in binary geothermal power plants to generate electricity for desalination. It was also found that the amount of electrical power that could be generated with tidal storage and from deep sea hydrothermal vents was of the order of several thousands of MW. Many locations with hot sea water were discovered (Figure 7), the best being at Los Cabos. As soon as the water table was reached, a temperature of 85°C was found at 50 meters from the sea shore. Alcocer et al., (2008) went on to claim that to have sea water at 90°C is a real advantage for thermal desalination; because this suggests temperatures around 150°C as one drills deeper.

Using satellite imagery, hundreds of anomalous "hot spots", where large amounts of hot water reach the surface through geological fractures, were identified. The information was corroborated by measuring the water temperature directly in the field, and obtaining samples of the water to determine its isotopic composition and to better understand its origin. Although some of these hot springs and wells were on-shore, many were under-sea, close to the coast and at very shallow depths. The most important of these shallow, hot sea-water vents were near Puertecitos, Bahia Conception and Ensenada. In those cases field measurements and sampling required some diving. Each hot spring was different; some had high amounts of dissolved gases; others had lower salinity than the surrounding sea water; others had high sulfur content. This information was very valuable for the group in charge of designing the thermal desalinating equipment where the availability and the quality of the hot sea water were very important.

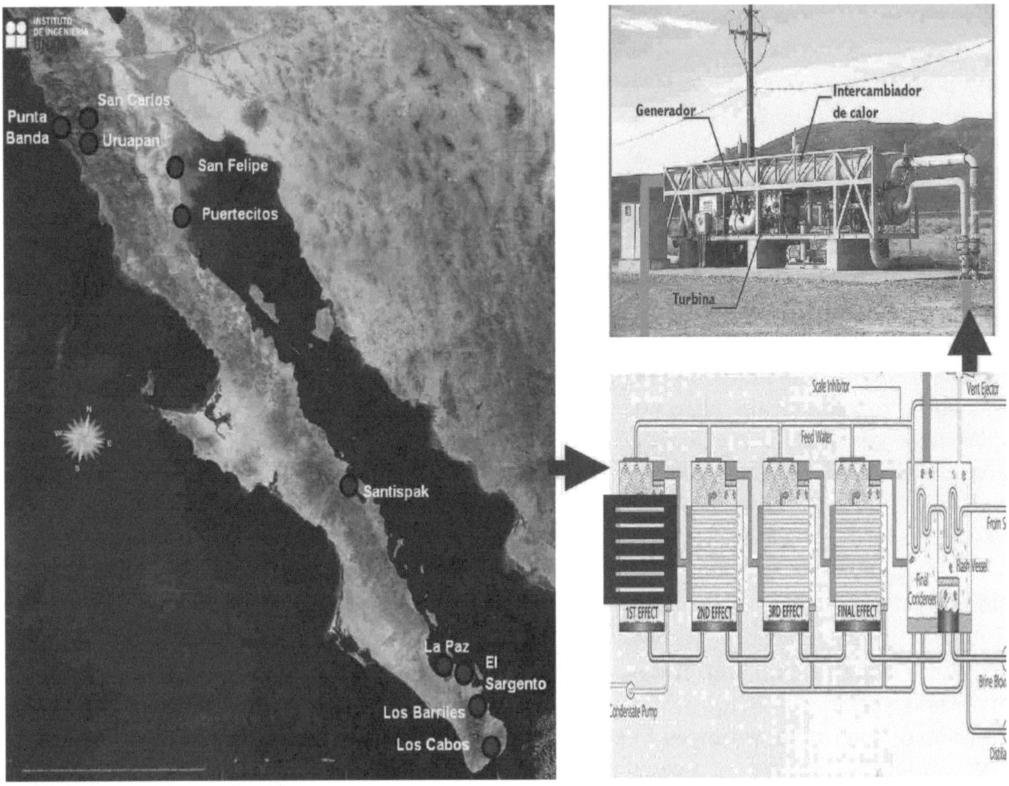

Fig. 7. *Left:* Locations with hot sea water at 50 m depths in the Baja Peninsula in Mexico (dots on map) with the Pacific Ocean on western side and the Gulf of California on the eastern side (Alcocer and Hiriart, 2008). *Lower right:* Schematic of multiple effects boiling/distillation (MEB) (or flashing) thermal desalination system with first effect highlighted in red. *Upper right:* Actual boiling or flashing effect (highlighted in black in previous schematic). (See also Table 2 for further information about MEB)

Hot water was used in a heat exchanger to heat clean seawater and then to decrease the pressure to produce instantaneous evaporation in a multistage set of chambers (Figure 7). The innovation introduced in the design was the use of hot sea-water to heat all the chambers, not just the first one as in a conventional Multi-Stage Flash (MSF) plant. The innovation can be considered as a combination of Multi-Effect Distillation/Boiling (MED or MEB) and Multi-Stage Flash (MSF), called "Multi-Flash with Heaters" (MFWH). Preliminary results indicated that for an initial temperature of 150°C, 4m^3 of sea water were required to produce 1m^3 of desalinated water. At an initial temperature of 80°C, 14 m^3 were required. For additional information on this topic see Rodrıguez et al. (1996).

3.4 The Kwinana desalinationp and wind farm in Perth, Western Australia

A larger scale seawater reverse osmosis (SWRO) plant located in Perth at Kwinana, with a capacity of 140,000 m^3/d, is the largest plant worldwide using renewable energy (BlurbWire 2010; Pankratz, 2008). The energy demand of the plant is 3.5 KWh/m^3 (185 GWh/a) which is met by wind energy. The plant's total energy consumption is offset by energy production from an 82 MW wind farm (Figure 8) with an expected power output of 272 GWh/a. The

Fig. 8. The Kwinana Seawater Reverse Osmosis Desalination (SWRO) Plant and Wind Farm in Perth, Western Australia . 8a (*Upper Left*) The SWRO Plant seaside location in Pert; 4b (*Upper Right*) The Emu Downs Wind Farm consisting of 48 Vestas wind turbines each with 1.65 MW generating capacity; 4c (*Lower*) The desalination plant, with 12 SWRO trains with a capacity of 160 megalitres per day and six BWRO trains delivering a final product of 144 megalitres per day. (Pankratz, 2008).

plant will be run continuously generating a constant base load electricity demand (Water Corporation, 2002). The reverse osmosis plant was the first of its kind in Australia and covers several acres in an industrial park near the suburb of Kwinana. The Wind Farm is a joint development between Stanwell Corporation and Griffin Energy. Construction of the project commenced in November 2005, and the project was commissioned in October 2006. Emu Downs Wind Farm consists of 48 Vestas wind turbines each with 1.65 MW generating capacity, a substation, interconnection to the main 132 kV electricity grid, administration and stores buildings, and a network of access roads. The wind farm is close to the coast, with a good quality wind resource that has increased wind speeds and reliability aligning with periods for peak power demand. Emu Downs Wind Farm is accredited under the Australian Government's Renewable Energy Electricity Act 2000 and as a Green Power Generator by the Sustainable Energy Development Authority. The wind farm contributes 270 GWh/year into the general power grid, offsetting the 180 GWh/year requirement from the desalination plant. The desalination plant, with 12 SWRO trains with a capacity of 160 megalitres per day and six BWRO trains delivering a final product of 144 megalitres per day, will have one of the world's lowest specific energy consumptions, due in part to the use of pressure exchanger energy recovery devices. The devices are isobaric chamber types which recover energy in the brine stream and deliver it to water going to the membrane feed at a net transfer efficiency at up to 98%. As a condition of its continued operation the Perth plant has a comprehensive environmental monitoring program, measuring the seawater intake and brine outfall. Excess water from the plant is stored in the hills dams.

3.5 Desalination using nuclear energy in Kazakhstan, India and Japan

The feasibility of integrated nuclear desalination plants has been proven with over 150 reactor-years of experience, chiefly in Kazakhstan, India and Japan (Stock Trading, 2010; Khamis, 2009; Misra, 2010). Large-scale deployment of nuclear desalination on a commercial basis will depend primarily on economic factors. Indicative costs are US$ 70-90 cents per cubic metre, much the same as fossil-fuelled plants in the same areas. One obvious strategy is to use power reactors which run at full capacity, but with all the electricity applied to meeting grid load when that is high and part of it to drive pumps for RO desalination when the grid demand is low. The BN-350 fast reactor at Aktau, in Kazakhstan, successfully supplied up to 135 MWe of electric power while producing 80,000 m^3/day of potable water over some 27 years, about 60% of its power being used for heat and desalination (Kadyrzhanov et al., 2007). In Japan, some ten desalination facilities linked to pressurised water reactors operating for electricity production have yielded 1000-3000 m^3/day each of potable water, and over 100 reactor-years of experience have accrued. MSF was initially employed, but MED and RO have been found more efficient there. The water is used for the reactors' own cooling systems. India has been engaged in desalination research since the 1970s. In 2002 a demonstration plant coupled to twin 170 MWe nuclear power reactors (PHWR) was set up at the Madras Atomic Power Station, Kalpakkam, in southeast India. This hybrid Nuclear Desalination Demonstration Project comprises a reverse osmosis (RO) unit with 1800 m^3/day capacity and a multi-stage flash (MSF) plant unit of 4500 m^3/day costing about 25% more, plus a recently-added barge-mounted RO unit. They incur a 4 MWe loss in power from the plant. In 2009 a 10,200 m^3/day MVC plant was set up at Kudankulam to supply fresh water for the new plant. It has four stages in each of four streams. An RO plant there supplies the plant's township. China Guangdong Nuclear Power has commissioned a 10,080 m^3/day desalination plant at its new Hongyanhe project at Dalian in the northeast. Much relevant experience comes from nuclear plants in Russia, Eastern Europe and Canada where district heating is a by-product. Large-scale deployment of nuclear desalination on a commercial basis will depend primarily on economic factors. The UN's International Atomic Energy Agency (IAEA) is fostering research and collaboration on the issue.

4. Environmental considerations and sustainability

Desalination of sea and brackish water requires large quantities of energy which normally results in a significant environmental impact if fossil fuels are used (e.g. CO_2 and SO_2 emissions, thermal pollution of seawater). The operating cost of different desalination techniques is also very closely linked to the price of energy. This makes the use of renewable energies associated with the growth of desalination technologies very attractive. Let us take a closer look at the environmental impacts that must be considered during utilization of geothermal resources as outlined by Rybach (2007), Lund (2007); Kagel, et al., (2005) and

Fridleifsson, et al. (2008). These include emission of harmful gases, noise pollution, water use and quality, land use, and impact on natural phenomena, as well as on wildlife and vegetation. The environmental advantages of renewable energy can be seen when comparing, for instance, a coal-fired power plant to a geothermal power plant; the former produces about 25 times as much carbon dioxide (CO_2) and sulfur dioxide (SO_2) emissions per MWh (i.e. 994kg vs. up to 40kg for CO_2, 4.71kg vs. up to 0.16kg for SO_2, respectively) (Lund, 2007; Fridleifsson, et al., 2008). However, in a geothermal power plant hydrogen sulfide (H_2S) also needs to be routinely treated and converted to elemental sulfur since about 0.08kg H_2S may be produced per MWh electricity generated. We can argue that this is still much better than oil-fired power plants and natural gas fired plants which produce 814 kg and 550 kg of H_2S per MWh, respectively.

Binary power plants and direct-use projects normally do not produce any pollutants, as the water is injected back into the ground after use without exposing it to the atmosphere. We can argue that the ready availability of inexpensive oil and natural gas reserves in such areas of the world as the Arabian Gulf may reduce the need for using renewable energy for desalination. However, looking at this more closely we see that this is non-sustainable since fossil fuels are non-renewable, and with a continually growing population there is an ever increasing demand on the use of fossil fuels for desalination. Take Saudi Arabia as a specific example; in 2008 total petroleum (i.e. oil and gas) production was 10.8 million bbl/d with internal oil consumption at 2.4 million bbl/d (i.e. about 25% of the total production) (U. S. Energy Information Administration, 2010). Most of the internal consumption was used for electricity generation and water desalination. The population is expected to increase from 30 million in 2010 to approximately 100 million by 2050 (U.S. Census Bureau, 2004). It has been estimated that by then 50% of the fossil fuel production will be used internally in the country for seawater desalination in order to provide fresh water for the people. This will reduce the state's income, increase pollution and is clearly non-sustainable. There are also concerns about the resulting political instability which could arise due to these effects (Cristo and Kovalcik, 2008). A possible solution to the environmental and sustainability problems is the increased use of renewable, including nuclear, energy sources for desalination (Lund, 2006, 2007; Fridleifsson, et al., 2008; Stock Trading, 2010)

5. Market potential, process selection and risk management

Kalogirou (2005) in a thorough study reviewed various renewable energy desalination systems is presented together with a review of a number of pilot systems erected in various parts of the world. The selection of the appropriate renewable energy desalination technology depends on a number of factors. These include, plant size, feed water salinity, remoteness, availability of grid electricity, technical infrastructure and the type and potential of the local renewable energy resource. Among the several possible combinations of desalination and renewable energy technologies, some seemed to be more promising in terms of economic and technological feasibility than others. However, their applicability strongly depended on the local availability of renewable energy resources and the quality of water to be desalinated. Kalogirou (2005) argued that the most popular combination of technologies is multiple effect boiling (MEB) with thermal collectors (Figure 9a) and reverse osmosis with photo voltaics (PV). PV is particularly good for small applications in sunny areas (Figure 9b). For large units, wind energy was reported to be more attractive as it requires less space than solar collectors. With distillation processes, large sizes were more attractive due to the relatively high heat loses elements in determining water costs when water is produced from desalination plants.

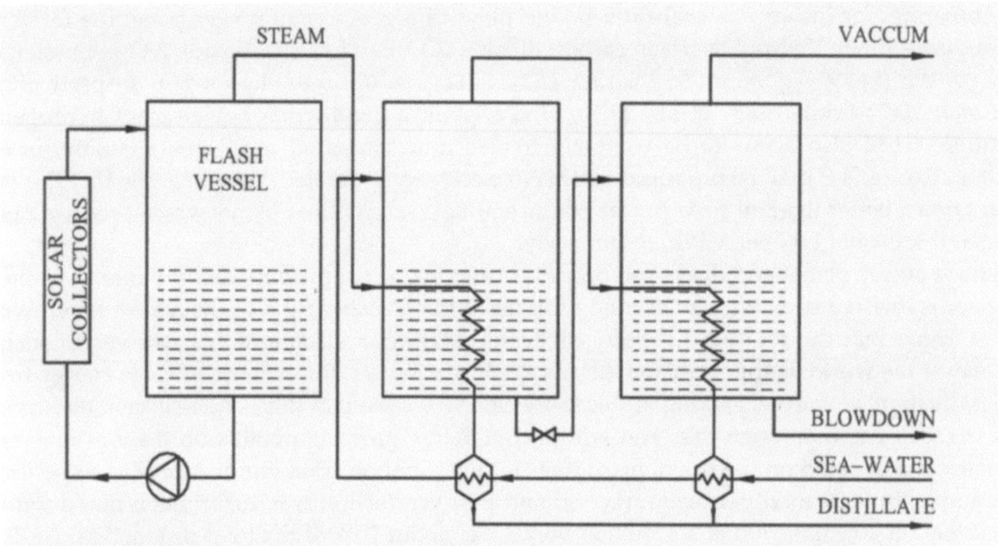

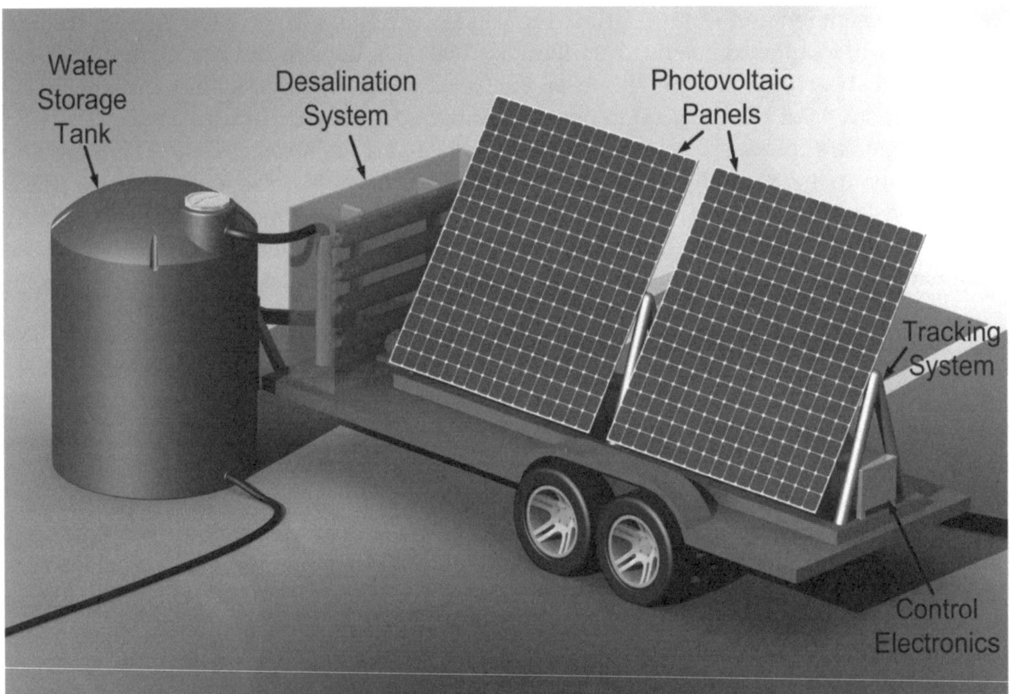

Fig. 9. a. (*Top*) Schematic of a multiple effect boiler (MEB) (Kalogirou (2005). 9b. (*Bottom*) A simple mobile solar powered reverse osmosis system (Dubowsky, 2010).

Many factors enter into the capital and operating costs for desalination: capacity and type of plants, plant location, feed water, labor, energy, financing, concentrate disposal, and plant reliability (USBR, 2003). For example, the price of desalted seawater is about 3 to 5 times the cost of desalting brackish water from the same size plant, due primarily to the higher salt content of the former. In any state or district, the economics of using desalination is not just the number of dollars per cubic meter of fresh water produced, but the cost of desalted water versus the other alternatives (e.g. superior water management by reducing consumption and improving water transportation). In many arid areas, the cost of alternative sources of water (i.e. groundwater, lakes and rivers) is already very high and often above the cost of desalting. Any economic evaluation of the total cost of water delivered to a customer must include all the costs involved. This includes the costs for environmental protection (such as brine or concentrate disposal), and losses in the storage and distribution system. The capital and operating costs for desalination plants have tended to decrease over the years due primary to improvements in technical efficiency (Reif, 2008). At the same time that desalting costs have been decreasing, the price of obtaining and treating water from conventional sources has tended to increase because of the increased levels of treatment being required to meet more stringent water quality standards. This rise in cost for conventionally treated water also is the result of an increased demand for water, leading to the need to develop more expensive conventional supplies, since the readily accessible water sources have already been used up.

Reif (2008) performed a profitability analysis and risk management of geothermal projects being implemented in Bavaria Germany. Reif's study concluded that the sensitive response of the project's rate of return to changes in the parameters of their computer simulations made it clear that geothermal projects are financially risky. For instance, every project faced the usual business risks, such as budget over-runs, increases in interest rates, and delays. Project management was used to limit these risks. It was recommended that the initiators of a plan must run profitability simulations in order to analyze varying scenarios before implementing the project. The results needed to be updated as the project progressed. Reif (2208) argued that reserves must always be planned for in the financing. In addition, business risks could also be limited by suitable structuring of the contracts with the partners in the project (e.g. drilling companies, power-plant supplier, and civil-engineering companies).

6. Concluding remarks

Water scarcity is an increasing problem around the world and there is a consensus that seawater desalination can help to alleviate this situation. Among the energy sources suitable to drive desalination processes, solar, wind, wave and geothermal energy, in addition to nuclear energy, are the most promising options, due to the ability to couple the availability of energy with water demand supply requirements in many world locations.

The cost for conventional desalination can be significant because of its intensive use of energy. However, the selection of a specific process should depend on a careful study of site conditions and the application at hand. Local circumstances will always play a significant role in determining the most appropriate process for an area. The best desalination system should be more than economically reasonable in the study stage. It should work when it is installed and continue to work and deliver suitable amounts of fresh water at the expected quantity, quality, and cost for the life of a project. Seawater desalination in itself is an expensive process, but the inclusion of renewable energy sources and the adaptation of desalination technologies to renewable energy supplies can in some cases be a particularly less expensive and economic way of providing water. The utilization of conventional energy sources and desalination technologies, notably in conjunction with cogeneration plants, is still more cost effective than solutions based on only renewable energies and, thus, is generally the first choice.

In closing, the world's water demands are rising considerably. Much research has been directed at addressing the challenges in using renewable energy to meet the power needs for desalination plants. Renewable energy technologies are rapidly emerging with the promise of economic and environmental viability for desalination. There is a need to accelerate the development of novel water production systems from renewable energies. These technologies will help to minimize environmental concerns. Our investigation has shown that there is great potential for the use of renewable energy in many parts of the world. Solar, wind, wave, geothermal and even nuclear sources could provide a viable source of energy to power both seawater and the brackish water desalination plants. Finally, it must be noted that part of the solution to the world's water shortage is not only to produce more water, but also to do it in an environmentally sustainable way and to use less of it. This is a challenge that we should well be able to meet.

7. References

Alcocer, S. M. and Hiriart G. (2008). An applied research program on water desalination with renewable energies. *Am. J. Environ. Sci.*, 4, 3, 204-211

Al-Hallaj, S.; Farid, M. M. and Tamimi, A. R. (1998). Solar desalination with a humidification-dehumidification cycle: performance of the unit. *Desalination*, 120, 273-280

BlurbWire (2010). Kiwana desalination plant (http://www.blurbwire.com/topics/Kwinana_Desalination_Plant) (accessed 8 August 2010)

Bouchekima B. (2003). A small solar desalination plant for the production of drinking water in remote arid areas of southern Algeria. *Desalination*, 159, 197–204

Bourouni K.; Martin R.and Tadrist, L. (1999a). The International Workshop on Desalination Technologies for Small and Medium Size Plants With Limited Environmental Impact, Rome , Italy (03/12/1998), vol. 122, no 2-3, pp. 301-313

Bourouni K.; Martin R.; Tadrist L. and Chaibi M.T. (1999b). Heat transfer and evaporation in geothermal desalination units, *Applied Energy*, 64, 1, 129-147

Bourouni. K, and Chaibi M. T. (2005). Application of geothermal energy for brackish water desalination in the south of Tunisia. *Proceedings World Geothermal Congress,* Antalya, Turkey, 24-29 April

Bourouni K.; Chaibi M. T. And Tadrist, L. (2001). Water desalination by humidification and dehumidification of air: state of the art. *Desalination*, 137, 167–176

Burgess, G. and Lovegrove, K. (2005). Solar thermal powered desalination: membrane versus distillation technologies (online: solar-thermal.anu.edu.au/wp-content/uploads/DesalANZSES05.pdf) (accessed 14 August 2010)

Cataldi, R., S. Hodgson and J. Lund (editors), 1999. Stories from a Heated Earth - Our Geothermal Heritage, Geothermal Resources Council, Davis, CA, 569 p.

Chapa, J. (2007) Wave energy: Aquabuoy 2.0 (http://www.inhabitat.com/2007/10/08/wave-energy-aquabuoy-20-wave-power-generator/) (accessed 16 August 2010)

Charcosset, C. (2009). A review of membrane processes and renewable energies for desalination. *Desalination*, 245, 214–231

Childs, W. D.; Dabiri, A. E.; Al-Hinai, H. A. and Abdullah, H. A. (1999) VARI-RO solar powered desalting study. *Desalination*, 125, 155-166

Cristo, M. W. and Kovalcik, M. P. (2008). Population pressure and the future of Saudi state stability. Master of Science Thesis, Naval Postgraduate School, Monterey CA, December, 77pp.

Davies P.A. and Paton C. (2005). The seawater greenhouse in the United Arab Emirates: Thermal modelling and evaluation of design options. *Desalination*, 173,103-111

Davies, P.A. (2005) Wave-powered desalination: resource assessment and review of technology, *Desalination*, 186, 97–109

Drude, B. C. (1967). Submarine units for reverse osmosis, *Desalination*, 2, 325–328

Dubowsky, S.; Wiesman, R.; Bilton, A.; Kelley, L. and Heller, R. (2010). Smart power and water for challenging environments. MIT current projects (http://robots.mit.edu/projects/KFUPM/index.html (accessed 20 August 2010)

Eltawil, M. A.; Zhengminga, Z. and Yuana, L. (2009). A review of renewable energy technologies integrated with desalination systems. *Renewable and Sustainable Energy Reviews*. 13, 9, 2245-2262

Fahrenbruch, A. and Bube, R. H. (1983). *Fundamentals of solar Cells*. Academic Press, Orlando FL, USA. 559pg

Fath, M. E. S. (1998). Solar desalination: A promising alternative for water provision with free energy, simple technology and a clean environment. *Desalination*, 116, 45-56

Fekraoui A. and Kedaid F. (2005). Geothermal resources and uses in Algeria: A country update report. *Proceedings World Geothermal Congress 2005*, Antalya, Turkey, 24-29 April.

Fernández J. L. and Norberto Chargoy N. (1990). Multistage, indirectly heated solar still. *Solar Energy*, 44, 4, 215-223

Fridleifsson, I. B.; Bertani, R.; Huenges, E.; Lund, J. W.; Ragnarsson, A. and Rybach, L. (2008). The possible role and contribution of geothermal energy to the mitigation of climate change. In: *IPCC Scoping Meeting on Renewable Energy Sources, Proceedings*, O. Hohmeyer and T. Trittin (Eds.) Luebeck, Germany, 20-25 January, 59-80

Glueckstern, P. (1982). Preliminary considerations of combining a large reverse osmosis plant with the Mediterranean–Dead Sea project, *Desalination*, 40, 143–156

Goosen M. F. A. and Shayya W. (1999). Water management, purification and conservation in arid climates. In: *Water Management, Purification and Conservation in Arid Climates: Volume I Water Management*. M. F. A. Goosen and W. H. Shayya (EDS.), Technomic Publishing Co., Lancaster, Pennsylvania, 1-6

Goosen, M. F. A.; Sablani, S.; Shayya, W. H.; Paton, C. and Al-Hinai, H. (2000). Thermodynamic and economic considerations in solar desalination", *Desalination*, 129, 63-89

Goosen, M. F. A.; Sablani, S.; Paton, C.; Perret, J.; Al-Nuaimi, A.; Haffar, J.; Al-Hinai, H. and Shayya, W. (2003). Solar energy desalination for arid coastal regions: development of a humidification-dehumidification seawater greenhouse, *Solar Energy*, 75, 413-419

Goosen, M. F. A.; Mahmoudi, H. and Ghaffour, N. (2010). Water desalination using geothermal Energy. *Energies*, 3, 1423-1442

Houcine, I.; Benjemaa, F.; Chahbani, M. H. and Maalej, M. (1999) Renewable energy sources for water desalting in Tunisia. *Desalination*, 125, 1–3,123–32

Huang, S. and Liu, J. (2010). Geothermal energy stuck between a rock and a hot place, *Nature*, 463, 293

International Desalination Association (IDA) Conference (2008). 360 Environmental Presentation, Gran Canaria, 2007 & Environmental Management Consultants Newsletter, Perth, 2008 (http://www.360environmental.com.au/).

Kadyrzhanov, K. K.; Lukashenko, S. N. and Lushchenko, V. N. (2007). Assessment of environmental impact of reactor facilities in Khazakstan. In: *Safety Related Issues of Spent Nuclear Fuel Storage*, J. D. B. Lambert and K. K. Kadyrzhanov (Eds), Springer. (http://www.springerlink.com/content/64270j0u525166u5/) (accessed 10 August 2010)

Kagel, A.; Bates, D. and Gawell, K. (2005). *A guide to geothermal energy and the environment*, Geothermal Energy Association, Washington, D. C., 75 p.

Kalogirou S. (2005). Seawater desalination using renewable energy sources. *Progress in Energy and Combustion Science*; 31, 242-281

Kedaid, F. Z. (2007). Database on the geothermal resources of Algeria. *Geothermics*, 36, 3, 265-275

Khamis, I. (2009). A global overview on nuclear desalination. *Int. J. of Nuclear Desalination*, 3, 4, 311-328

Koschikowski, J. and Heijman, B. (2008). Renewable energy drives desalination processes in remote or arid regions. *Membrane Technology*, 8, August, 8-9

Laboy, E.; Schaffner, F.; Abdelhadi, A. And Goosen, M.F.A. (Eds), (2009). *Environmental Management, Sustainable Development & Human Health,* Taylor & Frances (Balkema NDL) London. 35 Chapts, 594 pp ISBN: 978-0-415-46963-0

Lu, H.; Walton, J. C. and Swift, A. H. P. (2000). Zero discharge desalination. *Int Desalination Water Reuse Q*, 10, 3, 35–43

Lund, J. W. (2006). Chena Hot Springs, *Geo-Heat Center Quarterly Bulletin*, 27, 3 (September), Klamath Falls, OR, 2-4

Lund, J. W. (2007). Characteristics, development and utilization of geothermal resources, *GHC Bulletin*, June, 1-9

Magagna, D. and Muller, G. (2009). A wave energy driven RO stand-alone desalination system: initial design and testing. *Desalination and Water Treatment*, 7, 47–52

Mahmoudi, H.; Spahis, N.; Goosen, M.F.A.; Ghaffour, N.; Drouiche N. and Ouagued, A. (2010). Application of geothermal energy for heating and fresh water production in a brackish water greenhouse desalination unit: A case study from Algeria, *J. Renew Sustain Energy Rev*, 14, 1, 512-517

Mahmoudi, H.; Spahis, N.; Goosen, M.F.A.; Sablani, S.; Abdul-Wahab, S.; Ghaffour, N. and Drouiche, N. (2009a). Assessment of wind energy to power solar brackish water greenhouse desalination units: A case study from Algeria. *J. Renew Sustain Energy Rev*, doi:10.1016/j.rser.2009.03.001

Mahmoudi, H.; Ouagued, A. and Ghaffour, N. (2009b). Capacity building strategies and policy for desalination using renewable energies in Algeria, *J. Renew Sustain Energy Rev*; 13, 921–926

Mahmoudi, H. ; Abdul-Wahab, S. A.; Goosen, M. F. A. ; Sablani, S. S. ; Perret, J. and Ouagued, A. (2008). Weather data and analysis of hybrid photovoltaic-wind power generation systems adapted to a seawater greenhouse desalination unit designed for arid coastal countries, *Desalination*, 222, 119–27

Misra, B. M. (2010). Sustainable desalination technologies for the future. *Int. J. of Nuclear Desalination*, 4, 1, 37-48

Ophir, A. (1982). Desalination plant using low grade geothermal heat, *Desalination,* 40, 125–32

Pacenti, P.; de Gerloni, M.; Reali, M.; Chiaramonti, D.; Gärtner, G. O.; Helm, P. and Stöhr, M. (1999). Submarine seawater reverse osmosis desalination system, *Desalination*, 126, 213–218

Pankratz, T. (2008). Water desalination report (WDR), *MEDRC Workshop*, Muscat (www.waterdesalreport.com)

Paton, C. and Davies, A. (1996). The seawater greenhouse for arid lands. *Proc. Mediterranean Conference on Renewable Energy Sources for Water Production*, Santorini, Greece, 10–12 June

Popiel, C.; Wojtkowiak, J. and Biernacka, B. (2001). Measurements of temperature distribution in ground. *Exp Thermal Fluid Sci,* 25, 301–9

Reali, M.; de Gerloni, M. and Sampaolo, A. (1997). Submarine and underground reverse osmosis schemes for energy-efficient seawater desalination, *Desalination*, 109, 269–275

Reif, T. 2008 Profitability analysis and risk management of geothermal projects, *GHC Bulletin*, January, 1-4

Rodrıguez, G.; Rodrıguez, M., Perez, J. and Veza, J. (1996). A systematic approach to desalination powered by solar, wind and geothermal energy sources, In: *Proceedings of the Mediterranean conference on renewable energy sources for water production*. European Commission, EURORED Network, CRES, EDS, Santorini, Greece, 10–12 June, 20–5

Rybach, L. (2007). Geothermal sustainability, *Proceedings European Geothermal Congress*, Unterhaching, Germany, 5 p.

Sablani, S.; Goosen, M. F. A.; Paton, C.; Shayya, W. H. and Al-Hinai, H. (2003). Simulation of fresh water production using a humidification–dehumidification Seawater Greenhouse. *Desalination*, 159,283–288

Sadhwani, J. J. and Veza, J. M. (2008). Desalination and energy consumption in Canary Islands, *Desalination* 221, 143–150

Schwarzer K., Vieira M.E., Faber C. and Müller C. (2001) *Solar thermal desalination system with heat recovery. Desalination*, 137 (1-3), 23-29.

Serpen, U.; Aksoy, N. and Öngür, T. 2010. 2010 present status of geothermal energy in Turkey, *Proceedings of Thirty-Fifth Workshop on Geothermal Reservoir Engineering*, Stanford University, Stanford, California, 1-3 February SGP-TR-188

Soerensen, B. (1979). *Renewable Energy*. London: Academic Press.

Stefansson, V. (2005). World geothermal assessment. *Proceedings of the World Geothermal Congress*, Antalya, Turkey, 24-29 April (http://iga.igg.cnr.it)

Stock Trading (2010). How quickly will Saudi Arabia turn to nuclear vitality? (July 3rd, 2010 by admin) (http://www.profitablenicheinnovation.com/trading/day-trading/stock-trading-how-quickly-will-saudi-arabia-turn-to-nuclear-vitality) (accessed on 27 July 2010)

Szacsvay, T.; Hofer-Noser, P. and Posnansky, M. (1999). Technical and economic aspects of small-scale solar-pond-powered seawater desalination systems. *Desalination,* 122, 185–93

Tester, J. W.; Anderson, B. J.; Batchelor, A. S.; Blackwell, D. D.; DiPippo, R.; Drake, E. M.; Garnish, J.; Livesay, B.; Moore, M. C.; Nichols, K.; Petty, S.; Toksoz, M. N.; Veatch, R. W.; Baria, R.; Augustine, C.; Murphy, E.; Negraru, P. and Richards, M. (2007). Impact of enhanced geothermal systems on U. S. energy supply in the twenty-first century, *Phil Trans R Soc A,* 365, 1057-1094 (http://www1.eere.energy.gov/geothermal/future_geothermal.html) ISBN: 9780615134383

Tzen, E.; Theofilloyianakos, D and Karamanis, K. (2004). Design and development of a hybrid autonomous system for seawater desalination. *Desalination,* 166, 267–274

USBR (U.S. Bureau of Reclamation) (2003). Cost estimating procedures. In *Desalting Handbook for Planners,* 3rd ed.; Desalination and Water Purification Research and Development Program Report No. 72; United States Department of Interior, Bureau of Reclamation, Technical Service Center, US Government Printing Office: Washington DC, USA, Chapter 9, pp. 187–231

U.S. Census Bureau, (2004). Table A-4 Population by Region and Country 1950-2050, (http://www.census.gov/prod/2004pubs/wp-02.pdf)

U. S. Energy Information Administration/ International/ Country Briefs/ Saudi Arabia (2010). (http://www.eia.doe.gov/cabs/Saudi_Arabia/Profile.html) (Accessed 27 July 2010)

Wash Technology, (2009). Geothermal desalination: hot rocks key to producing low cost fresh water. 9 December

Water Corporation (2002) Perth Metropolitan Desalination Proposal, Perth

White, D. E. and Williams, D. L. (Eds.), (1975). Assessment of geothermal resources of the United States - 1975, U.S. Geological Survey Circular 727, U.S., Government Printing Office, 155 p.

Wright, J. D. (1982) Selection of a working fluid for an organic Rankine cycle coupled to a salt-gradient solar pond by direct-contact heat exchange, *J. Sol. Energy Eng.*, 104, 4, 286 293

Wright, M. (1998). *Nature of Geothermal Resources, in Geothermal Direct-Use Engineering and Design Guidebook,* John W. Lund,(Ed), Geo-Heat Center, Klamath Falls, OR, 27-69

6

Seawater Desalination: Trends and Technologies

Val S. Frenkel, Ph.D., P.E., D.WRE.
Kennedy/Jenks Consultants,
USA

1. Introduction

Figure 1 below provides information on our planet's available water resources which do not allow too many alternatives.

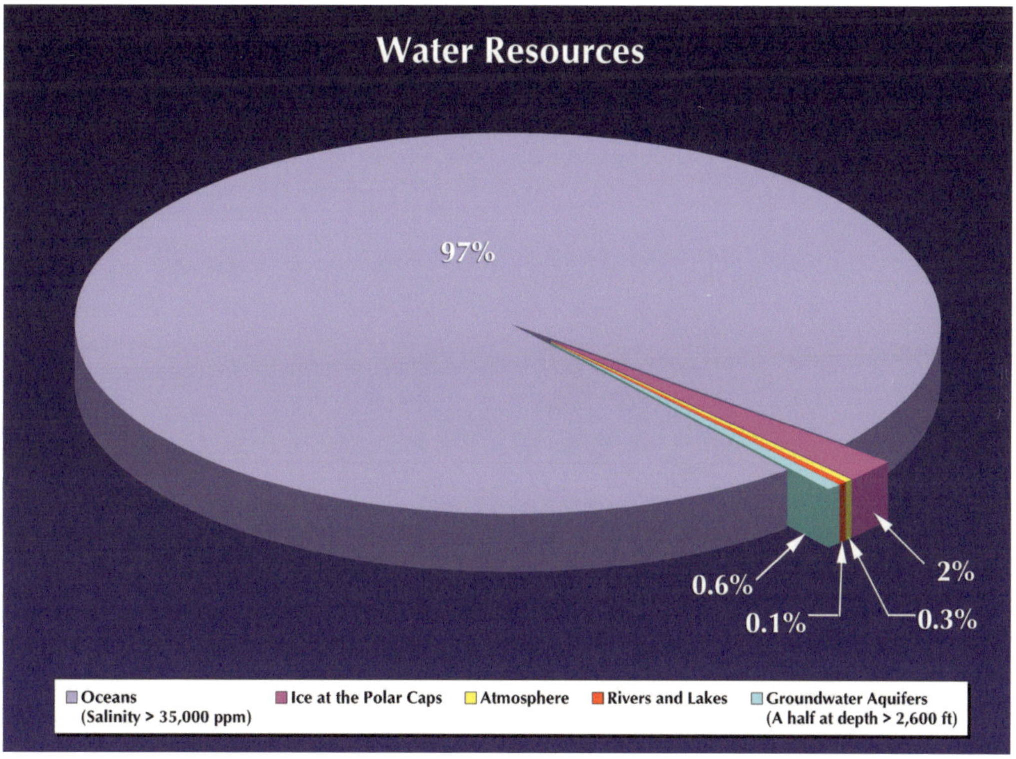

Fig. 1. Water Resources on the Earth

With 97% of available water represented by salty water with the Salinity Level > 35 g/l, the largest possible source of alternative water supply requires and will require desalination.

The conventional water treatment technologies have been known and widely used for centuries, and some, like media filtration, were applied thousands of years ago, while membranes were introduced to water treatment just in the second half of the 20th Century. Development of the first high pressure membrane, Reverse Osmosis (RO) was claimed at University of California in Los Angeles (UCLA) in 1962, and commercialized by the early 1970s.. The low pressure membranes, Microfitration (MF) and Ultrafiltration (UF) were commercialized for drinking water treatment just about one decade ago. Because they provide significant technical benefits and have become cost-competitive, membrane technologies are rapidly displacing and replacing traditional processes verified by the centuries.

The oldest desalination methods are based on evaporating water and collecting the condensate. The best known commercially applied thermal technologies are:

- **Multi Stage Flash (MSF)**
- **Multi Effect Distillation (MED**
- **Vapor Compression (VC)**

While MSF, MED, and VC use thermal power to separate water from the brine, **Electrodialisys Reversal (EDR)** uses high voltage current to remove Cations and Anions from the stream.

The newest commercial technology for Desalination is based on membrane treatment. **Reverse Osmosis (RO)** and **Brackish Water Reverse Osmosis (BWRO)** or **Sea Water Reverse Osmosis (SWRO)**, are the fastest growing desalination techniques with the greatest number of installations around the globe. Desalination by RO is beginning to dominate the current and future desalination markets. As seen in the chart below, the number of membrane desalination installations is close to 80% of all desalination facilities.

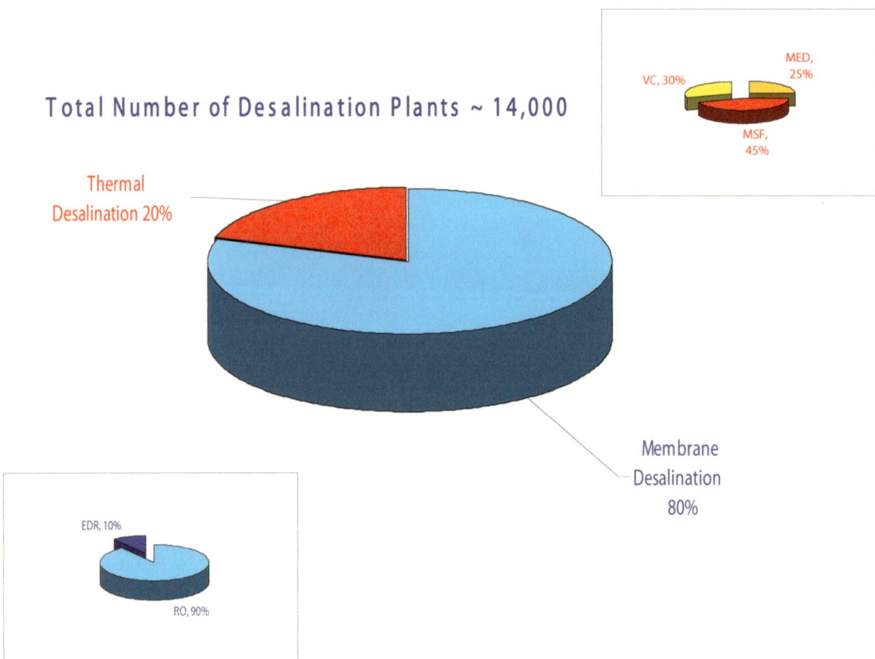

Fig. 2. Number of desalination plants worldwide. RO - Reverse Osmosis, EDR - Electro Dialysis Reversal, MSF - Multi Stage Flash, MED - Multi Effect Distillation, VC - Vapor Compression

The first RO desalination membranes were developed in the first half of the 20th Century. Desalination by RO entered the commercial market in the early 1970s when the membrane manufacturing process became efficient enough to produce desalted water that was competitive to thermal processes, and when the technological process for RO desalination was well established.
While leading in the number of installations, desalination by RO still provides only a comparable capacity to the thermal processes:

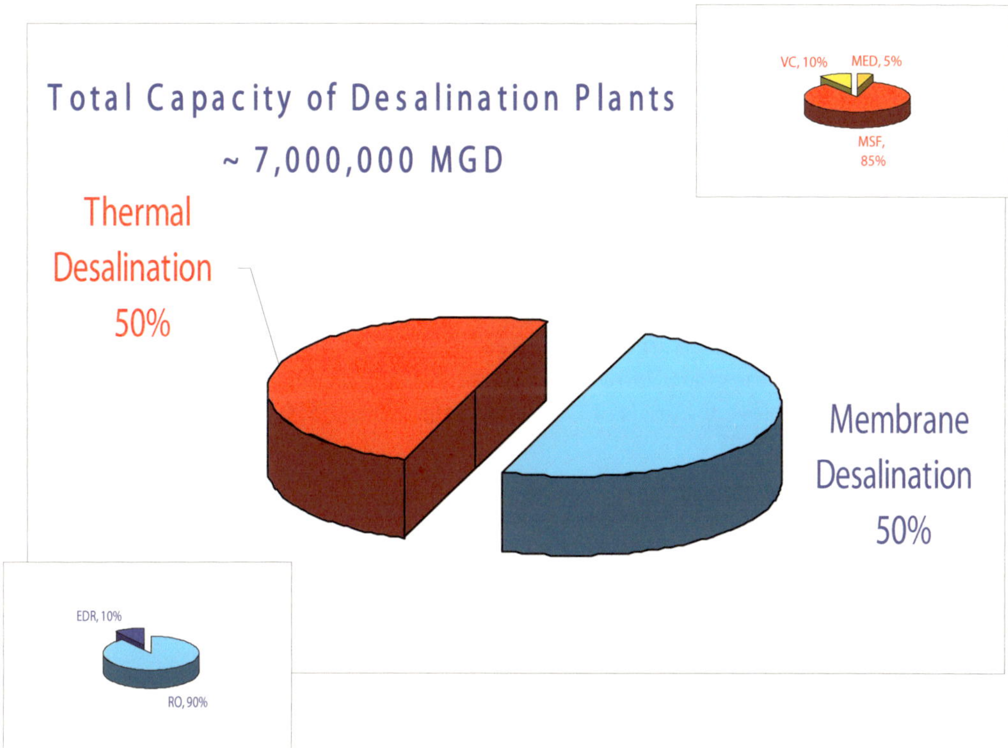

Fig. 3. Desalination Capacity Worldwide. RO - Reverse Osmosis, EDR - Electro Dialysis Reversal, MSF - Multi Stage Flash, MED - Multi Effect Distillation, VC - Vapor Compression

The lack of correlation between the number of installations and overall capacities can be explained by the development of membrane desalination. Thermal processes have been on the market for more than five decades and most of them provide relatively high capacities. However, this ratio is expected to change significantly because most of the desalination systems currently designed, constructed, and considered for construction are based on membrane technology. For example, the largest membrane desalination plant in the U.S. is the Tampa Bay SWRO, with a capacity of 25 MGD / 95,000 m3/day (and provision for up to 35 MGD / 130,000 m3/day expansion). The plant went into the operation in 2003. The newly considered Carlsbad desalination plant capacity 50 MGD / 190,000 m3/day is planning to use SWRO membrane technology. A much larger membrane desalination facility was commissioned in May 2005 in Israel, the Ashkelon SWRO, with a capacity of 44

MGD / 166,000 m3/day, which was expanded to 88 MGD / 330,000 m3/day at the end of 2005.

When different technologies were evaluated for these large desalination facilities, SWRO provided the most cost-effective solution for all considerations: capital expenditures, O&M, and cost per 1,000 gallons of treated water based on 20 – 30 years of operation.

As positive results, such as cost-effectiveness, emerge from large SWRO facilities in operation, they will provide more security and confidence in building SWRO plants with larger capacities.

2. Membrane technologies

Membranes are becoming a common commodity in water treatment, with four major membrane categories that depend on the membrane pore sizes in commercial use at the present time:

- Microfiltration (MF) - screens particles from 0.1 to 0.5 microns
- Ultrafiltration (UF) - screens particles from 0.005 to 0.05 microns
- Nanofiltration (NF) - screens particles from 0.0005 to 0.001 microns
- Reverse Osmosis (RO) - ranging molecular size down to 10 MWCO

The appropriate membrane treatment process for the removal of different constituents from water can be traced in the chart below. All four membrane categories are commonly used in water treatment to achieve the goals of Drinking Water Guidelines and Standards, as well as

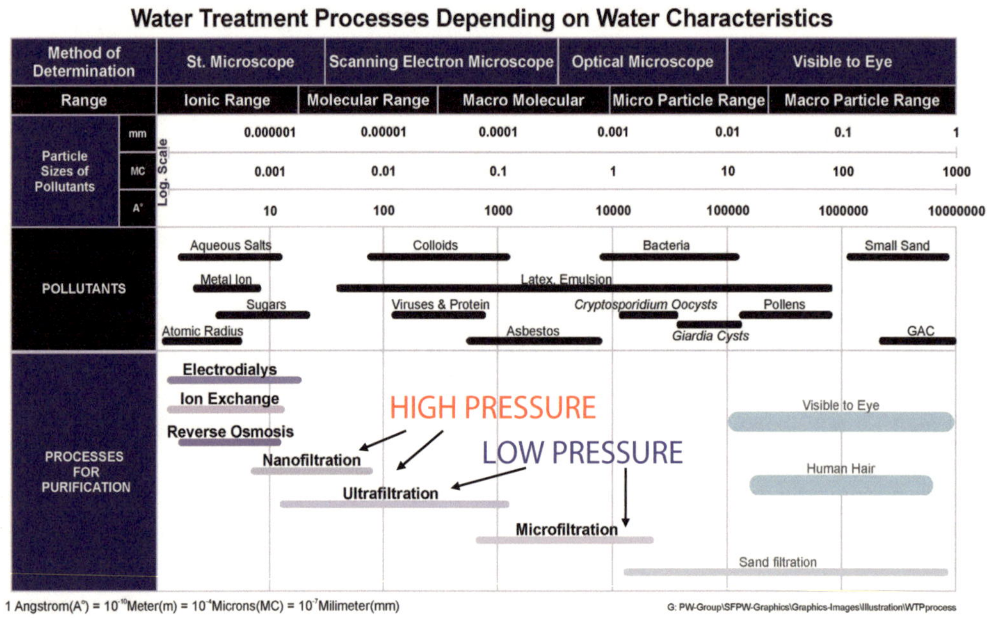

Fig. 4. Water Treatment Spectrum

to produce desalted and/or Ultra Pure Water (UPW) for different industrial and other needs, such as power plants make-up water, electronic ships manufacturing, food industry, pharmaceutical, medical, and others.

Water impurities depending on size and hydraulic properties:

- Suspended Solids (expressed as TSS, TVSS, Turbidity)
- Colloids (expressed as SDI)
- Dissolved Solids (expressed as TDS, TVDS)

Nature of water impurities:

- Mineral nature (non organic)
- Organic nature

Membrane Shape Type:

- Spiral Wound
- Hollow Fiber
- Flat Sheet

Membrane Type depending on driven pressure:

- Pressure Driven (MF, UF, NF and RO)
- Immersed, Vacuum Driven (MF only)

The first commercial use of membrane technology was desalination by RO, the process known decades ago and commercialized in the early 1960s.

3. Energy recovery

Implementation of efficient Energy Recovery Turbines (ERT) into the RO desalination technologies boosted growth of RO plants worldwide. There are three major types of ERT:

- Pelton Wheel
- Francis type
- Reversal pump

Recent developments in RO energy conservation brought the following technologies into the market:

- Double Work Exchanger Energy Recovery DWEER
- Hydraulic turbo-charger
- Pressure/Work Exchanger and others

From the ERT, the most popular and reliable was the first type, Pelton Wheel ERT, which can save up to 30% and higher of the energy consumed by high pressure RO pumps, represents the highest O&M expenditure for RO plant operation. Of the latest developments, DWEER and other systems can save up to 90-95% of the brine energy. For example, for high salinity water with the RO recovery of 40%, the overall energy savings can be as high as 50% or more of the energy for the entire plant operation.

4. Desalination statistics

Table 1 provides more detailed information and figures on the global production of desalinated water, by process and plant capacity.

Desalting process	Percentage	Capacity (×10^6 m^3/day)	Capacity (10^6 gal/day)	No. of plants
Unit capacity				
100-60 000 m^3 /day				
Multistage flash	44.4	10.02	2,204	1,244
Reverse osmosis	39.1	8.83	1,943	7,851
Multiple effect	4.1	0.92	202	682
Electrodialysis Reversal	5.6	1.27	279	1,470
Vapor compression	4.3	0.97	213	903
Membrane softening	2.0	0.45	99	101
Hybrid	0.2	0.05	11	62
Others	0.3	0.06	13	120
	100.0	22.57	4,965	**12,433**
Unit capacity				
500-60 000 m^3/ day				
Multistage flash	46.8	10.00	2,200	1,033
Reverse osmosis	37.9	8.10	1,782	3,835
Multiple effect	3.8	0.81	178	653
Electrodialysis Reversal	4.7	1.00	220	230
Vapor compression	4.2	0.90	198	486
Membrane softening	2.1	0.45	99	64
Hybrid	0.2	0.04	9	27
Others	0.23	0.05	11	11
EDI	0.05	0.01	2	97
	100.0	21.36	4,699	6,436
Unit capacity				
4000-60 000 m^3/day[1]				
Multistage flash	64.0	9.27	2,039	496
Reverse osmosis	25.7	3.72	818	613
Multiple effect	3.6	0.52	114	48
Electrodialysis Reversal	2.1	0.31	68	60
Vapor compression	1.9	0.28	62	42
Membrane softening	2	0.36	79	50
Hybrid	0	0.02	4	2
Others	0	0.00	0	0
	100.0	14.48	3,186	**1,311**

Table 1. Summary of worldwide desalination capacity to 1998, split by plant type and process capacity range. Source: 1998 IDA Worldwide Desalting Plants. Inventory Report No. 15. Wangnick Consulting GmbH.

Today, the desalination capacity of membranes using RO reaches close to 3,500,000 MGD / 14 000 000 000 m3/day total capacity, which is half of the entire desalination capacity worldwide. Membrane desalination is the fastest growing technology, and is expected to become the prevalent desalination technology for the 21st century.

Microfiltration and ultrafiltration technologies became commercial in the late 1980s -1990s. The major issues in membrane developments are:

- Increase membrane flux
- Decrease trans-membrane pressure
- Increase particles and salt rejection
- Extend membrane lifetime
- Improve operational process including back-wash technique and CIP cleaning

To address these issues, improve membrane performance, and bring membrane applications to a new level, the following membrane characteristics and parameters are subjects for current and future research and development:

- Improving pore shape, uniformity, and distribution
- Upgrading hydrophilic properties
- Increasing overall porosivity or pore density of membranes
- Developing more sophisticated and cost-effective membrane materials
- Improving the membrane manufacturing process

During the past 10 to 20 years, the availability, efficiency, and reliability of membrane systems have increased significantly, while the capital and operational costs of these

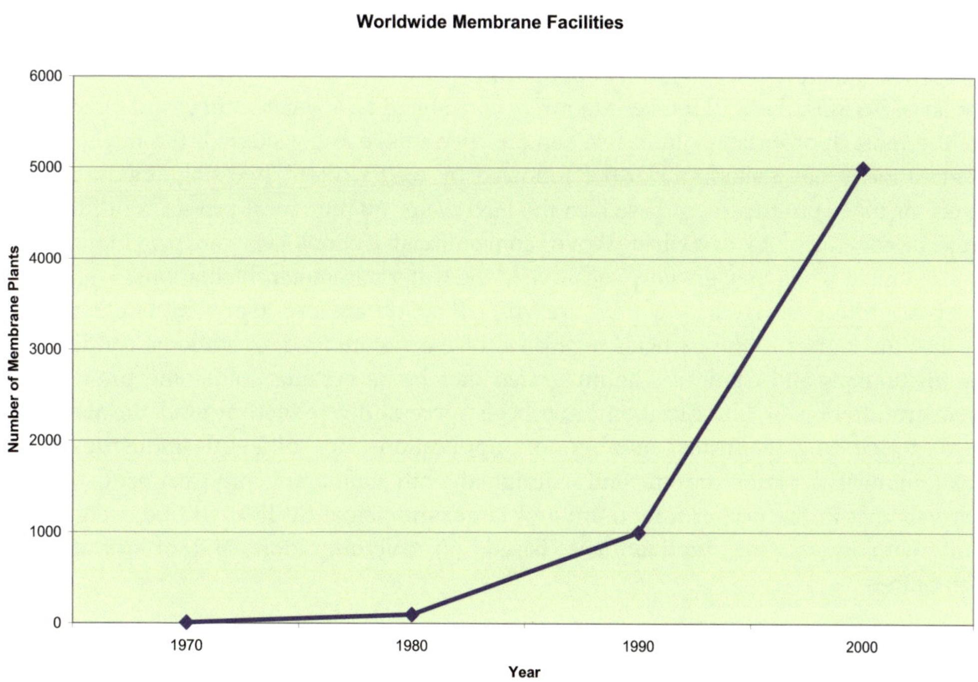

Fig. 5. Worldwide Membrane Facilities (not including Desalination)

systems have dropped considerably. These developments have resulted in worldwide exponential growth of membrane treatment plants. The latest AMTA information on the number of membrane plants worldwide (not including desalination systems) is provided in the following chart.

As a result of the growing membrane industry, membrane prices have trended lower during the last decade. Some membrane manufacturers supply membranes to membrane system integrators, while some suppliers also act as membrane integrators as well. When manufacturers are deciding whether or not to be system integrators, they evaluate many criteria, including competition with the professional system integrator companies. Since high pressure membranes such as RO and NF are no longer on the market, most of the RO/NF membrane manufacturers do not act as system integrators. Moreover, the industry has reached a consensus on the standard sizes for RO and NF membranes. The most widely used RO/NF elements are 2.5", 4" and 8" in diameter and 40" and 60" long. Currently, RO elements are sized 16", 17.5", 18 and 18.5" diameter in the commercialization process to increase the amount of active membrane area provided by each element.

The low pressure MF/UF membranes are a relatively new technology, and no industrial standard has yet been currently established. As the low pressure market grows, the low pressure membranes will likely develop their own across-the-industry standard in order to optimize MF/UF plant designs and reduce project costs. As a result of the current situation, most of the MF/UF membrane manufacturers are also acting as system integrators.

When comparing open sea intakes to beach wells, the latter have preference:

- Less turbidity and solids in the feed and less pre-treatment is required
- Less chemical consumption by pretreatment
- Less seasonal and daily fluctuation of water quality and temperature, which allows better and more reliable facility operation

For large SWROs, the well intakes are rarely considered as a water source, and open intakes are the most appropriate solution. When an open intake is considered, the most common pre-treatment may include a clarifier followed by media filters, two-stage media filters in series, or other pre-treatment based on the feed water quality, local project conditions, and project economics. As described above, conventional technologies can provide sufficient pre-treatment when they are very sensitive to the water parameter fluctuations.

The best prototype may be low pressure MF/UF membrane pre-treatment, which provides an absolute barrier to the particles regardless of the system load, operational conditions, or the fluctuations and changes. The integrated membrane systems containing pre-treatment by microfiltration or ultrafiltration have been successfully piloted around the world and have found a significant number of applications in different industries. Some small municipal plants are currently designed with membrane pre-treatment, and it is believed that in the near future, more and larger municipal SWROs will be designed and built with membrane pretreatment based on microfiltration and/or ultrafiltration membranes.

5. SWRO design considerations

Figure 6 below can be a guide on the major criteria to be considered when designing SWROs:

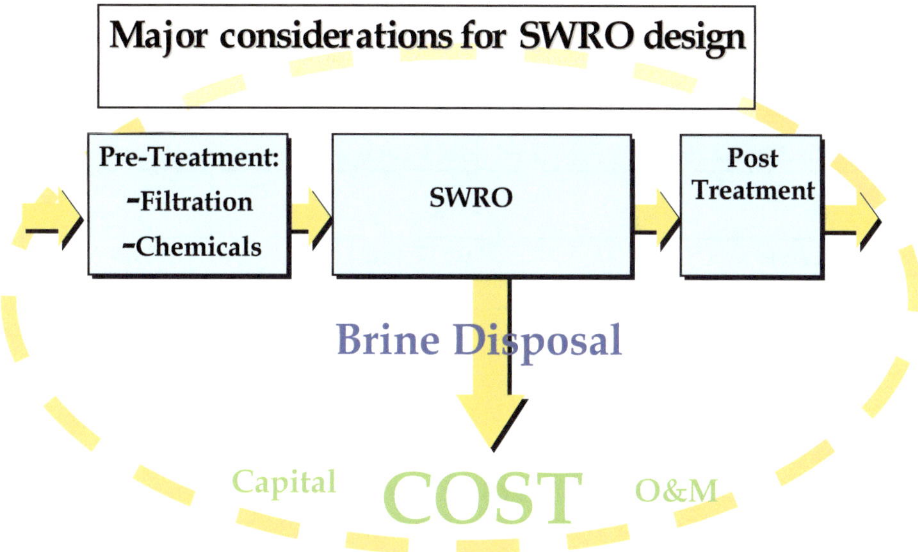

Fig. 6. Major considerations for SWRO design. The larger the SWRO, the lower the water cost and visa versa. For those who are considering SWRO, the tables below show comparable numbers for facilities of different sizes.

Parameter	Metric	US
Capacity	20,000 m3/day	5.3 MGD
CAPITAL TOTAL	$ 20 M	$ 20 M
Capital cost for 20 yrs at 6%	$ 0.27 /m3	$ 1.02 / 1000 gal
Energy cost at 4 KW-H/m3, $ 0.06 /KW-H	$ 0.24 /m3	$ 0.91 / 1000 gal
Chemicals + Labor	$ 0.21 /m3	$ 0.79 / 1000 gal
TOTAL WATER COST	$ 0.72 /m3	$ 2.73 / 1000 gal

Table 2. Typical SWRO – 5 MGD (based on Eilat SWRO) (TDS = 52,000 ppm)

Parameter	Value	US
Capacity	95,000 m3/day (135,000 m3/day)	25 MGD (35 MGD)
CAPITAL TOTAL (construction 20 months)	$ 110 M	$ 110 M
Capital cost for 20 yrs at 6%	$ 0.32 /m3	$ 1.21 / 1000 gal
Energy cost at 2.96 KW-H/m3, $ 0.04 /KW-H	$ 0.12 /m3	$ 0.45 / 1000 gal
Chemicals + Labor (12 people)	$ 0.219 /m3	$ 0.83 / 1000 gal
TOTAL O&M COST (Fixed $ 8.5M/yr for 30 yrs)	$ 0.659 /m3	$ 2.49 / 1000 gal

Table 3. The largest SWRO in the U.S., Tampa Bay SWRO (TDS = 26,000 ppm), 25 MGD, expansion to 35 MGD

Parameter	Metric	US
Capacity	165,000 m3/day (330,000 m3/day)	44 MGD (88 MGD)
CAPITAL TOTAL	$ 212 M	$ 212 M
Capital cost for 20 yrs at 6%	$ 0.17 /m3	$ 0.64 / 1000 gal
Energy cost at 4 KW-H/m3, $ 0.06 /KW-H	$ 0.24 /m3	$ 0.91 / 1000 gal
Chemicals + Labor	$ 0.117 /m3	$ 0.44 / 1000 gal
TOTAL WATER COST	$ 0.527 /m3	$ 1.99 / 1000 gal

Table 4. Largest SWRO in the world– Ashkelon, Israel 44 MGD (expansion to 88 MGD by 2005)

6. Summary

Membranes are becoming a commodity in the desalination and in the water treatment field, finding more applications and replacing traditional conventional technologies. Used in combination with different technologies, membranes may address removal of mineral and organic compounds in the water including volatile types such as endocrine disruptors (EDCs) (42 found in the U.S.), pharmaceutically-active compounds (PhACs), and personal care products (PCPs).

The major reasons membrane desalination is emerging as the dominant technology are:

- Absolute barrier for treatment/removal
- Product water is not affected by the feed water hydraulic and contaminant overloads, spikes, and fluctuations
- Less or no chemicals required
- Smaller footprint/layout
- Single step process
- Modular expandability (for future expansions)
- Less volume of discharged wastes (including sludge and chemicals)
- Simplicity of operation, process automation

7. References

[1] California Department Health Service (DHS) Surface Water Treatment Staff Guidance Manual, California, 1993.

[2] Frenkel, V., Gourgi, T. 1994. *Water Treatment Systems: Bed Filtration and Desalination by Reverse Osmosis (RO).*

[3] Frenkel, V., Pankratz, T., 2004. *Desalination Methods, Technology and Economics.* Desalination Conference, April 16, 2004, The Seminar Group, Santa Barbara, California.

[4] "The Guidebook to Membrane Desalination Technology. Reverse Osmosis, Nanofiltration and Hybrid Systems. Process, Design and Applications" by M. Wilf with chapters by C. Bartels, L. Awerbuch, M. Mickley, G. Pearce and N. Voutchkov, Balaban Desalination Publications, 2006.

[5] L. Stevens, J. Kowal, K. Herd, M. Wilf, W. Bates, Tampa Bay seawater desalination facility: start to finish, Proceedings of IDA Water Desalination Conference, Bahamas (2003).

7

Advanced Mechanical Vapor-Compression Desalination System

Jorge R. Lara, Omorinsola Osunsan and Mark T. Holtzapple
Texas A&M University
United States

1. Introduction

Vapor compression is a reliable and robust desalination technology that is attractive because of its capacity to treat large volumes of water with a wide range of salt concentrations. However, compared to other major desalination technologies such as reverse osmosis, mechanical vapor compression has had relatively high operating and capital costs. New innovative developments in compressor and evaporator designs make it possible to reduce energy consumption so it is a more competitive alternative. Texas A&M University has developed an advanced vapor-compression desalination system that operates at high temperatures. Advanced sheet-shell latent heat exchangers promote dropwise condensation allowing small temperature and pressure differentials between the saturated boiling liquid and the condensing steam, hence reducing the energy requirements. This newer system consists of a train of non-scaling evaporators arranged so feed water flows countercurrently, recovering heat from both the condensate stream and the concentrated discharge brine. A high-efficiency gerotor compressor provides the compression work required to return saturated steam to the initial stage of the evaporator train. An experimental investigation of hydrophobic copper plates described below shows that extraordinarily high heat transfer coefficients can be attained. The gerotor compressor is particularly advantageous for applications where either electricity or mechanical energy is available.

Extensive studies in dropwise condensation show that for low temperature differentials across the hydrophobic plate, heat transfer coefficients will increase with elevated steam pressures. According to the data described in this study, dropwise condensation of saturated steam and forced-convection boiling of saturated water separated by a thin hydrophobic copper plate result in ultra-efficient heat transfer. The forced convection in the water chamber is produced by a liquid jet ejector.

1.1 Advanced mechanical vapor-compression desalination system

Figure 1 shows the advanced mechanical vapor-compression desalination system. In this example, three evaporator stages are illustrated, but fewer or more could be employed (Holtzapple et al., 2010). The left-most evaporator is at the lowest pressure and the right-most evaporator is at the highest pressure. In the left-most evaporator, the vapor space above the boiling water is connected to the compressor inlet. The work added to the compressor causes the discharged steam to be superheated. The superheat is removed in the desuperheater.

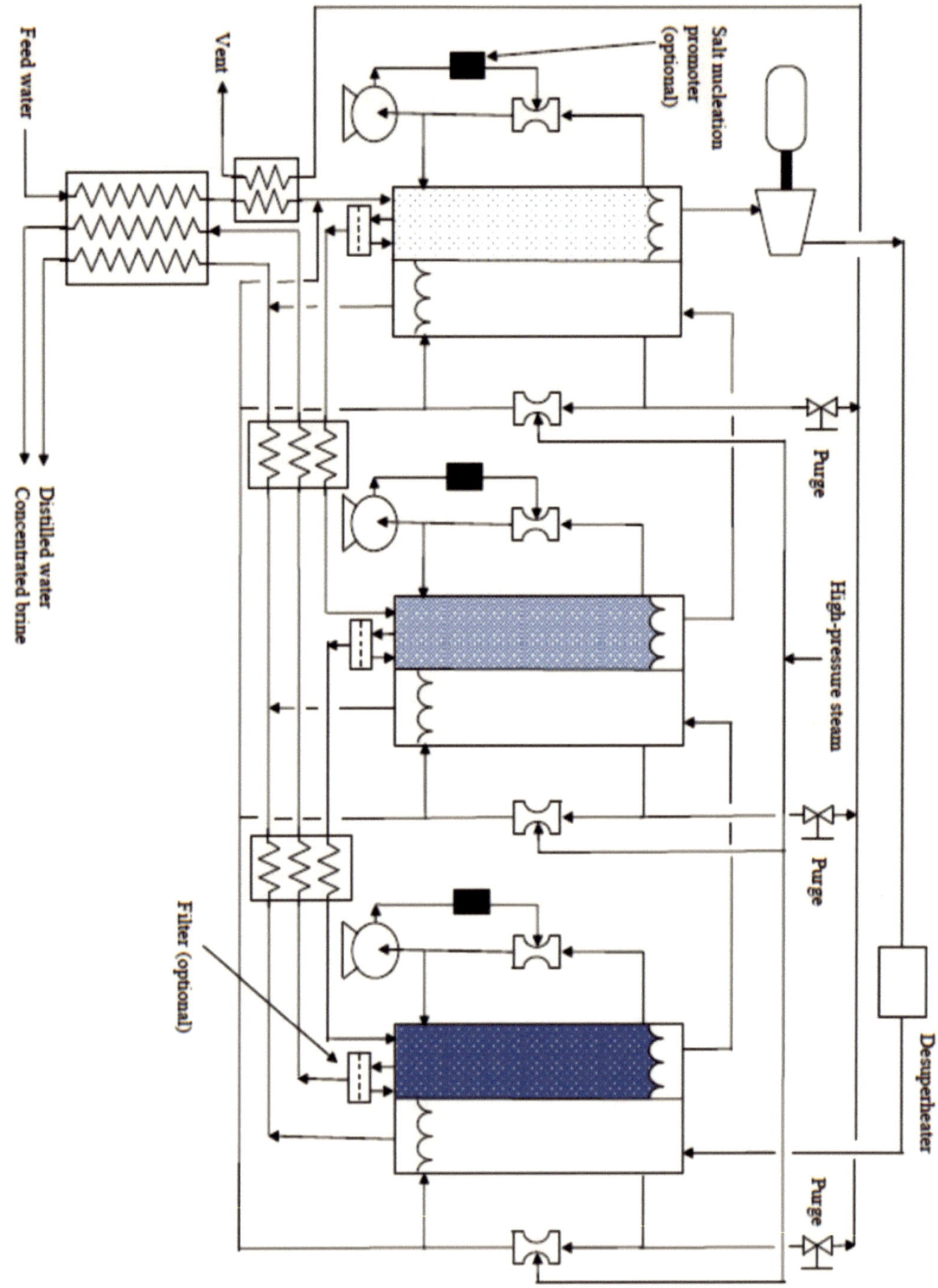

Fig. 1. Advanced mechanical vapor-compression desalination system.

The saturated high-pressure steam that exits the desuperheater enters the condensing side of the right-most evaporator. As this steam condenses, it evaporates water from the boiling side thereby producing steam that can be fed to the middle evaporator. In the middle evaporator, the steam condenses, which causes more steam to be produced on the boiling-water side. This steam then enters the left-most evaporator where it condenses and evaporates water from boiling side. The water evaporated from the boiling side enters the compressor, as previously described.

The evaporators are operated at elevated temperature and pressure, which accomplishes the following: (a) the physical size of the compressor is reduced, thereby reducing its cost and (b) in the evaporators, high heat transfer coefficients are obtained.

The primary disadvantage of operating at elevated temperature is that it promotes scaling on heat exchanger surfaces, primarily from salts with "reverse solubility," i.e., those salts in which the solubility decreases at elevated temperature. Examples of reverse solubility salts are calcium carbonate, magnesium carbonate, calcium sulfate, and magnesium sulfate. Commonly, to limit scaling, the maximum heat exchanger temperature is ~120°C; however, at this temperature and pressure, the compressor is physically large and heat transfer coefficients are poor. It is highly desirable to increase the operating temperature, which requires methods to address scale formation such as the following: (a) remove carbonates from the feed water by acidification and stripping the resulting carbon dioxide; (b) remove sulfates via ion exchange; (c) promote salt nucleation in the bulk fluid rather than on surfaces; (d) abrade heat exchanger surfaces with circulating "cleaning balls" commonly made from rubber; and (e) apply non-stick coatings to heat exchanger surfaces.

In the evaporators, the steam-side heat transfer coefficient improves up to 30% by inducing shearing steam on the condensing surface; the liquid-side heat transfer coefficient improves with forced-convection boiling. This can be accomplished using an internal jet ejector powered by a pump.

To preheat the feed to the evaporators, a sensible heat exchanger is employed, which exchanges thermal energy between the incoming feed water and the discharged distilled water and concentrated brine. As shown in Figure 1, the preheated feed water is fed to the left-most evaporator. In a countercurrent series manner, the brine exiting the left-most evaporator is directed to the middle evaporator and the brine exiting the middle evaporator is directed to the right-most evaporator. As the brine flows from left to right, it becomes ever more concentrated. In the left-most evaporator (lowest brine concentration), the pressure ratio between the condensing steam and boiling water is minimal. In the right-most evaporator (highest brine concentration), the pressure ratio between the condensing steam and boiling water is maximal.

Because noncondensable gases are present in the feed water, it will be necessary to purge them from the system. The purged gases exit with steam, which is sent to a heat exchanger that preheats the incoming feed to the left-most evaporator.

1.2 Mass and energy balance

The steam-side energy balance (Lara, 2005) is

$$q = m_s(H_s - H_c) = m_s h_{fg} \tag{1}$$

where

q = rate of heat transfer (W)

m_s = rate of steam flow (kg/s)
H_c = specific enthalpy of condensate (J/kg)
H_s = specific enthalpy of steam (J/kg)
h_{fg} = latent heat of evaporation (J/kg)
The saltwater-side energy balance is:

$$q = m_v H_v - m_f H_f + m_b H_b$$

$$= (m_f - m_b)\, H_v - m_f H_f + m_b H_b \tag{2}$$

where
m_f = rate of saltwater feed flow (kg/s)
m_b = rate of exiting brine flow (kg/s)
$m_v = m_f - m_b$ = rate of vapor flow to the next effect (kg/s)
H_v = specific enthalpy of vapor going to the next effect (J/kg)
H_f = specific enthalpy of saltwater feed (J/kg)
H_b = specific enthalpy of exiting brine (J/kg)
Using the boiling temperature as a reference, the enthalpy H_f can be calculated from the specific heat of saltwater C_{pf} (J/(kg·°C))

$$H_f = C_{pf}\,(T_b - T_f) \tag{3}$$

where
T_b = temperature of brine exiting latent heat exchanger (°C)
T_f = temperature of saltwater entering latent heat exchanger (°C)
At steady-state flow conditions in the evaporator, for seawater feed, the saltwater concentration in the right-most evaporator has been set to 7%. Under these circumstances, there is an appreciable boiling point elevation. The vapor leaving the evaporator solution is superheated by about 1.5 °C, which corresponds to the boiling point elevation. Using the boiling temperature as a reference (i.e., $H_b = 0$), the specific enthalpy H_v of the leaving vapor equals the latent heat of vaporization plus the sensible superheat. However, the sensible superheat is small so it is approximately true that H_v is the latent heat of vaporization, which is h_{fg}. With this simplifying assumption, the steady-state evaporator energy balance derived using Equations 1 to 3 becomes:

$$m_s\, h_{fg} = (m_f - m_b)\, h_{fg} - m_f\, C_{pf}\,(T_b - T_f) + 0 \tag{4}$$

1.3 Pressure drop in the heat exchanger

For two-phase flow inside horizontal tubes and channels (ASHRAE Fundamentals Handbook, 2001), the pressure gradient is the sum of frictional and momentum terms

$$\frac{dP}{dz} = \left(\frac{dP}{dz}\right)_{friction} + \left(\frac{dP}{dz}\right)_{momentum} \tag{5}$$

Detailed analysis of the pressure drop in the hydrophobic heat exchanger was performed by Lara (2005), and it was concluded that pressure drop in the sheet-shell heat exchanger is not a major issue.

As described below, the advanced mechanical vapor-compression desalination system has two key components: (1) hydrophobic heat exchanger, and (2) high-efficiency compressor.

2. Hydrophobic heat exchanger

An extensive experimental investigation on hydrophobic heat exchangers was performed (Lara & Holtzapple, 2010). The study shows dropwise condensation on the condensing surface and forced-convective boiling on the boiling surface make a very efficient heat transfer mechanism that delivers heat transfer coefficients of the order of 277 kW/(m^2·°C) for 0.2-mm-thick vertical copper plates coated with 2.54-μm-thick hydrophobic Ni-P-PTFE coating for steam at 827 kPa. The extraordinarily high heat transfer coefficient requires small pressure differentials between the condensing and the boiling chambers, hence the compression energy requirement of the system is small. Hydrophobic heat exchangers perform best at high pressure (Rose, 2002); therefore, the compressor must operate at high pressures and small compression ratios. The mechanical vapor-compression system described in this study uses an innovative gerotor compressor, which is now commercially available from StarRotor Corporation (Murphey et al., 2010).

During the experimental investigation, heat transfer coefficients were measured in vertical heat exchangers. Two different square, thin-sheet plate designs were tested. One had round-dimpled spacers, and the other had round-shaped vertical-grooved spacers. In both cases, the experimental plates were mounted in a sealed two-chamber apparatus with condensing saturated steam on one side and boiling liquid water on the other (Figure 2).

The liquid-side heat transfer mechanism employed either natural or forced convection pool boiling of saturated water. The steam-side heat transfer mechanism was condensing saturated steam with either filmwise or dropwise condensation.

2.1 Apparatus and procedure

The experimental apparatus is tailored to observe and manipulate key heat transfer variables. The apparatus (Figure 2) consists of two sections: (1) a boiling water chamber and (2) a condensing steam chamber. Both chambers are made of stainless steel 304 and are divided by the test plate. The whole assembly is bolted together. To prevent leakage, a gasket is placed between each side of the test plate and frame. Data were collected only after steady state was achieved.

High-pressure steam enters valve V1 into cyclone C1 where liquid is separated, thus ensuring the steam quality entering the apparatus is 1.0. Pressure regulator V2 sets the condenser pressure, which is measured by pressure gauge P. The steam enters the condenser, which has a 3.2-mm gap that is set by the thickness of the aluminum plate inserted into the condenser. At the bottom of the condenser, condensate flows into sight glass S2. By manually opening valve V4, the liquid level in sight glass S2 can be maintained constant. The drained liquid is collected in graduated cylinder G1 and is measured over a 90-s interval. (Note: This manual method of collecting condensate was more reproducible than steam traps.)

The rate of shearing steam flowing past the plate is regulated by valve V3. Cyclone C2 separates liquid entrained with the shearing steam. The collected liquid enters sight glass S3; by manually opening valve V5, the liquid level in sight glass S3 is kept constant. The drained liquid collected in graduated cylinder G2 is measured over a 90-s interval.

The steam exiting cyclone C2 enters heat exchanger HX2 where it condenses and is collected in graduated cylinder G3 over a 90-s interval. The amount of liquid collected in graduated cylinder G3 is compared to the amount of liquid collected in both graduated cylinders G1 and G2 so that the ratio R of each flow can be measured. Knowing the gap g (3.2 mm), the plate depth, and the steam density allows the velocity of the shearing steam v to be measured.

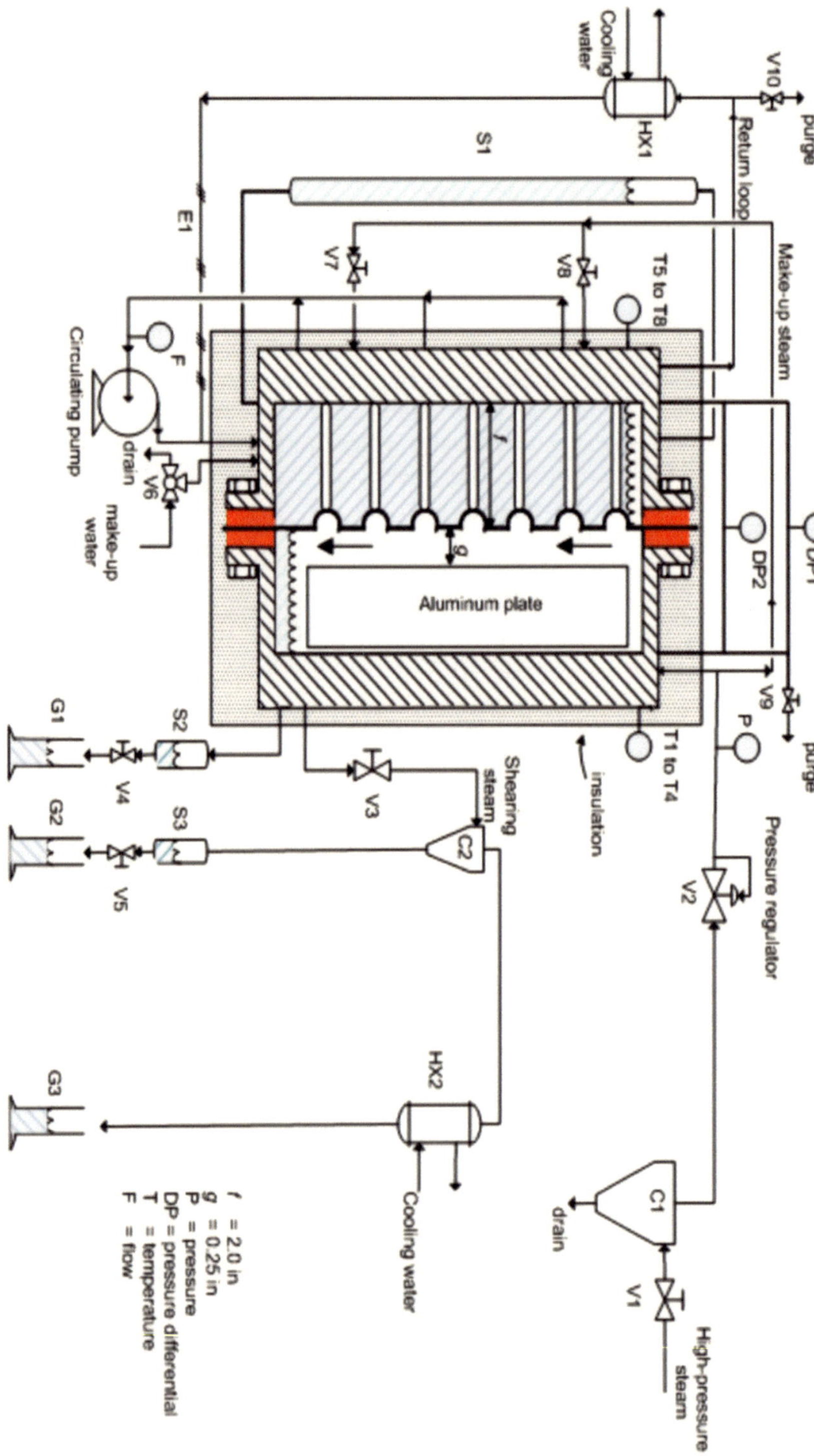

Fig. 2. Schematic of experimental apparatus (Lara & Holtzapple, 2010).

The boiling side is flooded with tap water; sight glass S1 ensures the liquid level is kept constant. If necessary, excess liquid can be drained or make up water added by manually opening three-way valve V6. The steam evaporated from the boiler enters heat exchanger HX1 where it condenses. The condensate is heated to saturation using electric resistance heater E1. If necessary, make-up steam can be added to the boiler by opening valves V7 and V8.
To induce forced convection in the boiler, a pump circulates the liquid. An all-metal flow meter measures the rate of circulating liquid. Knowing the gap f and the plate dimensions, the liquid velocity can be calculated.
To ensure that noncondensable gases are removed from the system, valves V9 and V10 allow a small stream to be purged to the atmosphere.
The temperature differential ΔT between the steam side and the liquid side is set by the amount of cooling water flowing through the heat exchanger HX1, the amount of make-up steam added through valves V7 and V8, and the heat added through resistance heater E1.
The differential pressure between the two chambers is measured using differential pressure gauges DP1 and DP2. One operates from 0 to 2 psid (0 to 13,800 Pa) and the other operates from 0 to 10 psid (0 to 69,000 Pa). The measured pressure differential ΔP between chambers and the steam pressure P allows ΔT to be determined using steam tables. Four thermocouples measure the temperatures in each quadrant of the condenser. Similarly, four thermocouples T5 to T8 measure the temperatures in each quadrant of the boiler. Because thermocouples are not particularly accurate, they were not used to measure ΔT across the test plate. Instead, their purpose was to ensure uniform temperatures in each quadrant of the boiler and of the condenser. Using steam tables, the thermocouple readings were found to be consistent with the readings taken by the pressure gauge P and differential pressure gauges DP1 and DP2.
Thermal losses from insulation are calculated by opening valves V7 and V8, which equalizes the pressures in both chambers with saturated steam so there is no temperature difference across the plate. The condensate is collected and used to determine the heat loss through the steam-side insulation. This collected steady-state condensate serves as the baseline, which is subtracted from the condensate collected in both graduated cylinders G1 and G2 during the experiments; the net condensate collected (m) is substituted in Equation 8 to calculate heat flux. This allows the heat transfer through the plate to be measured without interference from heat loss through the insulation.
Three different plate materials were tested: (1) 0.762-mm-thick naval brass 464 (2) 0.2-mm-thick copper, and (3) 0.127-mm-thick titanium grade 2.

2.2 Calculation of heat transfer coefficient

Neglecting fouling, the theoretical overall heat transfer coefficient can be calculated using

$$U = \frac{1}{\frac{1}{h_{cond}} + \frac{1}{h_{boiling}} + \left(\frac{x}{k}\right)} \tag{6}$$

where

U = overall heat transfer coefficient (kW/(m^2·°C))
x = plate thickness (m)
k = thermal conductivity (kW/(m·°C))
h_{cond} = condensation heat transfer coefficient (kW/(m^2·°C))
$h_{boiling}$= boiling heat transfer coefficient (kW/(m^2·°C))

Measured heat transfer coefficients U are obtained from

$$U = \left(\frac{q}{\Delta T} \right) \tag{7}$$

and

$$q = (mh_{fg})/A \tag{8}$$

where

q = heat flux (kW/m^2)
m = net condensate collected from the apparatus (kg/s)
h_{fg} = latent heat of condensation (kJ/kg)
A = effective heat transfer area (m^2)
ΔT = temperature differential across the plate (°C)

Dropwise condensation performs best at higher pressures and small temperature differentials across the plate (Rose, 2002). Moreover, high operating pressures increase steam density which allows mechanical vapor-compression (MVC) desalination systems to use smaller compressors.

2.3 Heat transfer enhancement techniques

Active and passive heat transfer enhancement techniques for heat exchangers have been investigated intensively (Bergles, 2002). Fourth-generation heat transfer technology involves simultaneous application of several techniques to produce an enhancement larger than the individual techniques operating separately. Dropwise condensation has been studied for the past 60 years (Rose, 2002). Experiments with brass tubes show dropwise condensation has heat transfer coefficients 1.6–28.6 times greater than filmwise condensation (Ma et al., 2002). The experiments reported herein enhanced heat transfer by simultaneously employing the following: (1) passive electroless Ni-P-PTFE thin-hydrophobic coating to promote dropwise condensation on the steam side and to inhibit fouling in the boiling side, (2) active forced convection circulating saturated liquid in the boiling chamber, and (3) active shearing steam on the condensing surface. This study measures the heat transfer with forced-convective boiling (liquid side) and dropwise condensation (steam side). The following factors were investigated: (a) saturated steam temperature; (b) temperature differential ΔT across the plate; (c) shearing steam; (d) shearing liquid.

2.4 Summary of experimental results

The first plate was round-dimpled 0.762-mm-thick naval brass (k = 116 W/(m·°C)), which was roughened via sand-blasting on the liquid side to promote nucleation. The condensing metal surface was either bare (filmwise condensation) or coated with 0.635-μm-thick layer of Ni-P-PTFE (dropwise condensation).

Shearing steam on the condensing surface enhanced the overall heat transfer coefficient by 1.6 times and forced liquid convection increased it by additional 1.4 times. Interestingly, excessive shearing steam reduced the overall heat transfer coefficient. Presumably, this occurred because a film formed that increased the thermal resistance across the plate and disrupted the dropwise condensation mode. Without coating, the best operating point delivered U = 16.5 kW/(m^2·°C) (saturated steam T = 166 °C, P = 722 kPa, ΔT = 0.2 °C). With 0.635-μm Ni-P-PTFE hydrophobic coating, the best operating point delivered an overall heat transfer coefficient U = 99.4 kW/(m²·°C) (saturated steam T = 166 °C, P = 722 kPa, ΔT = 0.2 °C, shearing steam v = 0.16 m/s, $R \approx 1$ kg shearing steam/kg condensate, saturated liquid side v = 1.57 m/s).

The second plate was round-dimpled 0.2-mm-thick copper (k = 400 W/(m·°C)). The plate surfaces in both chambers were modified with 0.635-μm Ni-P-PTFE hydrophobic layer. Experiments on this plate were performed under two different conditions in the saturated liquid chamber: (1) forced convection and (2) forced convection with PTFE boiling stones as a dynamic nucleation agent. For the first condition, the best operating point delivered an overall heat transfer coefficient U = 159 kW/(m²·°C) (saturated steam T = 166 °C, P = 722 kPa, ΔT = 0.2 °C, shearing steam v = 0.4 m/s, $R \approx 1$ kg shearing steam/kg condensate, saturated liquid side v = 1.57 ft/s). For the second condition, the best operating point delivered an overall heat transfer coefficient U = 182 kW/(m²·°C) (saturated steam T = 166 °C, P = 722 kPa, ΔT = 0.2 °C, steam velocity v = 0.49 m/s, $R \approx 1$ kg shearing steam/kg condensate, saturated liquid velocity v = 1.57 m/s).

The third round-dimpled plate was made of grade-2 bare 0.12-mm-thick titanium (k = 21.9 W/(m·°C)). The best design point delivered U = 77.8 kW/(m²·°C) (saturated steam T = 166 °C, P = 722 kPa, ΔT = 0.2 °C, steam velocity v = 0.15 m/s, R = 1.5 kg shearing steam/kg, saturated liquid velocity v = 1.57 m/s).

The fourth plate was vertical-grooved 0.2-mm-thick copper (k = 400 W/(m·°C)) coated with 0.635-μm Ni-P-PTFE hydrophobic coating. The best design point delivered U = 192 kW/(m²·°C) (saturated steam T = 166 °C, P = 722 kPa, ΔT = 0.2 °C, steam velocity v = 0.16 m/s, $R \approx 0.43$ kg shearing steam/kg condensate, saturated liquid velocity v = 1.57 m/s). The last experiment was performed on the copper plate previously described but with a modified chemistry for the coating. Lead-free 2.5-μm-thick hydrophobic Ni-P-PTFE delivered 21% better heat transfer coefficient. For this case, the best design point was U = 240 kW/(m²·°C) (saturated steam T = 166°C, P = 722 kPa, ΔT = 0.2 °C, steam velocity v = 0.23 m/s, $R \approx 0.6$ kg shearing steam/kg condensate, saturated liquid velocity v = 1.57 m/s).

2.5 Experimental results of vertical groove plates

Figure 3 shows the overall heat transfer coefficients corresponding to different temperature differences across the vertical groove plate at different constant saturated steam pressures (Lara & Holtzapple, 2010). Forced convection is imposed in the saturated liquid side with $v_{\text{sat liq}}$ = 1.57 m/s. Optimal shearing steam ratio (R) (Figure 4) was employed.

The following empirical equations describe each of the curves shown in Figure 3:

$$U = 61.1(\Delta T)^{-0.9153} \qquad (P = 722\ \text{kPa}) \tag{9}$$

$$U = 39.8(\Delta T)^{-0.8214} \qquad (P = 653\ \text{kPa}) \tag{10}$$

$$U = 25.9(\Delta T)^{-0.7715} \qquad (P = 446\ \text{kPa}) \tag{11}$$

Equations 9 to 11 can be used to calculate the heat flux:

$$q = U\,\Delta T = 61.1(\Delta T)^{1-0.9153} = 61.1(\Delta T)^{0.0847} \qquad (P = 722\ \text{kPa}) \tag{12}$$

$$q = U\,\Delta T\ 39.8(\Delta T)^{1-0.8214} = 39.8(\Delta T)^{0.1786} \qquad (P = 653\ \text{kPa}) \tag{13}$$

$$q = U\,\Delta T = 25.9(\Delta T)^{1-0.7715} = 25.9(\Delta T)^{0.2285} \qquad (P = 446\ \text{kPa}) \tag{14}$$

Figure 5 presents this information graphically.

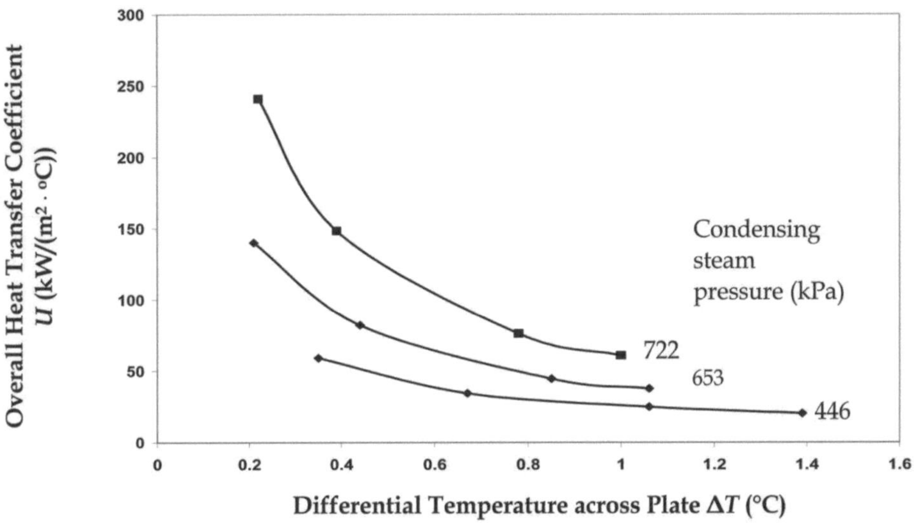

Fig. 3. Overall heat transfer coefficients corresponding to different ΔT. Copper 0.2-mm-thick plate with round-shaped vertical grooves and fully coated with 2.54-μm-thick lead-free Ni-P-PTFE hydrophobic coating. Forced-convection saturated pool boiling ($v_{sat\ liq}$ = 1.57 m/s). Optimal shearing steam on the condensing surface (see Figure 4). Condensing steam at different pressures. (Lara & Holtzapple, 2010)

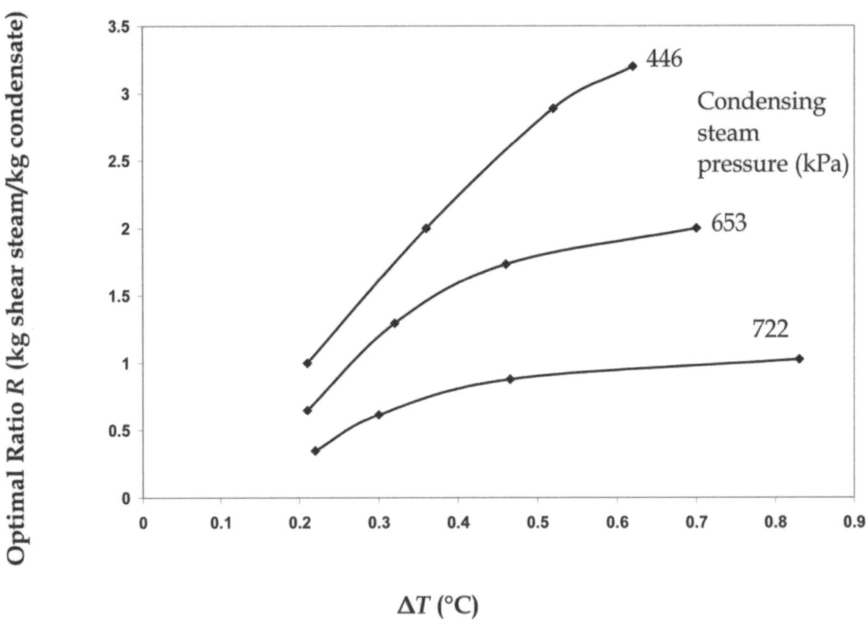

Fig. 4. Optimal shearing steam ratio R corresponding to saturated steam pressure P and temperature differential across the test plate ΔT. Experiments conducted on 0.20-mm-thick copper plate with round-shaped vertical grooves and fully coated with 2.54-μm-thick lead-free Ni-P-PTFE hydrophobic coating. Forced convection in the saturated pool boiling side ($v_{sat\ liq}$ = 1.57 m/s). (Lara & Holtzapple, 2010)

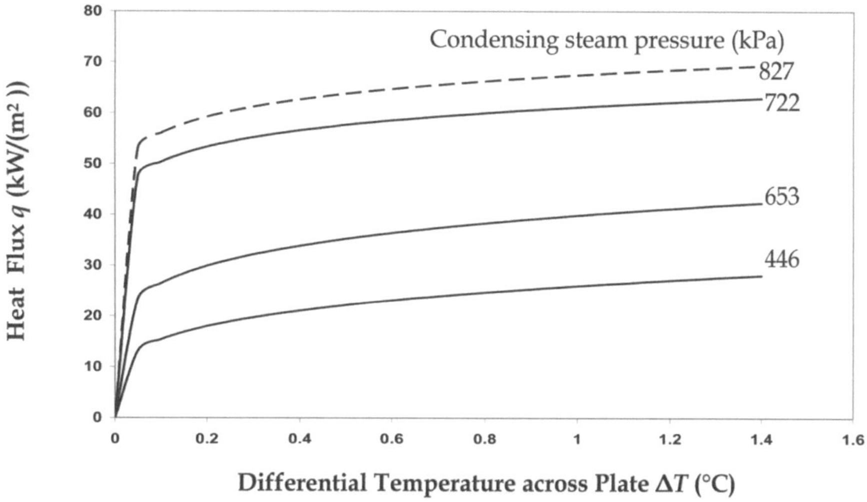

Fig. 5. Heat flux across the plate corresponding to different ΔT. Forced convection in saturated pool boiling. R is the optimal corresponding value (Figures 4). Smooth curves were calculated using Equations 12 to 14. Dashed line is a projection to desired operating pressure using Equations 15 to 19. (Lara & Holtzapple, 2010)

At P = 166 kPa, the design point (U = 240 kW/(m^2·°C)) requires shear velocity v = 0.23 m/s and the flow ratio R = 0.6 kg shearing steam/kg condensate.

Previously, Figure 3 showed heat transfer coefficient U as a function of ΔT for a constant P. Figure 6 shows the same data where U is a function of P for a given ΔT. The following correlations were used to construct Figure 6:

$$U = 0.461\,(P)^{1.978} \qquad (\Delta T = 0.22\ °C) \tag{15}$$

$$U = 0.733\,(P)^{1.827} \qquad (\Delta T = 0.40\ °C) \tag{16}$$

$$U = 1.131\,(P)^{1.686} \qquad (\Delta T = 0.70\ °C) \tag{17}$$

$$U = 1.315\,(P)^{1.637} \qquad (\Delta T = 0.85\ °C) \tag{18}$$

$$U = 1.935\,(P)^{1.51} \qquad (\Delta T = 1.40\ °C) \tag{19}$$

3. Gerotor compressor

Injecting liquid water into the compressor allows the compression to be nearly isothermal, which minimizes energy consumption. Conventional centrifugal compressors do not allow water injection because the high-speed blades can be damaged from the impact with the droplets. In contrast, a gerotor positive-displacement compressor operates at lower speeds and has robust components that can tolerate liquid injection. Other advantages follow: (1) less expensive, (2) can be easily sized to the specific compression needs, and (3) efficient over a wide range of operating conditions. Gerotor compressors are available from StarRotor Corporation (Murphey et al., 2010) and are a key component of the MVC system because of its low energy consumption and low capital cost.

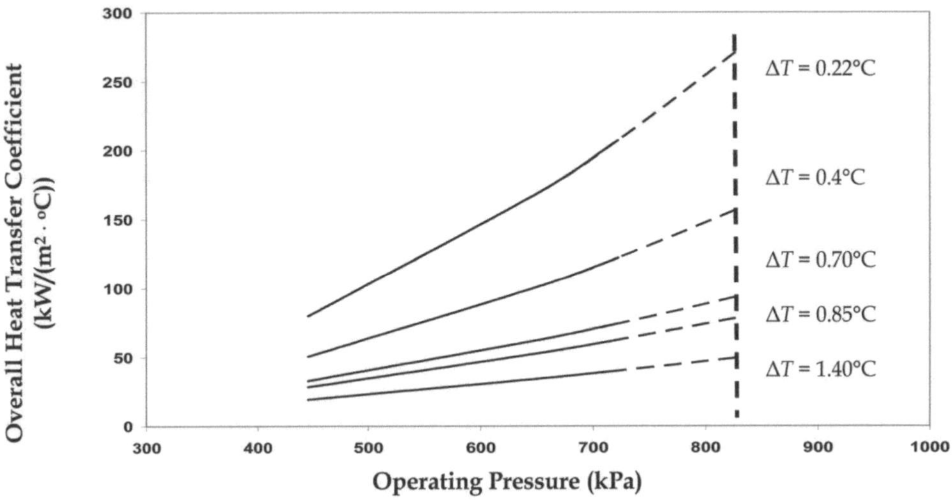

Fig. 6. Overall heat transfer coefficient related to operating pressure. Copper plate 0.20-mm thick with round-shape vertical grooves coated with lead-free 2.54-μm Ni-P-PTFE hydrophobic coating. Force-convection shearing steam on the condensing surface and forced convective saturated pool boiling ($v_{\text{sat liq}} = 1.57$ m/s). Smooth curves were determined using Equations 15 to 19. Solid line is interpolation. Dashed line is extrapolation. (Lara & Holtzapple, 2010)

For the case of liquid water injection, the compressor work W is evaluated (Lara, 2005) as

$$W = \frac{(1+x)H_2^{vap} - \left(H_1^{vap} + xH_1^{liq}\right)}{\eta_c} \tag{20}$$

where the amount of liquid water injected is

$$x = \frac{S_1^{vap} - S_2^{vap}}{S_2^{vap} - S_1^{liq}} \tag{21}$$

A 25-kW gerotor compressor has been reported to have an isentropic efficiency of 84 – 86% over a three-fold range in speed (1200 – 3600 rpm) (Murphey et al., 2010).
The presence of salt lowers the vapor pressure of water according to the following formula (Emerson & Jamieson, 1967), which is valid for 100 to 180 °C.

$$\log_{10}\left(\frac{P}{P_0}\right) = -2.1609 \times 10^{-4} S - 3.5012 \times 10^{-7} S^2 \tag{22}$$

where

P = actual vapor pressure above the salt solution at temperature T (kPa)
P_o = vapor pressure above pure water at temperature T (kPa)
S = salinity (g salt/kg solution)

Using this relationship, the required compression ratio can be calculated as a function of salt concentration, condenser temperature, and heat exchanger ΔT. Figure 7 shows the variation of compression ratio as function of salinity and ΔT.

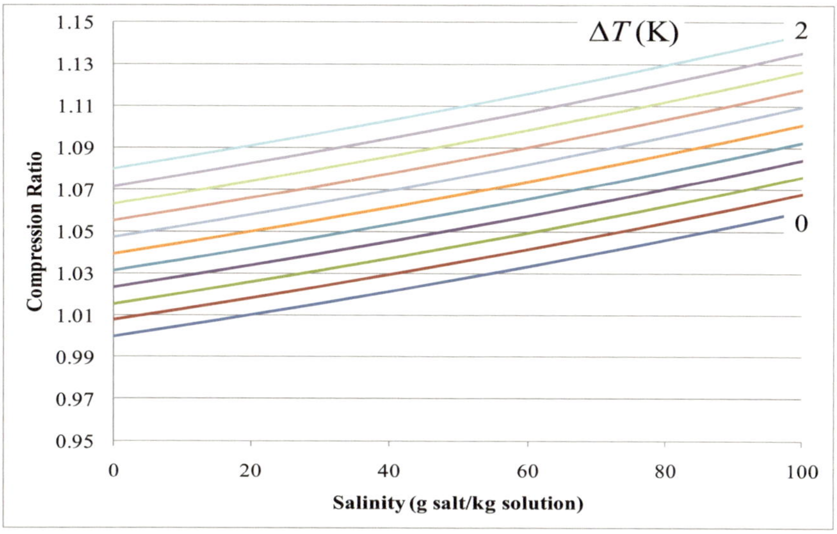

Fig. 7. Compression ratio as a function of salinity and ΔT across the heat exchanger. Operating point is typical of a seawater desalination system. P_{cond} = 0.06895 MPa, T_{cond} = 362.7 K, ΔT in 0.2 K increments.

4. Approach temperature in sensible heat exchangers

Compressor work (W) enters the system and exits as thermal energy in the distillate $\left(m_s C_{ps}\left(T_s - T_f\right)\right)$ and the brine $\left(m_b C_{pb}\left(T_b - T_f\right)\right)$ (Lara, 2005). Therefore, the energy balance is

$$W = m_b C_{pb}(T_b - T_f) + m_s C_{ps}(T_s - T_f) \tag{23}$$

where

- m_s = rate of distillate flow (kg/s)
- m_b = rate of exiting brine flow (kg/s)
- C_{pb} = specific heat of brine (J/(kg·K)
- C_{ps} = specific heat of distillate (J/(kg·K)
- T_s = temperature of distillate exiting desalination system (°C)
- T_b = temperature of brine exiting desalination system (°C)
- T_f = temperature of entering saltwater (°C)

Letting $\Delta T = (T_b - T_f) = (T_s - T_f)$

$$W = m_b C_{pb} \Delta T + m_s C_{ps} \Delta T \tag{24}$$

and using the following relationships:

$$\text{total mass balance: } m_s + m_b = m_f$$

$$\text{salt mass balance: } m_f x_f = m_b x_b$$

where

x_b = brine concentration

x_f = entering saltwater concentration

the following equation is derived:

$$\Delta T = \left(\frac{1}{\left(\frac{x_b}{x_f} - 1 \right)} C_{pb} + C_{ps} \right)^{-1} \left(\frac{W}{m_s} \right) \tag{25}$$

This ΔT represents the temperature rise of both the exiting distillate and brine. In addition, it is the approach temperature in the sensible heat exchangers.

4. Desalination plant cost analysis

A cost analysis for a 37,850 m^3/day seawater desalination plant is described below. The cost of the distilled water (US $/$m^3$) is the sum of (a) capital costs and (b) operating costs. The analysis described is for seawater (35,000 ppm TDS) and brackish water (~1200 ppm TDS).

The major pieces of equipment required for the advanced mechanical vapor-compression desalination system and the operating conditions considered for the capital investment follow:

1. **Hydrophobic latent heat exchanger:** P_{steam} = 827 kPa; T_{steam} = 172 °C; ΔT = 0.22 °C; U = 277 kW/($m^2 \cdot$°C); A = 16,607 m^2. This area is divided equally among 10 stages.
2. **Sensible heat exchanger (plate-and-frame):** T_{in} = 21.1 °C; T_{out} = 171 °C; U = 31 kW/($m^2 \cdot$°C); A = 16,467 m^2.
3. **Gerotor compressor:** W = 3187 kW; P_{in} = 570.2 kPa; T_{in} = 159.7 °C; P_{out} = 827 kPa; T_{out} = 172 °C; $\eta_{compressor}$ = 85%; volumetric flow rate of steam after Stage 10 = 13.84 m^3/min.
4. **Electric motor:** 3366 kW; totally enclosed; η_{motor} = 96%.
5. **Pump:** 900 kW; 0.6 m^3/s; 1400 kPa; η_{pump} = 80%.
6. **Degassing unit:** D = 0.35 m; 7.68 kW; air flow = 0.4 m^3/s; column height = 3 m; packing height = 2.4 m.
7. **Brine injection well:** A cost of $1,880,363 is estimated. This cost will vary depending on local regulations.

The approach temperature of the sensible heat exchanger was optimized to minimize operating costs for a given interest rate, steam cost, and sensible heat exchanger cost. Depending upon the scenario, the approach temperature varied from 0.37 to 1.3°C, which is larger than the temperature difference in the latent heat exchanger. To elevate the temperature of the water entering the latent heat exchanger to saturation, it is necessary to inject steam.

Table 1 shows different variables used to calculate the cost of product water for different scenarios. The base case is shown in bold.

The total capital cost of equipment was multiplied by a Lang factor of 3.68 to estimate the fixed capital investment (FCI). (Note: This desalination system is assumed to be sold as a

packaged unit, which has a lower Lang factor than a field-erected plant.) The capital cost of the purchased equipment for seawater desalination is given in Table 2.
The operating cost includes insurance, maintenance, labor, debt service, electricity, and steam. The annual maintenance and insurance costs were assumed to be 4% and 0.5% of the FCI, respectively. Labor cost was assumed to be \$500,000/yr. To determine the debt service, the fixed capital investment was amortized using the ordinary annuity equation

$$PV = R\frac{(1+i)^N - 1}{(1+i)^N} \tag{26}$$

where PV is the present value of the bond, R is the yearly cost of the bond, i is the annual interest rate, and N is the lifetime of the project (30 years). The annual operating cost is given in Table 3.
Table 4 shows the cost per m³ of drinking water for different bond interest rates.

Variable	Units	
Production	m^3/day	37,850
Inlet salt concentration	%	3.5 (seawater), 0.15 (brackish water)
Outlet salt concentration	%	7 (seawater), 1.5 (brackish water)
Latent heat exchanger cost	\$/m^2	108, **215**,323
Sensible heat exchanger cost	\$/m^2	161, **215**, 269
Steam cost	\$/1000 kg	7.7, **15.4**, 30.8
Electricity cost	\$/kWh	**0.05**, 0.10, 0.15, 0.20
Interest rate	%	**5**, 10, 15, 20
Bond	years	30

Table 1. Variables used for the different cases evaluated. (Base case is in bold.)

Equipment	Purchase Cost (\$)
Latent heat exchanger	3,573,820
Sensible heat exchanger	3,543,779
Compressor	1,761,942
Centrifugal pump	474,121
Degassing unit	15,308
Electric motor (totally enclosed)	56,336
Brine injection well	1,880,363
Total Equipment Cost	11,305,668
Lang Factor	3.68
Fixed Capital Investment (FCI)	41,604,858

Table 2. Capital cost of a desalination plant equipment that treats 37,850 m^3/day of seawater.

	Cost ($/yr)	Cost ($/m³)
Electricity ($0.05/kWh)	1,850,550	0.13
Steam ($15.4/1000 kg)	235,177	0.01
Labor	500,000	0.04
Maintenance (0.04 x FCI)	1,664,194	0.12
Insurance (0.005 x FCI)	208,024	0.01
Total annual operating cost	4,457,945	0.32

Table 3. Annual operating cost for a 37,850 m³ per day seawater desalination plant.

Interest Rate[a]	5%	10%	15%	20%
	Cost ($/m³)			
Debt service	0.19	0.30	0.40	0.53
Electricity ($0.05/kWh)	0.13	0.13	0.13	0.13
Steam ($15.4/1000 kg)	0.01	0.02	0.02	0.03
Labor	0.04	0.04	0.04	0.04
Maintenance (0.04 x FCI)	0.12	0.11	0.10	0.10
Insurance (0.005 x FCI)	0.01	0.01	0.01	0.01
Total	0.51	0.61	0.73	0.84

Table 4. Water costs ($/m³) for seawater feed at varying interest rates.

	Water Cost ($/m³)							
Electricity ($/kWh)	0.05		0.10		0.15		0.20	
Feed water % Interest	Seawater	Brackish water	Seawater	Brackish water	Seawater	Brackish water	Seawater	Brackish water
5%	0.51	0.42	0.65	0.47	0.79	0.53	0.92	0.58
10%	0.62	0.50	0.75	0.56	0.89	0.62	1.00	0.67
15%	0.73	0.61	0.86	0.67	0.99	0.72	1.13	0.77
20%	0.84	0.72	0.97	0.77	1.10	0.83	1.24	0.88

Table 5. Cost of water ($/m³) from seawater and brackish water at varying interest rates and electricity costs using base-case assumptions (i.e., latent and sensible heat exchanger = $215/m². Steam = $15.4/1000 kg).

Table 5 shows the cost of water for both seawater and brackish water at varying interest rates and electricity costs. In this case, a cost of $215/m² for the latent and sensible heat exchanger area was considered. Steam cost was $15.4/1000 kg. The debt service for both seawater and brackish water feed increases with the interest rate and is the major contributor to the cost of water at a fixed electricity cost. The debt service and the electricity cost are the dominant costs.

Figure 8 shows the costs of product water when all cost variables change while the unitary cost for the sensible heat exchanger is held constant at $161/m². This is the lower bound of the unitary cost of the sensible heat exchanger.

Figure 9 shows the costs of product water when all cost variables change while the unitary cost for the sensible heat exchanger is held constant at $215/m². This is the mid bound of the unitary cost of the sensible heat exchanger.

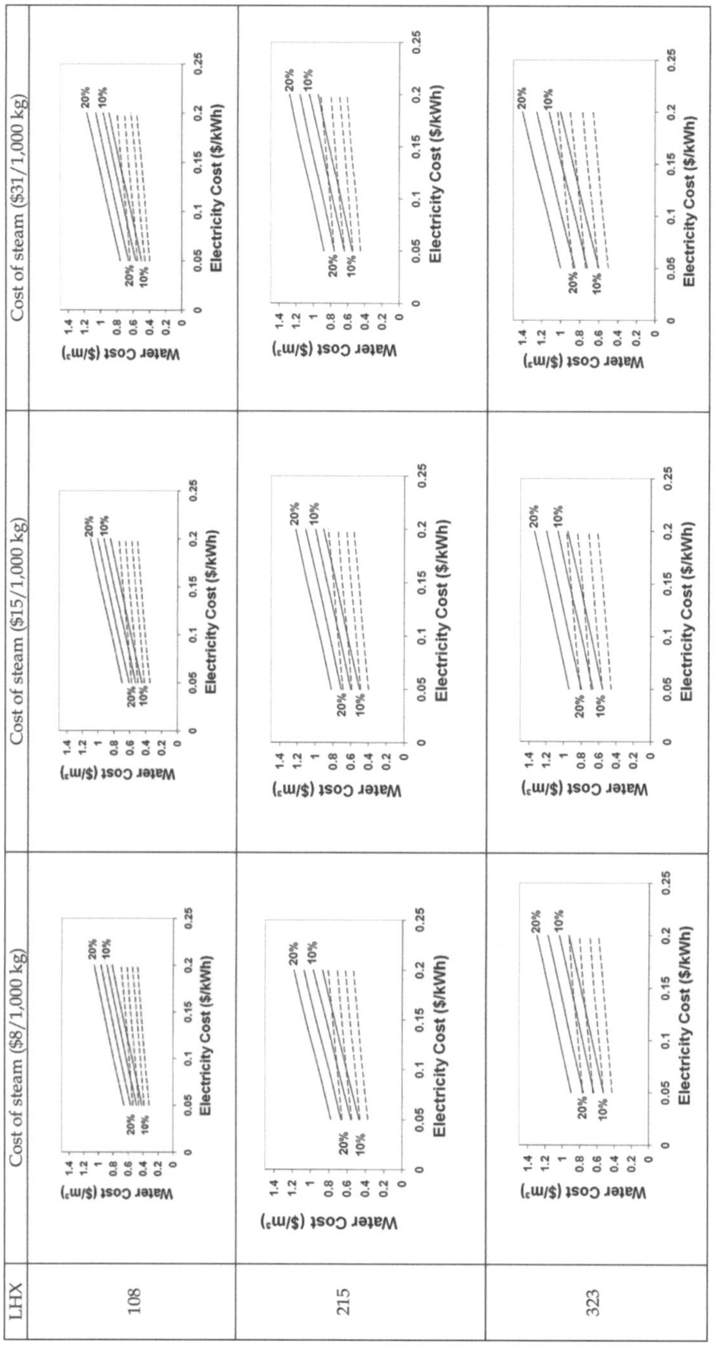

Fig. 8. Cost of water ($/m^3$) for different costs of steam ($/1,000 kg) and different costs of latent heat exchanger (LHX) unit area ($/m^2$) when the unitary cost of sensible heat exchanger area is held at $161/m^2$. Lines indicate different interest rate for debt service. Solid line is for seawater (35,000 ppm TDS), dotted line is for brackish water (~1,200 ppm TDS).

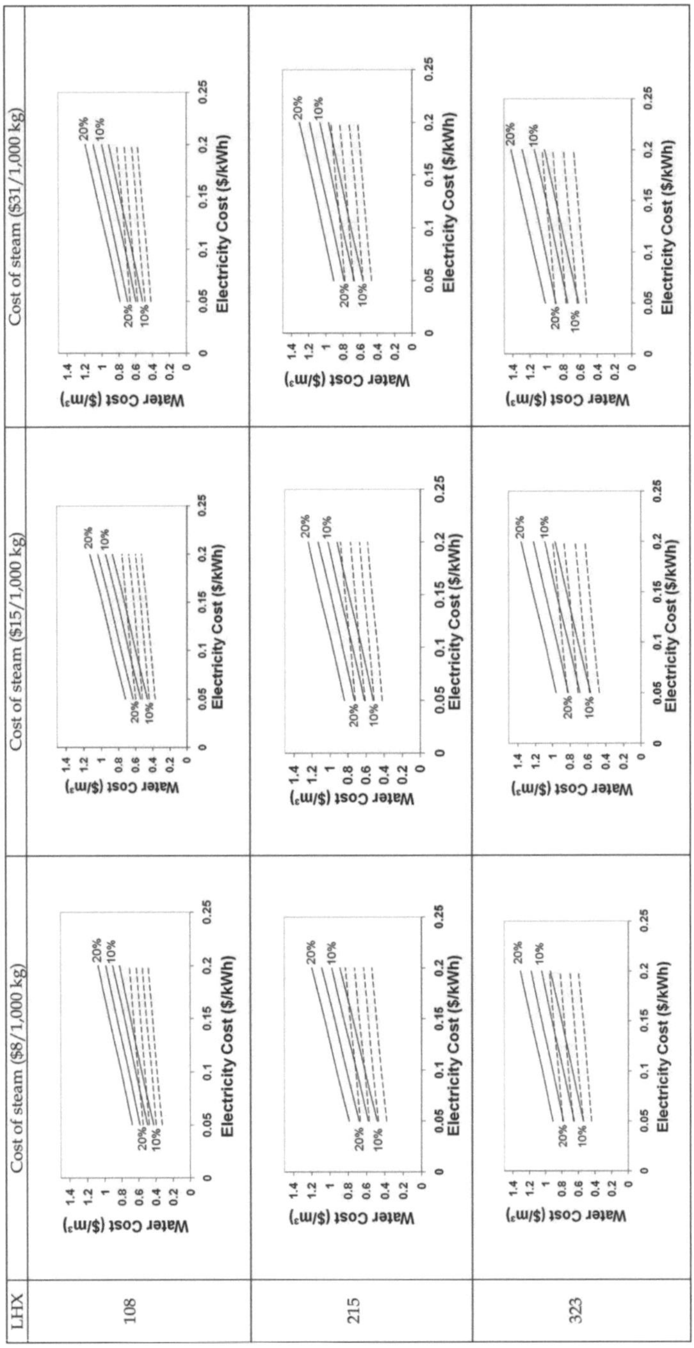

Fig. 9. Cost of water ($/m^3$) for different costs of steam ($/1,000 kg) and different costs of latent heat exchanger (LHX) unit area ($/m^2$) when the unitary cost of sensible heat exchanger area is held at $215/m^2$. Lines indicate different interest rate for debt service. Solid line is for seawater (35,000 ppm TDS), dotted line is for brackish water (~1,200 ppm TDS).

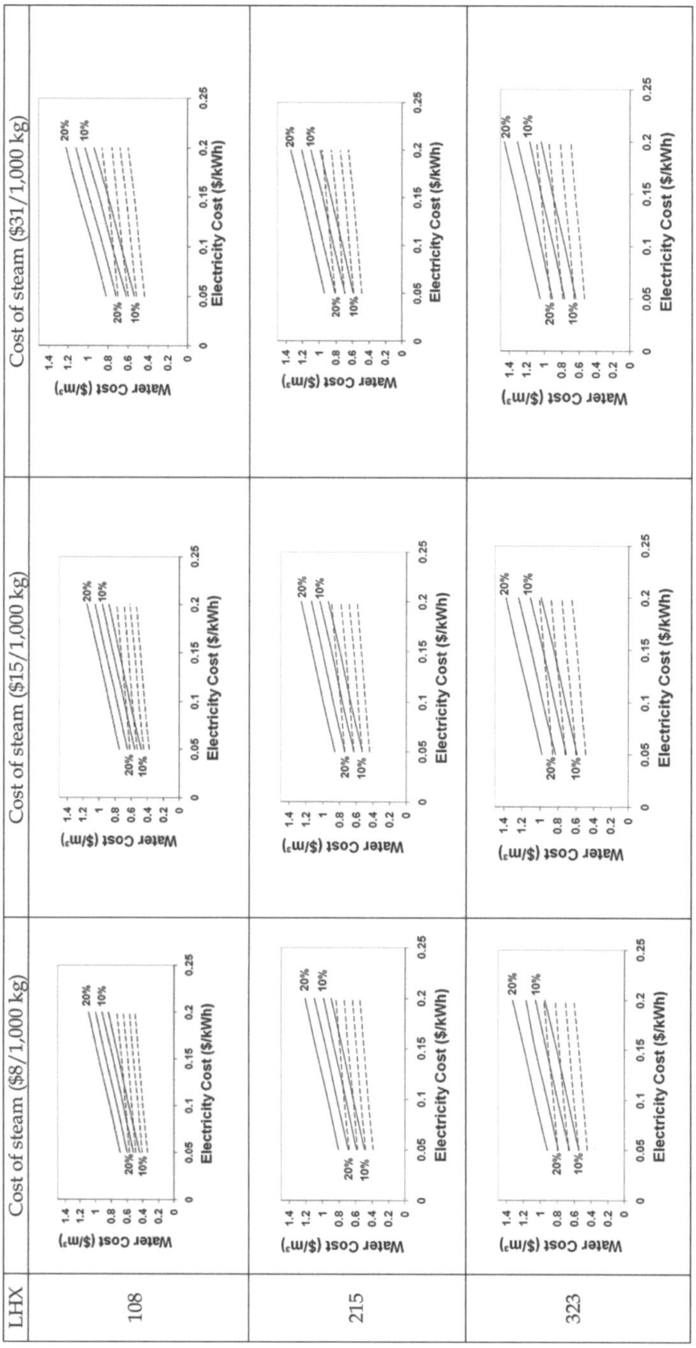

Fig. 10. Cost of water ($/m^3$) for different costs of steam ($/1,000 kg) and different costs of latent heat exchanger (LHX) unit area ($/m^2$) when the unitary cost of sensible heat exchanger area is held at $269/m^2$. Lines indicate different interest rate for debt service. Solid line is for seawater (35,000 ppm TDS), dotted line is for brackish water (~1,200 ppm TDS).

Figure 10 shows the costs of product water when all cost variables change while the unitary cost for the sensible heat exchanger is held constant at $269/m². This is the upper bound of the unitary cost of the sensible heat exchanger.

5. Conclusion

Traditionally, mechanical vapor-compression desalination systems are more energy intensive than reverse osmosis and require higher capital and operation costs. The present study describes recent developments in latent heat exchangers and gerotor compressors that make mechanical vapor-compression a competitive alternative to treat high-TDS waters with a robust, reliable, yet economical technology. Using base-case assumptions, fresh water can be produced at $0.51/m³ from seawater and at $0.42/m³ from brackish water (electricity $0.05/kWh, 5% interest, 30-year bond).

6. Legal notice

This desalination technology has been licensed to Terrabon, Inc. The information, estimates, projections, calculations, and assertions expressed in this paper have not been endorsed, approved, or reviewed by any unaffiliated third party, including Terrabon, Inc., and are based on the authors' own independent research, evaluation, and analysis. The views and opinions of the authors expressed herein do not state or reflect those of such third parties, and shall not be construed as the views and opinions of such third parties.

7. References

American Society of Heating, Refrigerating and Air-Conditioning Engineers, ASHRAE Fundamentals Handbook, Atlanta, GA, 2001.

Bergles, A. E. ExHFT for fourth generation heat transfer technology, *Experimental Thermal and Fluid Science*, 26 (2002) 335-344.

Emerson, W. H. and Jamieson, D. T. Some physical properties of seawater in various concentrations, *Desalination*, 3 (1967) 213.

Holtzapple, M. T., Lara, J. R. Watanawanavet, S. Heat exchanger system for desalination. Patent Disclosure. Texas A&M University, College Station Texas 77843, Sept 2010.

Lara, J. R., An Advanced Vapor-Compression Desalination System. PhD. Dissertation., Texas A&M University. Dec 2005.

Lara, J. R., Holtzapple, M. T. Experimental Investigation of Dropwise Condensation on Hydrophobic Heat Exchangers. Department of Chemical Engineering Texas A&M University, 3122 TAMU, College Station, TX 77843-3122, February 2010.

Lara, J. R., Noyes, G., Holtzapple M. T. An investigation of high operating temperatures in mechanical vapor-compression desalination, *Desalination*, 227 (2008) 217-232.

Ma, X., Chen, D., Xu, J., Lin, C., Ren, Z. Long, Influence of processing conditions of polymer film on dropwise condensation heat transfer, *International Journal of Heat and Mass Transfer*, 45 (2002) 3405–3411.

Murphey, M., Rabroker, A., Holtzapple, M. T. 30-hp Desalination Compressor, Final Report, StarRotor Corporation, 1805 Southwood Dr., College Station, TX 77840.

Rose, J. W. Dropwise condensation theory and experiment: a review, *Journal of Power and Energy*, 16 (2002) 115-128.

8

Renewable Energy Opportunities in Water Desalination

Ali A. Al-Karaghouli and L.L. Kazmerski
National Renewable Energy Laboratory
Golden, Colorado, 80401,
USA

1. Introduction

Desalination is a water-treatment process that separates salts from saline water to produce potable water or water that is low in total dissolved solids (TDS). Globally, the total installed capacity of desalination plants was 61 million m^3 per day in 2008 [1]. Seawater desalination accounts for 67% of production, followed by brackish water at 19%, river water at 8%, and wastewater at 6%. Figure 1 show the worldwide feed-water percentage used in desalination. The most prolific users of desalinated water are in the Arab region, namely, Saudi Arabia, Kuwait, United Arab Emirates, Qatar, Oman, and Bahrain [2].

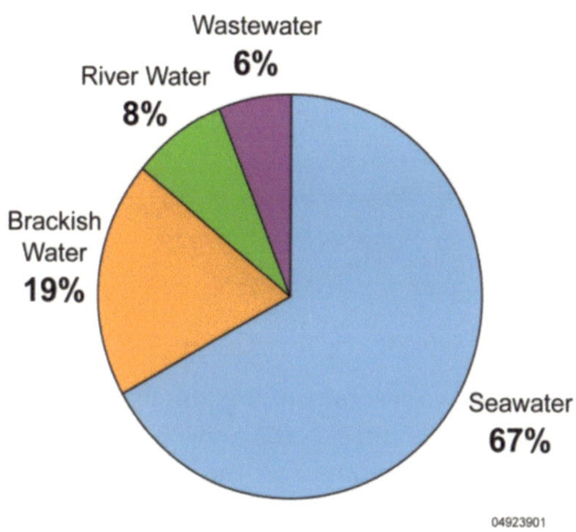

Fig. 1. Worldwide feed-water percentage used in desalination (http://desaldata.com/).

Desalination can be achieved by using a number of techniques. Industrial desalination technologies use either phase change or involve semi-permeable membranes to separate the solvent or some solutes. Thus, desalination techniques may be classified into two main categories [3]:

- Phase-change or thermal processes—where base water is heated to boiling. Salts, minerals, and pollutants are too heavy to be included in the steam produced from boiling and therefore remain in the base water. The steam is cooled and condensed. The main thermal desalination processes are multi-stage flash (MSF) distillation, multiple-effect distillation (MED), and vapor compression (VC), which can be thermal (TVC) or mechanical (MVC).

- Membrane or single-phase processes—where salt separation occurs without phase transition and involves lower energy consumption. The main membrane processes are reverse osmosis (RO) and electrodialysis (ED). RO requires electricity or shaft power to drive a pump that increases the pressure of the saline solution to the required level. ED also requires electricity to ionize water, which is desalinated by using suitable membranes located at two oppositely charged electrodes.

All processes require a chemical pre-treatment of raw seawater—to avoid scaling, foaming, corrosion, biological growth, and fouling—as well as a chemical post-treatment.

The two most commonly used desalination technologies are MSF and RO systems. As the more recent technology, RO has become dominant in the desalination industry. In 1999, about 78% of global production capacity comprised MSF plants and RO accounted for a modest 10%. But by 2008, RO accounted for 53% of worldwide capacity, whereas MSF consisted of about 25%. Although MED is less common than RO or MSF, it still accounts for a significant percentage of global desalination capacity (8%). ED is only used on a limited basis (3%) [4]. Figure 2 shows the global desalination plant capacity by technology in 2008.

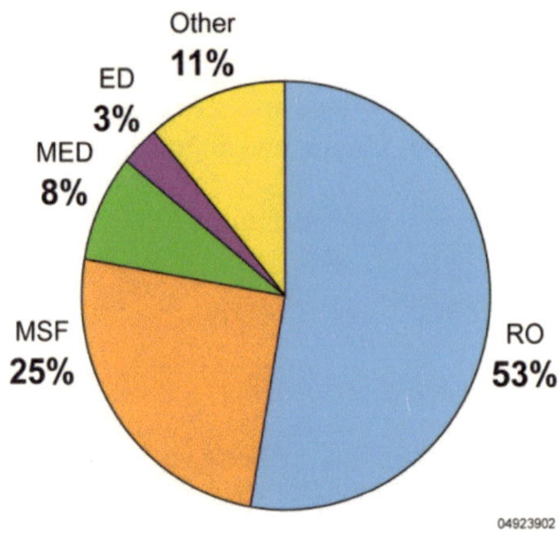

Fig. 2. Global desalination plant capacity by technology, 2008 (http://desaldata.com/).

The cost of water desalination varies depending on water source access, source water salinity and quality, specific desalination process, power costs, concentrate disposal method, project delivery method, and the distance to the point of use. Power costs in water desalination may account for 30% to 60% of the operational costs; thus, slight variations in power rates directly impact the cost of treated water.

Using renewable energy sources in water desalination has many advantages and benefits. The most common advantage is that they are renewable and cannot be depleted. They are a clean energy, not polluting the air, and they do not contribute to global warming or greenhouse gas emissions. Because their sources are natural, operational costs are reduced and they also require less maintenance on their plants. Using these resources in water desalination in remote areas also represents the best option due to the very high cost of providing energy from the grid. And implementing renewable energy in these areas will foster socioeconomic development. Renewable energy can be used for seawater desalination either by producing the thermal energy required to drive the phase-change processes or by producing electricity required to drive the membrane processes. The major sources of alternative energy discussed here are solar, wind, and geothermal.

This chapter provides insight into various aspects of desalination and how renewable energy resources can be coupled to desalination systems. A brief outline of the technical side of the main desalination processes is followed by an assessment of their respective advantages and disadvantages. The chapter then provides a general economic assessment of the conventional process versus desalination processes coupled with renewable energy. This analysis includes a range of cost estimates of competing processes as stated in the literature and how they compare to alternative sources of water supply.

2. Main desalination technologies

The two major types of desalination technologies used around the world can be broadly classified as either phase change (thermal) or membrane, and both technologies need energy to operate. Within these two types are sub-categories (processes) using different techniques, as shown below and in Figure 3:

- Phase-change processes, include:
  - Multi-stage flash distillation (MSF)
  - Multi-effect distillation (MED)
  - Vapor compression (VC)—thermal (TVC) and mechanical (MVC)
  - Other processes include solar still distillation, humidification-dehumidification , membrane distillation, and freezing.
- Membrane technology, include:
  - Reverse osmosis (RO)
  - Electrodialysis (ED and EDR).

Three other membrane processes that are not considered desalination processes, but that are relevant, are: microfiltration (MF), ultrafiltration (UF), and Nanofiltration (NF). The ion-exchange process is also not regarded as a desalination process, but is generally used to improve water quality for some specific purposes, e.g., boiler feed water [5].

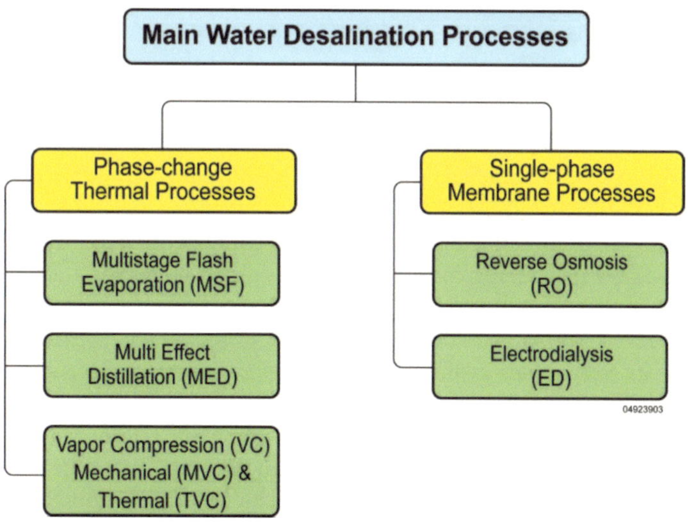

Fig. 3. Main desalination technologies.

2.1 Phase-change or distillation processes

Distillation processes mimic the natural water cycle as saline water is heated, producing water vapor, which, in turn, is condensed to form fresh water. The processes typically used include MSF, MED, and VC. Currently, about 25% of the world's desalination capacity is based on the MSF distillation principle. However, other distillation technologies, such as MED and VC distillation, are rapidly expanding and are anticipated to have a more important role in the future as they become better understood and more accepted. These processes require thermal or mechanical energy to cause water evaporation. As a result, they tend to have operating cost advantages when low-cost thermal energy is available [6].

2.1.1 Multi-stage flash distillation (MSF)

In MSF, seawater feed is pressurized and heated to the plant's maximum allowable temperature. When the heated liquid is discharged into a chamber maintained at slightly below the saturation vapor pressure of the water, a fraction of its water content flashes into steam. The flashed steam is stripped of suspended brine droplets as is passes through a mist eliminator and condenses on the exterior surface of the heat-transfer tubing [7]. The condensed liquid drips into trays as hot fresh-water product. Figure 4 is a diagram of a typical MSF unit.

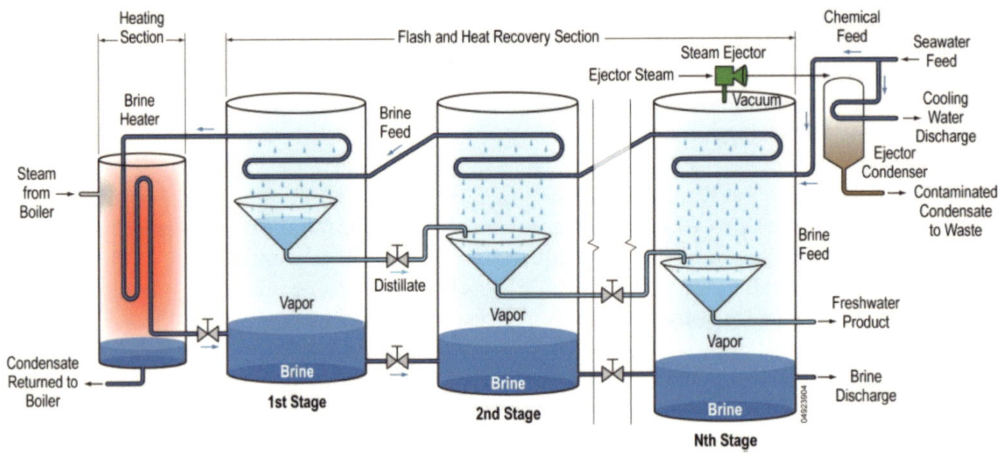

Fig. 4. Diagram of typical MSF unit (modified from [7])

2.1.2 Multi-effect distillation (MED)

MED units operate on the principle of reducing the ambient pressure at each successive stage, allowing the feed water to undergo multiple boiling without having to supply additional heat after the first stage. In this unit, steam and/or vapor from a boiler or some other available heat source (such as renewable sources or waste energy) is fed into a series of tubes, where it condenses and heats the surface of the tubes and acts as a heat-transfer surface to evaporate saline water on the other side. The energy used for evaporation of the saline water is the heat of condensation of the steam in the tube. The evaporated saline water—now free of a percentage of its salinity and slightly cooler—is fed into the next, lower-pressure stage where it condenses to fresh-water product, while giving up its heat to evaporate a portion of the remaining seawater feed [8]. Figure 5 is a diagram of an MED unit.

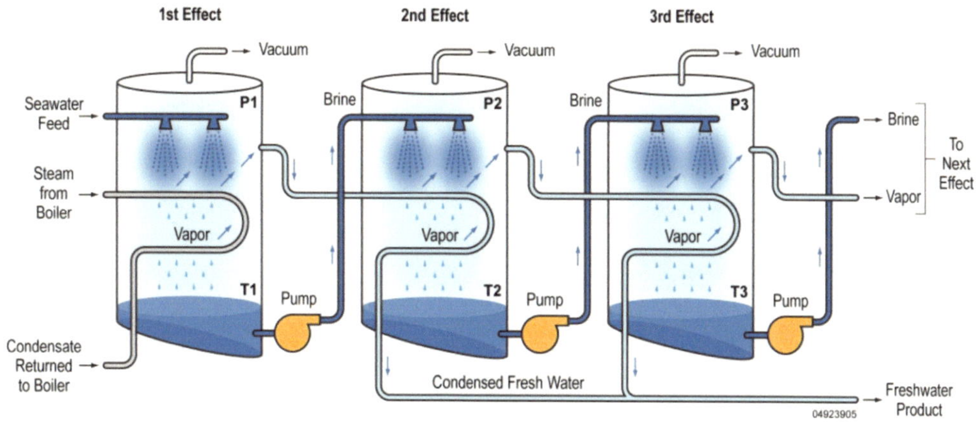

Fig. 5. Diagram of an MED unit (modified from [7])

2.1.3 Vapor-compression distillation

The VC distillation process is generally used for small- and medium-scale seawater desalting units. The heat for evaporating the water comes from the compression of vapor, rather than from the direct exchange of heat from steam produced in a boiler. Two primary methods are used to condense vapor so as to produce enough heat to evaporate incoming seawater: a mechanical compressor or a steam jet [9]. The mechanical compressor (MVC) is usually electrically driven, allowing the sole use of electrical power to produce water by distillation (Fig. 6a). With the steam jet-type of VC unit, also called a thermo compressor (TVC), a Venturi orifice at the steam jet creates and extracts water vapor from the main vessel by creating a lower ambient pressure in the main vessel. The extracted water vapor is compressed by the steam jet. This mixture is condensed on the tube walls to provide the thermal energy (heat of condensation) to evaporate the seawater being applied on the other side of the tube walls in the vessel (Fig. 6b). MVC units typically range in size up to about 3,000m^3/day, whereas TVC units may range in size up to 20,000 m^3/day; they are often used for resort and industrial applications.

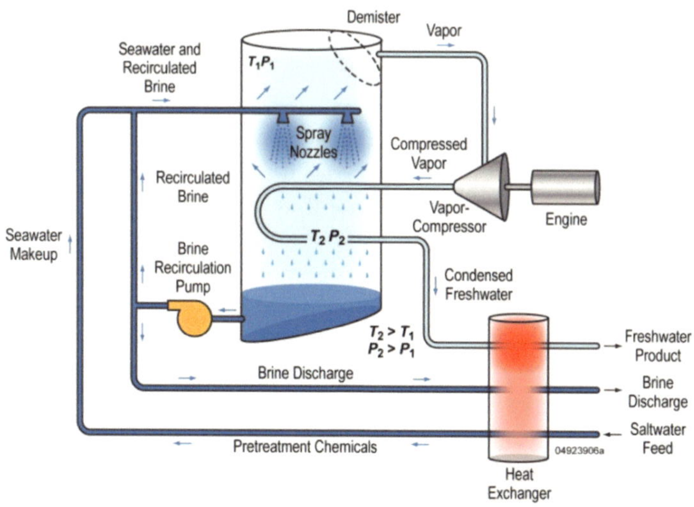

Fig. 6a. Diagram of a mechanical vapor-compression plant.

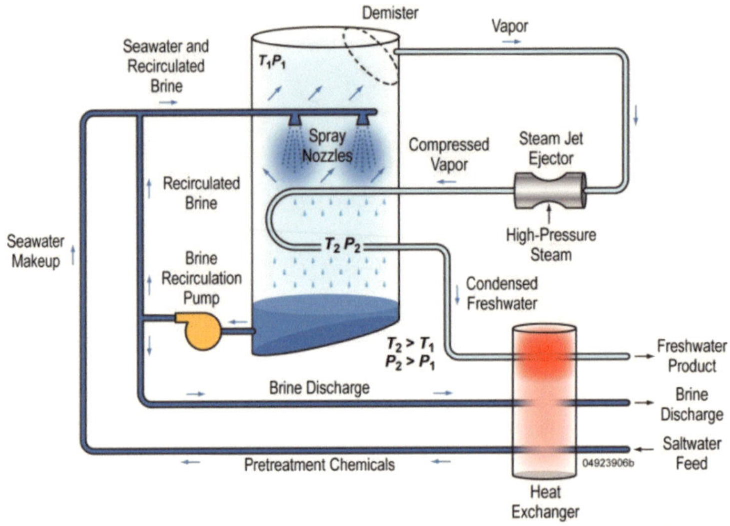

Fig. 6b. Diagram of a thermal vapor-compression plant (modified from [7])

2.2 Membrane processes

Membranes and filters can selectively permit or prohibit the passage of certain ions, and desalination technologies have been designed around these capabilities. Membranes play an important role in separating salts in the natural processes of dialysis and osmosis. These natural principles have been adapted in two commercially important desalting processes: electrodialysis (ED) and reverse osmosis (RO). Although they have typically been used to desalinate brackish water, versions are increasingly being applied to seawater, and these two approaches now account for more than half of all desalination capacity. A growing number of desalination systems are also adding filtration units prior to the membranes to remove contaminants that affect long-term filter operation. The filtration systems include microfiltration, nanofiltration, and ultrafiltration.

2.2.1 Reverse osmosis (RO)

RO technology description

Reverse osmosis is a form of pressurized filtration in which the filter is a semi-permeable membrane that allows water, but not salt, to pass through. A typical RO system consists of four major subsystems (see Fig. 7): pretreatment system, high-pressure pump, membrane module, and post-treatment system [10].

Feed-water pretreatment is a critical factor in operating an RO system because membranes are sensitive to fouling. Pretreatment commonly includes sterilizing feed water, filtering, and adding chemicals to prevent scaling and bio-fouling. Using a high-pressure pump, the pretreated feed water is forced to flow across the membrane surface. RO operating pressure ranges from 17 to 27 bars for brackish water and from 55 to 82 bars for seawater. Part of the feed water—the product or permeate water—passes through the membrane, which removes the majority of the dissolved solids [11]. The remainder, together with the rejected salts, emerges from the membrane modules at high pressure as a concentrated reject stream (brine). The energy efficiency of seawater RO depends heavily on recovering the energy

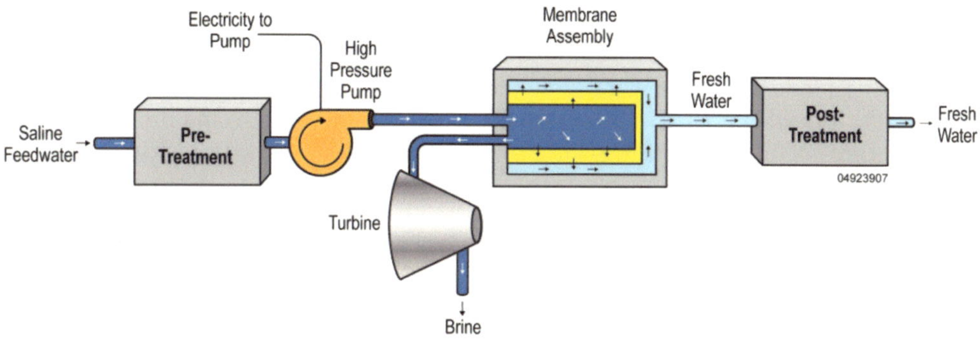

Fig. 7. Major subsystems in a reverse-osmosis system.

from the pressurized brine. In large plants, the reject brine pressure energy is recovered by a turbine—commonly a Peloton-wheel turbine—recovering 20% to 40% of the consumed energy.

The RO membrane is semi-permeable, possessing a high degree of water permeability, but presents an impenetrable barrier to salts. It has a large surface area for maximum flow and is extremely thin so that it offers minimal resistance to water flow; but it is also sturdy enough to withstand the pressure of the feed stream [12].

Polymers currently used for manufacturing RO membranes are based on either cellulose acetates (cellulose diacetate, cellulose triacetate, or combinations of the two) or polyamide polymers. Two types of RO membranes commonly used commercially are spiral-wound (SW) membranes and hollow-fiber (HF) membranes. Other configurations, including tubular and plate-frame designs, are sometimes used in the food and dairy industries. Seawater membrane elements are most commonly manufactured from a cellulose diacetate and triacetate blend or a thin-film composite usually made from polyamide, polysulphone, or polyurea polymers.

A typical industrial SW membrane is about 100–150 cm long and 20–30 cm in diameter. An HF membrane is made from both cellulose acetate blends and non-cellulose polymers such as polyamide. Millions of fibers are folded to produce bundles about 120 cm long and 10-20 cm in diameter. SW and HF membranes are used to desalt both seawater and brackish water. The decision of which to use is based on factors such as cost, feed-water quality, and product-water capacity. The main membrane manufacturers are in the United States and Japan [13].

The post-treatment system consists of sterilization, stabilization, and mineral enrichment of the product water. Because the RO unit operates at ambient temperature, corrosion and scaling problems are diminished compared to distillation processes. However, effective pretreatment of the feed water is required to minimize fouling, scaling, and membrane degradation. In general, the selection of proper pretreatment and proper membrane maintenance are critical for the efficiency and life of the system.

RO technology deployment

RO units are available in a wide range of capacities due to their modular design. Large plants are made of hundreds of units that are accommodated in racks. A typical maximum plant capacity is 128,000 m^3 /day, and very small units (down to 0.1 m^3/day) are also used for marine purposes, houses, or hotels. PV power is used for small-size RO units especially in remote places due to initial-cost benefits [14].

Numerous RO plants have been installed for both seawater and brackish-water applications. The process is also widely used in manufacturing, agriculture, food processing, and pharmaceutical industries. The worldwide total installed capacity of RO units in the United States is 32%, followed by 21% in Saudi Arabia, 8% in Japan, and 8.9% in Europe. Some 23% of RO units are manufactured in the United States, 18.3% in Japan, and 12.3% in Europe [14].

2.2.2 Electrodialysis

Electrodialysis (ED) is an electrochemical separation process that uses electrical currents to move salt ions selectively through a membrane, leaving fresh water behind. ED is a low-cost method for desalinating brackish water. Due to the dependence of energy consumption on the feed-water salt concentration, the ED process is not economically attractive for desalinating seawater. In the ED process, ions are transported through a membrane by an electrical field applied across the membrane. An ED unit consists of the following basic components: pretreatment system, membrane stack, low-pressure circulation pump, power supply for direct-current (rectifier or PV system), and post-treatment system.

The principle of ED operation is as follows: electrodes (generally constructed from niobium or titanium with a platinum coating) are connected to an outside source of direct current (such as a battery or PV source) in a container of salt water. The electrical current is carried through the solution, with the ions tending to migrate to the electrode with the opposite charge. Positively charged ions migrate to the cathode and negatively charged ions migrate to the anode. Salinity of the water is removed as water passes through ion-selective membranes positioned between the two electrodes (see Fig. 8). These membranes consist of flat-sheet polymers subjected to special treatment in which micro-sized cracks or crevices are produced in the plastic film surface. These devices permit the transport of ions, while ion-exchange sites incorporated into the membrane's polymer matrix promote membrane selectivity. Anion-permissible membranes allow anions to pass through to the positively charged electrode, but reject cations. Conversely, cation permissible membranes allow cations to pass through to the negatively charged electrode, but reject anions.

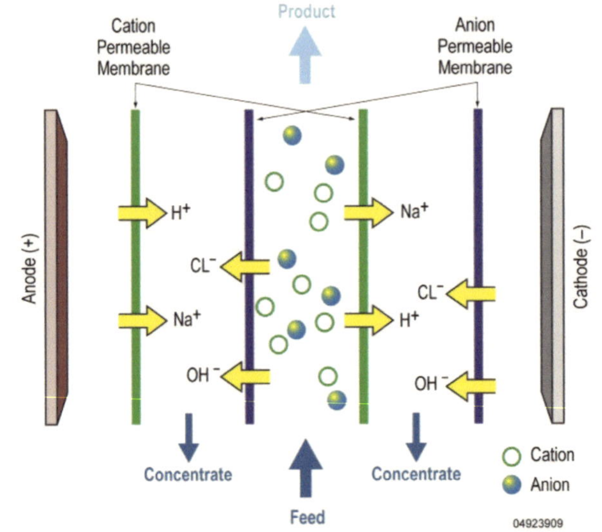

Fig. 8. Ion exchange in electrodialysis unit.

Between each pair of membranes, a spacer sheet is placed to permit water flow along the face of the membrane and to induce a degree of turbulence. One spacer provides a channel that carries feed water (and product water), whereas the next spacer carries brine. By this arrangement, concentrated and diluted solutions are created in the spaces between the alternating membranes. ED cells can be stacked either horizontally or vertically. In practice, several membrane pairs are used between a single pair of electrodes, forming an ED stack. Feed water passes simultaneously in parallel paths through all the cells, providing a continuous flow of product water and brine out of the stack (see Fig. 9). Stacks on commercial ED plants contain a large number of cell pairs, usually several hundred [15].

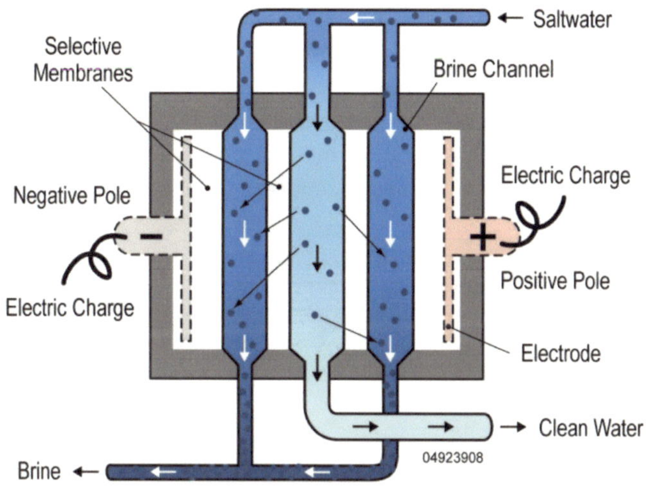

Fig. 9. Diagram of electrodialysis unit (modified from [7])

A modification to the basic ED process is electrodialysis reversal (EDR). An EDR unit operates on the same general principle as a standard ED plant, except that the product and brine channels are both identical in construction. In this process, the polarity of the electrodes changes periodically in time, reversing the flow through the membranes. This inhibits deposition of inorganic scales and colloidal substances on the membranes without the addition of chemicals to the feed water. This development considerably enhances the viability of this process because the process is now self-cleaning. In general, EDR requires minimum feed-water pretreatment and minimum use of chemicals for membrane cleaning [16].

ED and EDR technology deployment

The ED process is usually only suitable for brackish water with a salinity of up to 12,000 ppm TDS. With higher salinity, the process rapidly becomes more costly than other desalination processes because the power consumption is directly proportional to the salinity of the water to be desalinated. ED has been in commercial use since 1954, more than ten years before RO. Since then, this process has seen widespread applications, especially for the production of potable water. Due to its modular structure, ED is available in a wide range of sizes, from small capacities (down to 2 m^3/d) to large capacities (145,000 m^3/day). ED is widely used in the United States, which has 31% of the total installed capacity. In

Europe, the ED process accounts for 15% and the Middle East has 23% of the total installed capacity. The EDR process was developed in the early 1970s. Today, the process is used in about 1,100 installations worldwide. Typical industrial users of EDR include power plants, semiconductor manufacturers, the pharmaceutical industry, and food processors. The installed PV-ED units are only of small capacity and are used in remote areas.

3. Desalination with renewable energy systems

Using desalination technologies driven by renewable energy resources is a viable way to produce fresh water in many locations today. As the technologies continue to improve—and as fresh water and cheap conventional sources of energy become scarcer—using renewable energy technology in desalination will become even more attractive. The selection of the appropriate renewable energy desalination technology depends on a number of factors, including plant size, feed-water salinity, remoteness, availability of grid electricity, technical infrastructure, and the type and potential of the local renewable energy resources. Figure 10 shows the possible combination of renewable energy systems with desalination units.

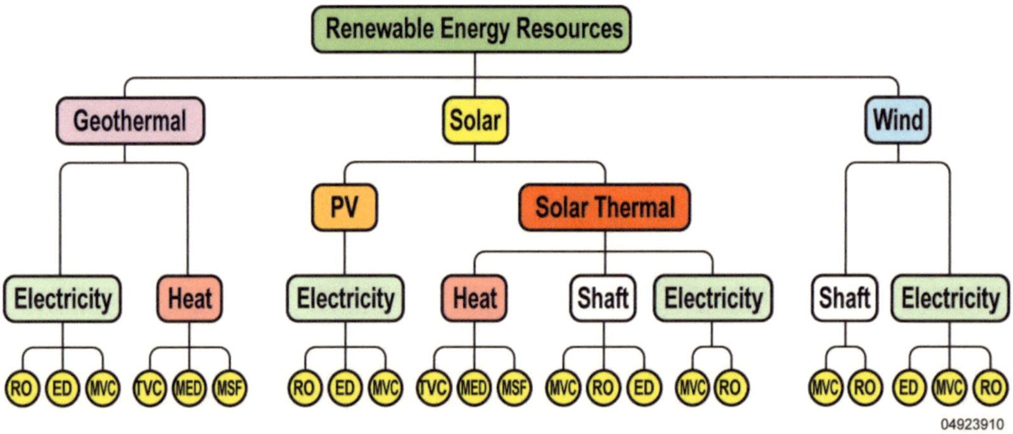

Fig. 10. Possible combinations of renewable energy systems with desalination units.

Proper matching of standalone power-supply desalination systems has been recognized as being crucial if the system is to provide a satisfactory supply of power and water at a reasonable cost. Standalone renewable energy systems for electricity supply are now a proven technology and economically promising for remote regions, where connection to the public electric grid is either not cost effective or feasible, and where water scarcity is severe. Solar thermal, solar PV, wind, and geothermal technologies could be used as energy suppliers for desalination systems.

Table 1 presents the most promising combinations of renewable energy resources with desalination technologies. According to this table, solar energy—both solar thermal and solar PV—can be used to drive MSF, MED, RO, and ED. Wind energy can drive VC, RO, and ED. Geothermal energy reservoirs with moderate temperature can drive MSF and MED units, while geothermal high pressure reservoirs can be utilized to drive mechanically driven desalination units by shaft power or by producing electricity to drive VC, RO, and ED units.

RE Resource	Desalination Process				
	MSF	MED	VC	RO	ED
Solar thermal	◙	◙			
Solar PV				◙	◙
Wind			◙	◙	◙
Geothermal	◙	◙	◙	◙	◙

Table 1. Possible combination of RE resources with desalination units

3.1 Solar-assisted desalination systems

Solar energy can drive the desalination units by either thermal energy and electricity generated from solar thermal systems or by PV systems. The cost distribution of solar distillation is dramatically different from that of RO and MSF. The main cost is in the initial investment. However, once the system is operational, it is extremely inexpensive to maintain and the energy has minimal or even no cost. Solar-assisted desalination systems are divided into two parts: solar thermal-assisted systems and solar photovoltaic-assisted systems.

3.1.1 Solar thermal-assisted systems

Solar thermal energy can be harnessed directly or indirectly for desalination. Collection systems that use solar energy to produce distillate directly in the solar collector are called direct-collection systems, whereas systems that combine solar energy collection devices with conventional desalination units are called indirect systems. In indirect systems, solar energy is used either to generate the heat required for desalination and/or to generate electricity used to provide the required electric power for conventional desalination plants such as MED and MSF plants. Direct solar desalination requires large land areas and has a relatively low productivity. However, it is competitive with indirect desalination plants in small-scale production due to its relatively low cost and simplicity.

3.1.1.1 Direct solar thermal desalination

Direct systems are those where the heat collection and distillation processes occur in the same equipment. Solar energy is used to produce the distillate directly in the solar still. The method of direct solar desalination is mainly suited for small production systems, such as solar stills, and it is used in regions where the freshwater demand is low. This device has low efficiency and low water productivity due to the ineffectiveness of solar collectors to convert most of the energy they capture, and to the intermittent availability of solar radiation. For this reason, direct solar thermal desalination has so far been limited to small-capacity units, which are appropriate in serving small communities in remote areas having scarce water. Solar-still design can generally be grouped into four categories: (1) basin still, (2) tilted-wick solar still, (3) multiple-tray tilted still, and (4) concentrating mirror still.

The basin still consists of a basin, support structure, transparent glazing, and distillate trough. Thermal insulation is usually provided underneath the basin to minimize heat loss. Other ancillary components include sealants, piping and valves, storage, external cover, and a reflector (mirror) to concentrate light. Single basin stills have low efficiency, generally below 45%, and low productivity (4–6 liter/m^2/day) due to high top losses. Double glazing can potentially reduce heat losses, but it also reduces the transmitted portion of the solar radiation [17]. On a much smaller scale, a solar micro-desalination unit [18] may be used in remote areas and is capable of producing about 1.5 liter/day.

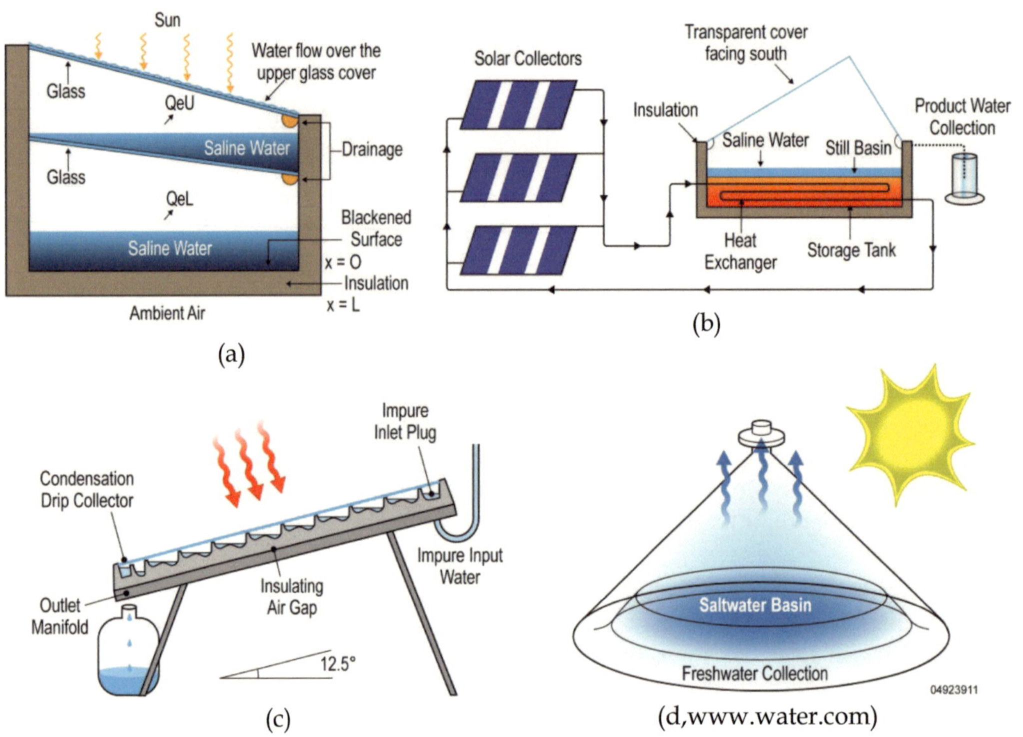

Fig. 11. Diagrams of various solar stills: (a) double-basin solar still, (b) single-basin solar still coupled with flat plate collector, (c) multi-steps tilted solar still, (d) micro-solar solar still.

A tilted-wick solar still uses the capillary action of fibers to distribute feed water over the entire surface of the wick in a thin layer. This allows a higher temperature to form on this thin layer. Insulation in the back of wick is essential. A cloth wick needs frequent cleaning to remove sediment built-up and regular replacement of wick material due to weathering and ultraviolet degradation. Uneven wetting of the wick can result in dry spots that reduce efficiency [19].

In a multiple-tray tilted still, a series of shallow horizontal black trays are enclosed in an insulated container with a transparent glazing on top. The feed-water supply tank is located above the still, and the vapor condenses and flows down to the collection channel and finally to the storage. The construction of this still is fairly complicated and involves many components that are more expensive than simple basin stills. Therefore, the slightly better efficiency it delivers may not justify its adoption [20].

The concentrating mirror solar still uses a parabolic mirror for focusing sunlight onto an evaporator vessel. The water is evaporated in this vessel exposed to extremely high temperature. This type of still entails high construction and maintenance costs [21].

Figure 11 shows four various types of solar stills.

3.1.1.2 Indirect solar thermal desalination

Indirect solar thermal desalination methods involve two separate systems: the collection of solar energy by a solar collecting system, coupled to a conventional desalination unit.

Processes include humidification-dehumidification (HD), membrane distillation (MD), solar pond-assisted desalination, and solar thermal systems such as solar collectors, evacuated-tube collectors, and concentrating collectors (CSP) systems driving conventional desalination processes such as MSF and MED.

Humidification-dehumidification process

These units consist of a separate evaporator and condenser to eliminate the loss of latent heat of condensation. The basic idea in humidification–dehumidification (HD) process is to mix air with water vapor and then extract water from the humidified air by the condenser. The amount of vapor that air can hold depends on its temperature. Some advantages of HD units are the following: low-temperature operations, able to combine with renewable energy sources such as solar energy, modest level of technology, and high productivity rates. Two different cycles are available for HD units: HD units based on open-water closed-air cycle, and HD units based on open-air closed-water cycle. These two options are described below. Figure12a shows an open-water closed-air cycle. In the process, seawater enters the system, is heated in the solar collector, and is then sprayed into the air in the evaporator. Humidified air is circulated in the system and when it reaches the condenser, a certain amount of water vapor starts to condense. Distilled water is collected in a container. Some of the brine can also be recycled in the system to improve the efficiency, and the rest is removed [22].

Figure12b shows an open-air closed-water cycle, which is used to emphasize recycling the brine through the system to ensure a high utilization of the salt water for freshwater production. As air passes through the evaporator, it is humidified. And by passing through condenser, water vapor is extracted [23].

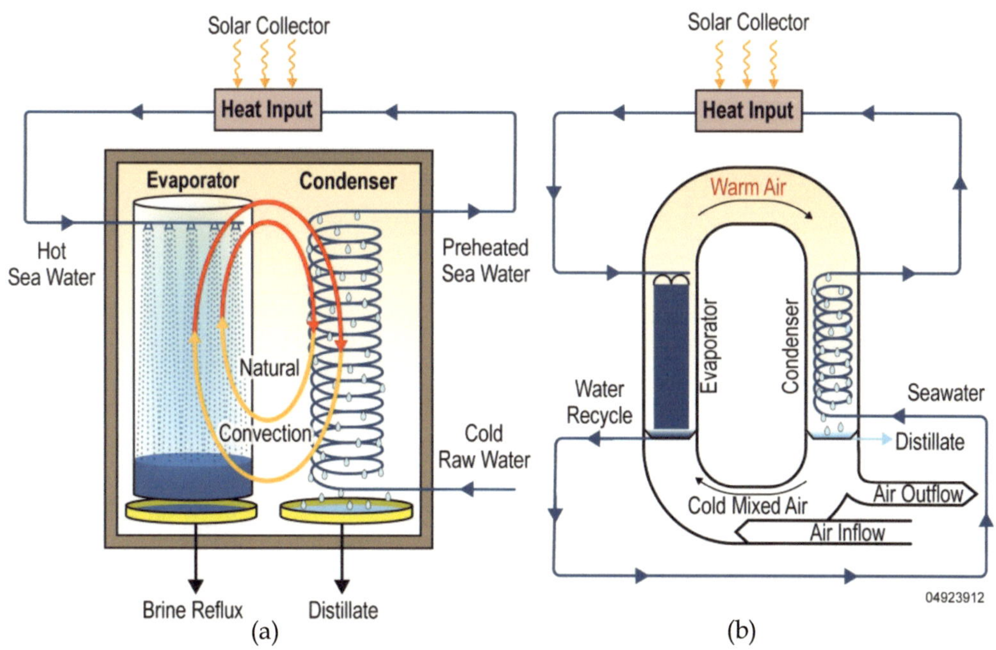

Fig. 12. Humidification-dehumidification systems: (a) open-water closed-air cycle, and (b) open-air closed-water cycle (modified from [64].

Membrane distillation

Membrane distillation (MD) is a separation/distillation technique, where water is transported between "hot" and a "cool" stream separated by a hydrophobic membrane, permeable only to water vapor, which excludes the transition of liquid phase and potential dissolved particles. The exchange of water vapor relies on a small temperature difference between the two streams, which results in a vapor pressure difference, leading to the transfer of the produced vapor through the membrane to the condensation surface. Figure 13 is a typical schematic diagram of the process. In the MD process, the seawater passes through the condenser usually at about 25ºC and leaves at a higher temperature, and then it is heated to about 80°C by an external source such as solar, geothermal, or industrial waste [24]. The main advantages of membrane distillation lie in its simplicity and the need for only small differentials to operate. However, the temperature differential and the recovery rate determine the overall efficiency for the process. Thus, when it is run with a low temperature differential, large amounts of water must be used, which adversely affects its overall energy efficiency.

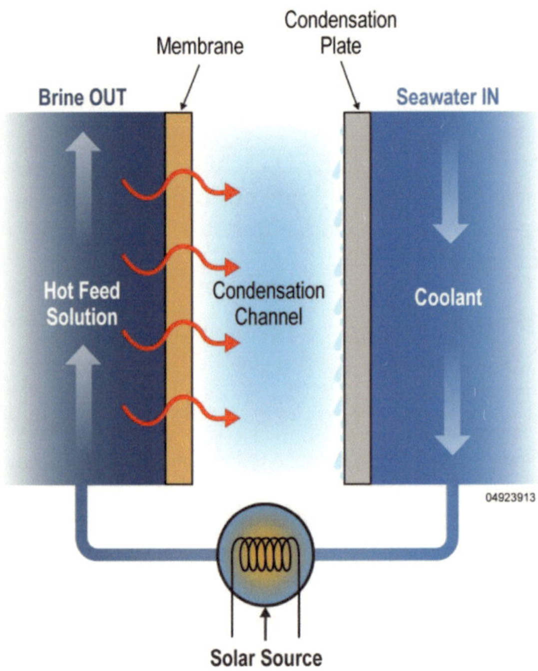

Fig. 13. Schematic of the membrane distillation process.

Membrane desalination is a promising process, especially for situations where low-temperature solar, geothermal, waste, or other heat is available. MD was introduced commercially on a small scale during the 1980s, but it has not demonstrated large-scale commercial success due to the high cost and problems associated with membranes. Therefore, more intensive research and development is needed, both in experimentation and modeling, focusing on key issues such as long-term liquid/vapor selectivity, membrane aging and fouling, feed-water contamination, and heat-recovery optimization. Scale-up studies and realistic assessment of the basic working parameters on real pilot plants, including cost and long-term stability, are also considered to be necessary [25].

Solar ponds-assisted desalination

Salinity-gradient solar ponds are a type of heat collector, as well as a mean of heat storage. Hot brine from a solar pond can be used as a heat source for MSF or MED desalination units. Solar ponds can store heat because of their unique chemically stratified nature. A solar pond has three layers: (1) upper or surface layer, called the upper convection zone, (2) middle layer, which is the non-convection zone or salinity-gradient zone, and (3) lower layer, called the storage zone or lower convection zone. Salinity increases with depth from near pure water at the surface to the bottom, where salts are at or near saturation. Salinity is relatively constant in the upper and lower convection zones, and increases with depth in the non-convection zone. Saline water is denser than fresh water; therefore, the water at the bottom of the pond is more dense (has a higher specific gravity) than water at the surface. The solar pond system is able to store heat because circulation is suppressed by the salinity-related density differences in the stratified water. Convection of hot water to the surface is repressed by the salinity (density) gradient of the non- convection zone. Thus, although solar energy can penetrate the entire depth of the pond, it cannot escape the storage zone [26]. Figure 14 show how a typical MSF unit operates using solar pond brine as a heat source.

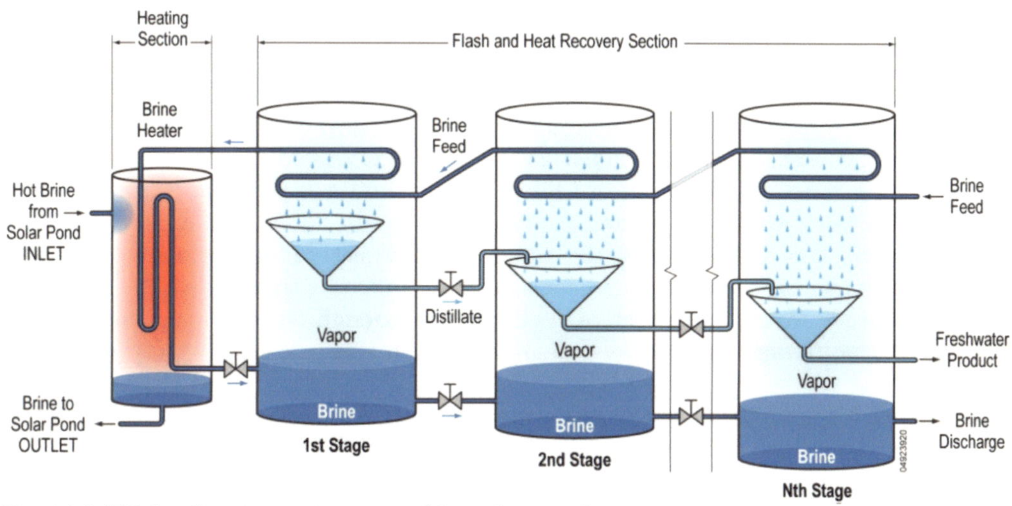

Fig. 14. MSF desalination unit operated by solar pond.

Concentration solar thermal desalination

Concentrating solar thermal power technologies are based on the concept of concentrating solar radiation to provide high-temperature heat for electricity generation within conventional power cycles using steam turbines, gas turbines, or Stirling and other types of engines. For concentration, most systems use glass mirrors that continuously track the position of the sun. The four major concentrating solar power (CSP) technologies are parabolic trough, Fresnel mirror reflector, power tower, and dish/engine systems. Debate continues as to which of these is the most effective technology [27]. Figure 15 shows diagrams of these systems.

Parabolic trough. Parabolic trough power plants consist of large parallel arrays of parabolic trough solar collectors that constitute the solar field. The parabolic collector is made of

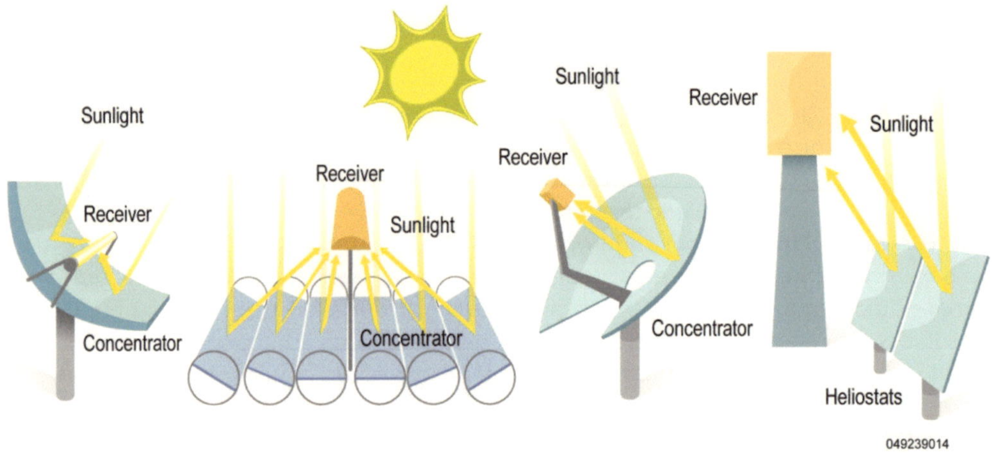

(a) (b) (c) (d)

Fig. 15. Solar concentrating systems, (a) parabolic trough, (b) Fresnel lenses, (c) dish engine, and (d) power tower.

reflectors, each of which focuses the sun's radiation on a receiver tube that absorbs the reflected solar energy. The collectors track the sun so that the sun's radiation is continuously focused on the receiver. Parabolic troughs are recognized as the most proven CSP technology, and at present, experts indicate the cost to be 10 US cents/kWh or less.

Fresnel mirror reflector. This type of CSP is broadly similar to parabolic trough systems, but instead of using trough-shaped mirrors that track the sun, flat or slightly curved mirrors mounted on trackers on the ground are configured to reflect sunlight onto a receiver tube fixed in space above these mirrors. A small parabolic mirror is sometimes added atop the receiver to further focus the sunlight. As with parabolic trough systems, the mirrors change their orientation throughout the day so that sunlight is always concentrated on the heat-collecting tube.

Dish/Stirling engine systems and concentrating PV (CPV) systems. Solar dish systems consist of a dish-shaped concentrator (like a satellite dish) that reflects solar radiation onto a receiver mounted at the focal point. The receiver may be a Stirling or other type of engine and generator (dish/engine systems) or it may be a type of PV panel that has been designed to withstand high temperatures (CPV systems). The dish is mounted on a structure that tracks the sun continuously throughout the day to reflect the highest percentage of sunlight possible onto the thermal receiver. Dish systems can often achieve higher efficiencies than parabolic trough systems, partly because of the higher level of solar concentration at the focal point. Dish systems are sometimes said to be more suitable for stand-alone, small power systems due to their modularity. Compared with ordinary PV panels, CPV has the advantage that smaller areas of PV cells are needed; because PV is still relatively expensive, this can mean a significance cost savings.

Power tower. A power tower system consists of a tower surrounded by a large array of heliostats, which are mirrors that track the sun and reflect its rays onto the receiver at the top of the tower. A heat-transfer fluid heated in the receiver is used to generate steam, which, in turn, is used in a conventional turbine generator to produce electricity. Some power towers use water/steam as the heat-transfer fluid. Other advanced designs are experimenting with molten nitrate salt because of its superior heat-transfer and energy-storage capabilities. Power towers also reportedly have higher conversion efficiencies than parabolic trough systems. They are projected to be cheaper than trough and dish systems, but a lack of commercial experience means that there are significant technical and financial risks in deploying this technology now. As for cost, it is predicted that with higher efficiencies, 7-8 cents/kWh may be possible. But this technology is still in its early days of commercialization.

CSP systems coupled with desalination plant

The primary aim of CSP plants is to generate electricity, yet a number of configurations enable CSP to be combined with various desalination methods. When compared with photovoltaics or wind, CSP could provide a much more consistent power output when combined with either energy storage or fossil-fuel backup. There are different scenarios for using CSP technology in water desalination [28], and the most suitable options are described below.

Parabolic trough coupled with MED desalination unit. Figure 16 shows a typical parabolic trough configuration combined with a MED system, where steam generated by the trough (superheated to around 380°C) is first expended in a non-condensing turbine and then used in a conventional manner for desalination. The steam temperature for the MED plant is around 135°C; therefore, there is sufficient energy in the steam to produce electricity before it is used in the MED plant. It is important to emphasize that water production is the main purpose of the plant—electricity is a byproduct. Although conventional combined-cycle

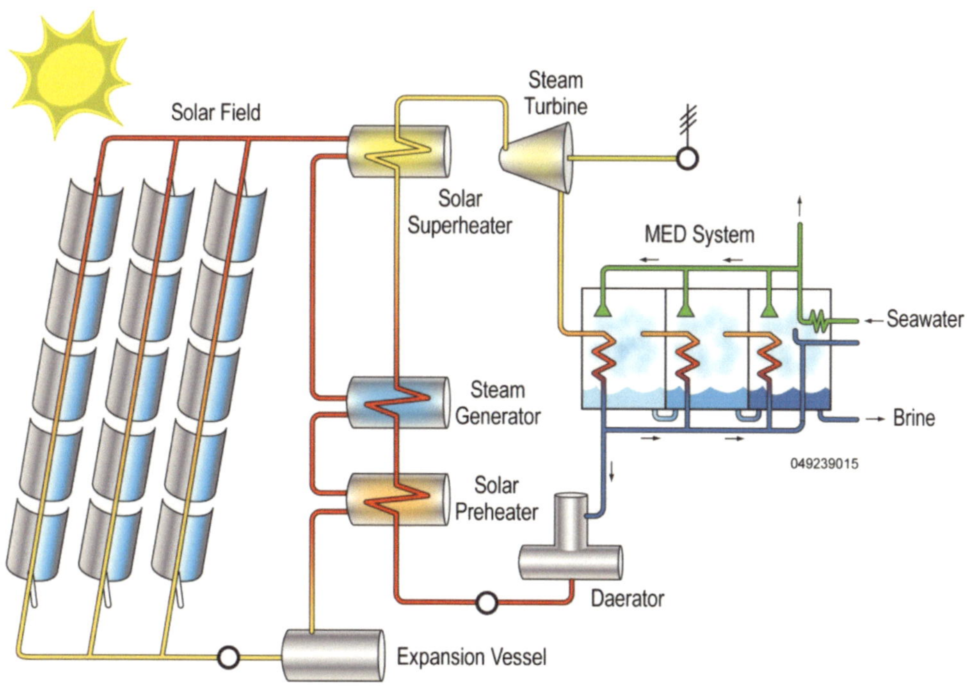

Fig. 16. Parabolic trough power plant with oil steam generator and MED desalination (Source: Bechtel Power)

(CC) power plants can be configured in a similar manner for desalination, a fundamental difference exists in the design approach for solar and for fossil-fuel-fired plants. The fuel for the solar plant is free; therefore, the design is not focused primarily on efficiency but on capital cost and capacity of the desalination process. In contrast, for the CC power plant, electricity production at the highest possible efficiency is the ultimate goal [29].

Parabolic trough coupled with RO desalination unit. In this case, as in MED, the steam generated by the solar plant can be used through a steam turbine to produce the electric power needed to drive the RO pumps. As an alternative for large, multi-unit RO systems, the high-pressure seawater can be provided by a single pump driven by a steam turbine. This arrangement is similar to the steam-turbine-driven boiler feed pumps in a fossil-fuel power plant. Often, MED and RO are compared in terms of overall performance, and specifically for energy consumption. Based on internal studies by Bechtel [30], one can conclude that in specific cases, the CSP/RO combination (see Fig. 17) requires less energy than a similar CSP/MED combination.

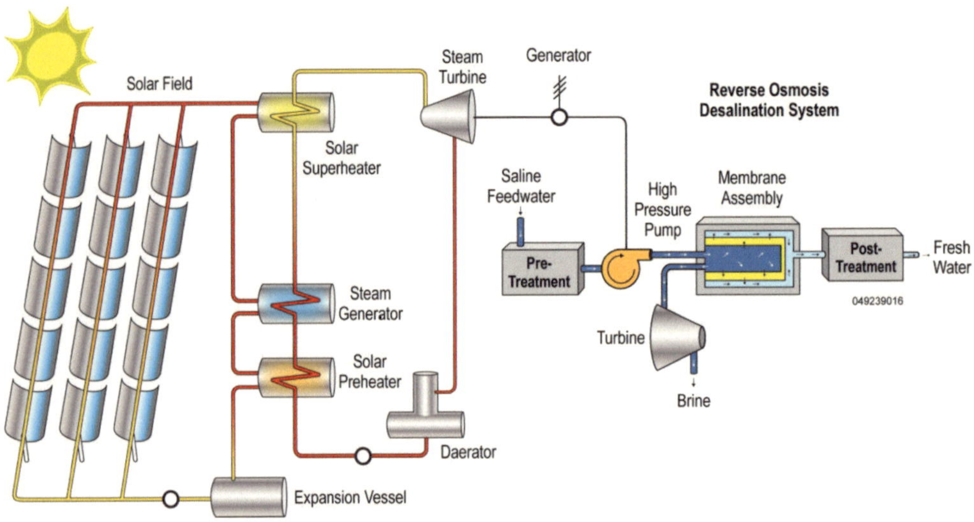

Fig. 17. Parabolic trough coupled with seawater RO desalination unit (modified from Bechtel Power)

However, an analysis presented in [31] suggests that, for several locations, CSP/MED requires 4% to 11% less input energy than CSP/RO. Therefore, before any decision can be made on the type of desalination technology to be used, we recommend that a detailed analysis be conducted for each specific location, evaluating the amount of water, salinity of the input seawater, and site conditions. It appears that CSP/MED provides slightly better performance at sites with high salinity such as in closed gulfs, whereas CSP/RO appears to be more suitable for low-salinity waters in the open ocean.

One additional advantage of the RO system is that the solar field might be located away from the shoreline. The only connection between the two is the production of electricity to drive the RO pumps and other necessary auxiliary loads.

3.1.1.3 Solar thermal applications

Although the strong potential of solar thermal energy to seawater desalination is well recognized, the process is not yet developed at the commercial level. The main reason is that the existing technology, although demonstrated as technically feasible, cannot presently compete, on the basis of produced water cost, with conventional distillation and RO technologies. However, it is also recognized that there is still potential to improve desalination systems based on solar thermal energy.

Among low-capacity production systems, solar stills and solar ponds represent the best alternative in low fresh water demands. For higher desalting capacities, one needs to choose conventional distillation plants coupled to a solar thermal system, which is known as indirect solar desalination [32]. Distillation methods used in indirect solar desalination plants are MSF and MED. MSF plants, due to factors such as cost and apparent high efficiency, displaced MED systems in the 1960s, and only small-size MED plants were built. However, in the last decade, interest in MED has been significantly renewed and the MED process is currently competing technically and economically with MSF [33]. Recent advances in research of low-temperature processes have resulted in an increase of the desalting capacity and a reduction in the energy consumption of MED plants providing long-term operation under remarkable steady conditions [34]. Scale formation and corrosion are minimal, leading to exceptionally high plant availabilities of 94% to 96%.

Many small systems of direct solar thermal desalination systems and pilot plants of indirect solar thermal desalination systems have been implemented in different places around the world [35]. Among them are the de Almería (PSA) project in 1993 and the AQUASOL project in 2002. Study of these systems and plants will improve our understanding of the reliability and technical feasibility of solar thermal technology application to seawater desalination. It will also help to develop an optimized solar desalination system that could be more competitive against conventional desalination systems. Table 2 presents several of the implemented indirect solar thermal pilot systems.

Plant Location	Year of Commission	Water Type	Capacity (L/hr)	RES Installed Power	Unit Water Cost (US$/m^3)
Almeria, Spain, CIEMAT	1993	SW	3000	2.672 m^2 solar collector area	3.6-4.35
Hazeg, Sfax, Tunisia	1988	BW	40-50	80 m^2 solar collector area	25.3
Pozo Izquierdo, Gran Canaria, SODESA Project	2000	SW	25	50 m^2 solar collector area	-
Sultanate of Oman, MEDRC Project	2002	SW	42	5.34 m^2 solar collector area	-
AQUASOL Project	2002	SW	3000	14 cells of parabolic concentrator	-

SW: seawater, BW: brackish water

Table 2. Solar thermal distillation plants

On a commercial basis, CSP technology will take many years until it becomes economic and sufficiently mature for use in power generation and desalination.

3.2 Solar PV desalination

General description of a PV system

A photovoltaic or solar cell converts solar radiation into direct-current (DC) electricity. It is the basic building block of a PV (or solar electric) system. An individual PV cell is usually quite small, typically producing about 1 or 2 watts of power. To boost the power output, the solar cells are connected in series and parallel to form larger units called modules. Modules, in turn, can be connected to form even larger units called arrays. Any PV system consists of a number of PV modules, or arrays. The other system equipment includes a charge controller, batteries, inverter, and other components needed to provide the output electric power suitable to operate the systems coupled with the PV system. PV systems can be classified into two general categories: flat-plate systems and concentrating systems. CPV system have several advantages compared to flat-plate systems: CPV systems increase the power output while reducing the size or number of cells needed; and a solar cell's efficiency increases under concentrated light.
Figure 18 is a schematic diagram of a PV solar system that has everything needed to meet a particular energy demand, such as powering desalination units.

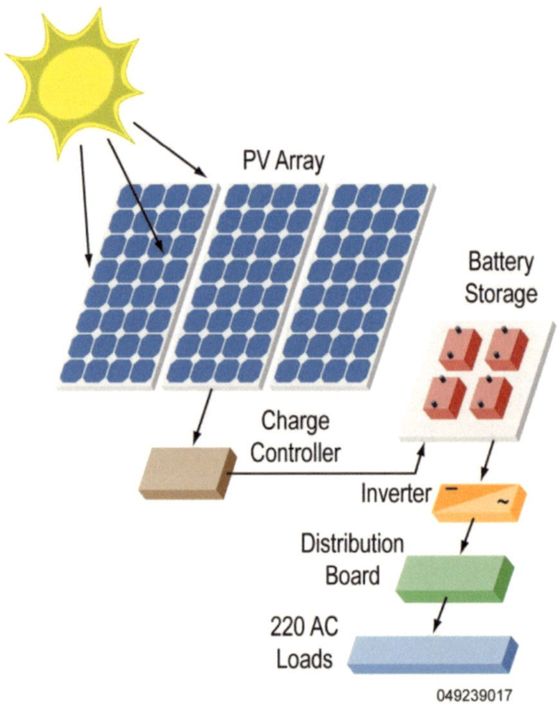

Fig. 18. Schematic of a typical photovoltaic system.

Typical PV system driving RO-ED units

PV is a rapidly developing technology, with costs falling dramatically with time, and this will lead to its broad application in all types of systems. Today, however, it is clear that PV/RO and PV/ED will initially be most cost competitive for small-scale systems installed in remote areas where other technologies are less competitive. RO usually uses alternating current (AC) for the pumps, which means that DC/AC inverters must be used. In contrast, ED uses direct current for the electrodes at the cell stack, and hence, it can use the energy supply from the PV panels without major modifications. Energy storage is again a concern, and batteries are used for PV output power to smooth or sustain system operation when solar radiation is insufficient.

PV/RO systems applications

PV-powered reverse osmosis is considered one of the most promising forms of renewable-energy-powered desalination, especially when it is used in remote areas. Therefore, small-scale PV/RO has received much attention in recent years and numerous demonstration systems have been built. Figure 19 is a schematic diagram of a PV/RO system. Two types of PV/RO systems are available in the market: brackish-water (BWRO) and seawater (SWRO) PV/RO systems. Different membranes are used for brackish water and much higher recovery ratios are possible, which makes energy recovery less critical [36].

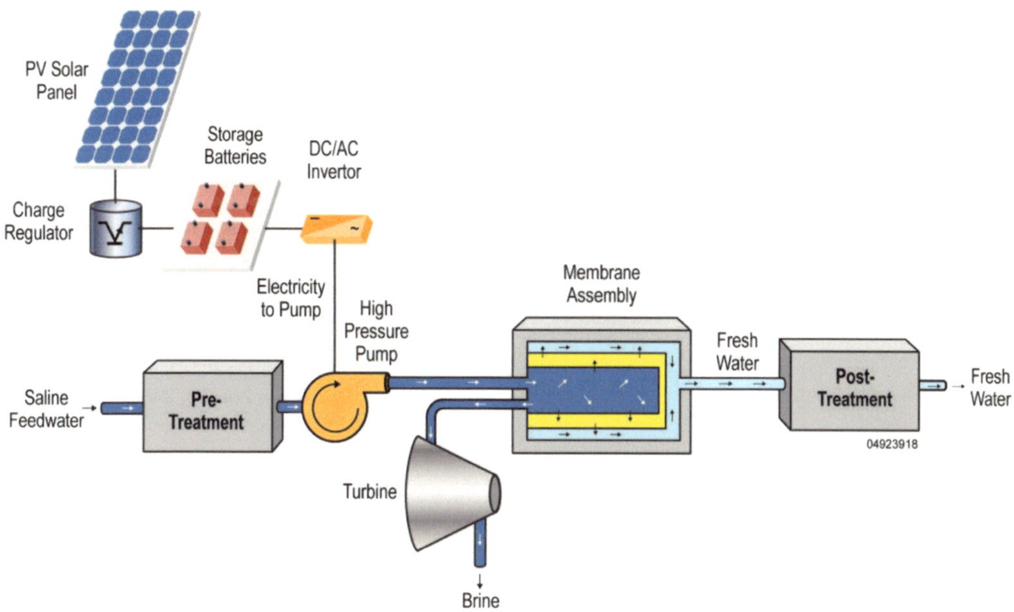

Fig. 19. Schematic of a PV/RO system.

Brackish water PV/RO systems

Brackish water has a much lower osmotic pressure than seawater; therefore, its desalination requires much less energy and a much smaller PV array in the case of PV/RO. Also, the lower pressures found in BWRO systems permit the use of low-cost plastic components. Thus, the total cost of water from brackish water PV/RO is considerably less than that from seawater, and systems are beginning to be offered commercially [37]. Table 3 presents information on installed brackish water PV/RO systems [38–42]. Many of the early PV/RO demonstration systems were essentially a standard RO system, which might have been designed for diesel or mains power, but powered from batteries charged by PV. This approach generally requires a rather large PV array for a given flow of product because of poor efficiencies in the standard RO systems and batteries. Large PV arrays and the regular replacement of batteries typically make the cost of water from such systems rather high.

Location	Feedwater (ppm)	Capacity (m^3/day)	PV (kWp)	Batteries (kWh)	Energy Consumption (kWh/m^3)	Water Cost (US$/$m^3$)	Year
Sadous, Riyadh, SA	5,800	15	10.08	264			1994
Magan, Isreal	4,000	3	3.5+0.6 wind	36		11.6	1997
Elhamarawien, Egypt	3,500	53	19.8+0.64 control	208	0.89		1986
Heelafar Rahab Oman	1,000	5	3.25	9.6		6.25	1995
White Cliffs, Australia	3,500	0.5	0.34	none	2-8		
Solar flow, Australia	5,000	0.4	0.12	none	1.86	10–12	
Hassi-Kheba, Algeria	3,200	0.95	2.59			10	
INETI, Lisbon, Portugal	5,000	0.1–0.5	0.05–0.15	none			2000
Conception del Oro, Mexico	3,000	0.71	2.5	none	6.9		1982
Thar desert, India	5,000	1	0.45		1 kWh/kg salt		1986
Perth, Australia	BW	0.4–0.7	1.2		4-5.8		1989
Gillen Bore, Australia	1,600	1.2	4.16	none			1996
Wano Road, Australia	BW		6				
Kasir Ghilen, Tunis	5,700	50				7.25	2006
Coite-Pedreias, Brazil	BW	0.25	1.1	9.6	3–4.7	14.9	
Mesquite, Nevada	3,500	1.5	0.4		1.38	3.6	2003
N. Jawa, Indonesia	BW	12	25.5				
Univ. of Almeria, Spain	BW	2.5	23.5				

Table 3. Brackish water RO plants driven by PV power

Seawater PV/RO application systems

The osmotic pressure of seawater is much higher than that of brackish water; therefore, its desalination requires much more energy, and, unavoidably, a somewhat larger PV array. Also, the higher pressures found in seawater RO systems require mechanically stronger components. Thus, the total cost of water from seawater PV/RO is likely to remain higher than that from brackish water, and systems have not yet passed the demonstration stage. Table 4 shows some of the installed seawater PV/RO plants [38–42].

Location	Feedwater (ppm)	Capacity (m^3/day)	PV (kWp)	Batteries (kWh)	Energy Consumption (kWh/m^3)	Water Cost (US$/$m^3$)	Year
Lampedusa, Italy	SW	40	100	880	5.5	9.5	1990
Jeddah, S. Arabia	42,800	3.2	8				1981
St. Luice, FL	32,000	0.64	2.7		13		1995
Doha, Qatar	35,000	5.7	11.2	none	10.6		
Cress, Laviro, Greece	36,000	< 1	4+ 0.9 wind	44		33	2001
ITC Canaries Island, Spain	SW	3	4.8	19	5.5	13	1998
Crest, UK	SW	0.5 L/h	1.54	none		4.2	2003
Vancouver, Canada	SW	0.5–1.0	0.48				
Ponta Libeco, Italy	SW		9.8				1993

Table 4. Seawater RO plants driven by PV power.

PV/ED applications

ED uses DC for the electrodes; therefore, the PV system does not include an inverter, which simplifies the system. Figure 20 shows a schematic diagram of a PV-powered ED system. Currently, there are several installations of PV/ED technology worldwide. All PV/RD applications are of a standalone type, and several interesting examples are discussed below.

In the city of Tanote, in Rajasthan, India, a small plant was commissioned in 1986 that features a PV system capable of providing 450 peak watts (W_p) in 42 cell pairs. The ED unit includes three stages, producing 1 m^3/d water from brackish water (5000 ppm TDS). The unit energy consumption is 1 kWh/kg of salt removed [43]. A second project is a small experimental unit in Spencer Valley, New Mexico (USA), where two separate PV arrays are used: two tracking flat-plate arrays (1000 W_p power, 120 V) with DC/AC inverters for pumps, plus three fixed arrays (2.3 kW_p, 50 V) for ED supply. The ED design calls for 2.8 m^3/d product water from a feed of about 1000 ppm TDS. This particular feed water contains uranium and radon, apart from alpha particles. Hence, an ion-exchange process is required prior to ED. Unit consumption is 0.82 kWh/m^3 and the reported cost is 16 US$/$m^3$ [44-45]. A third project is an unusual application in Japan, where PV technology is used to drive an ED plant fed with seawater, instead of the usual brackish water of an ED system [46]. The solar field consists of 390 PV panels with a peak power of 25 kW_p, which can drive a 10 m^3/d ED unit. The system, located on Oshima Island (Nagasaki), has been operating since 1986. Product-water quality is reported to be below 400 ppm TDS, and the ED stack is provided with 250 cell pairs.

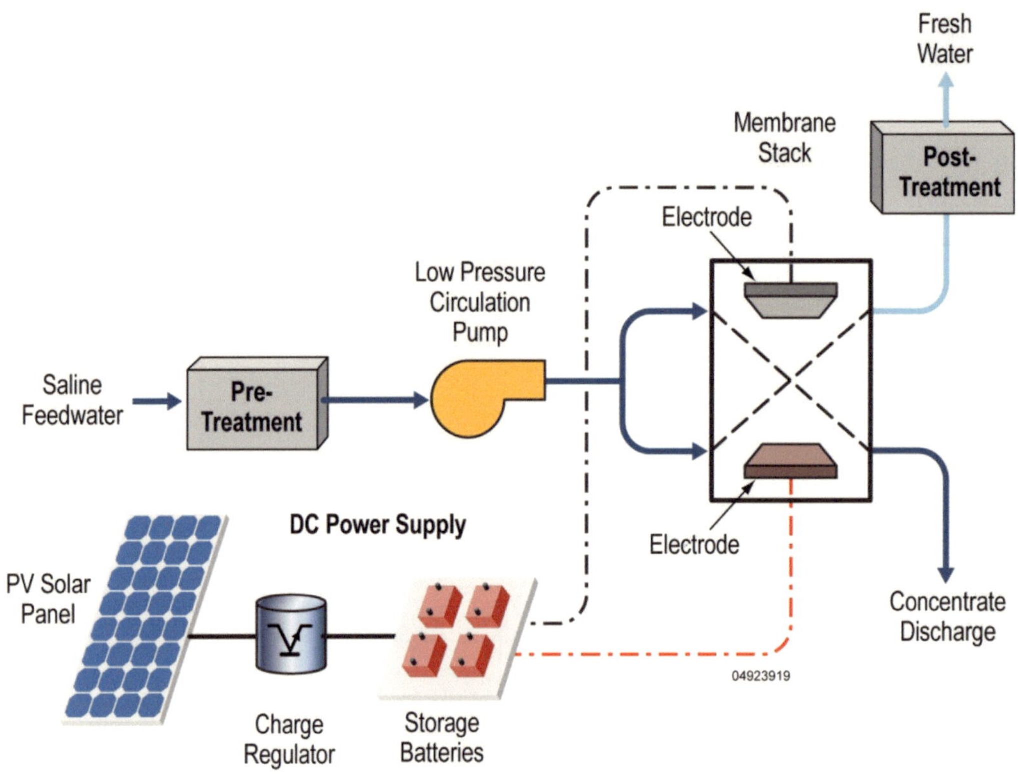

Fig. 20. Shows a schematic diagram of a PV-powered ED system.

3.3 Desalination systems driven by wind

Wind turbines can be used to supply electricity or mechanical power to desalination plants. Like PV, wind turbines represent a mature, commercially available technology for power production. Wind turbines are a good option for water desalination especially in coastal areas presenting a high availability of wind energy resources. Many different types of wind turbines have been developed. A distinction can be made between turbines driven mainly by drag forces versus those driven mainly by lift forces. As shown in Fig. 21, a distinction can also be made between turbines with axes of rotation parallel to the wind direction (horizontal) and with axes perpendicular to the wind direction (vertical). The efficiency of wind turbines driven primarily by drag forces is low compared with the lift-force-driven type. Therefore, all modern wind turbines are driven by lift forces. The most common types are the horizontal-axis wind turbine (HAWT) and the vertical-axis wind turbine (VAWT).

Wind-driven desalination has particular features due to the inherent discontinuous availability of wind power. For standalone systems, the desalination unit has to be able to adapt to the energy available; otherwise, energy storage or a backup system is required. Wind energy is used to drive RO, ED, and VC desalination units. A hybrid system of wind/PV is usually used in remote areas. Few applications have been implemented using wind energy to drive a mechanical vapor compression (MVC) unit. A pilot plant was installed in 1991 at Borkum, an island in Germany, where a wind turbine with a nominal power of 45 kW was coupled to a 48 m^3/day MVC evaporator. A 36-kW compressor was

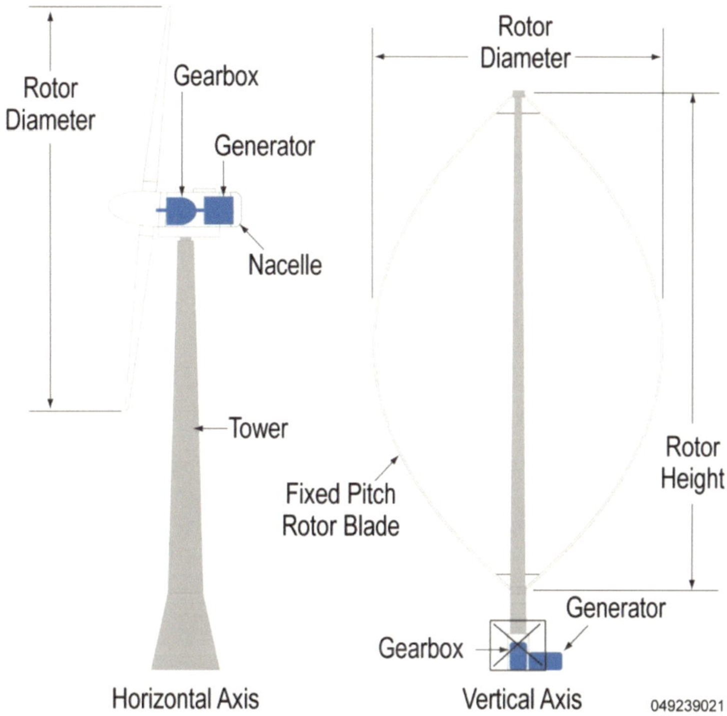

Fig. 21. Presents the horizontal and vertical wind turbine configurations.

required. The experience was followed in 1995 by another larger plant at the island of Ru¨gen. Additionally, a 50 m^3/day wind MVC plant was installed in 1999 by the Instituto Tecnologico de Canarias (ITC) in Gran Canaria, Spain, within the Sea Desalination Autonomous Wind Energy System (SDAWES) project [47]. The wind farm is composed of two 230-kW wind turbines, a 1500-rpm flywheel coupled to a 100-kVA synchronous machine, an isolation transformer located in a specific building, and a 7.5- kW uninterruptible power supply located in the control dome. One of the innovations of the SDAWES project, which differentiates it from other projects, is that the wind generation system behaves like a mini power station capable of generating a grid similar to conventional ones without the need to use diesel sets or batteries to store the energy generated.

Regarding wind energy and RO combinations, a number of units have been designed and tested. As early as 1982, a small system was set at Ile du Planier, France [48], which as a 4-kW turbine coupled to a 0.5-m^3/h RO desalination unit. The system was designed to operate via either a direct coupling or batteries. Another case where wind energy and RO were combined is that of the Island of Drenec, France, in 1990 [48]. The wind turbine, rated at 10 kW, was used to drive a seawater RO unit. A very interesting experience was gained at a test facility in Lastours, France, where a 5-kW wind turbine provides energy to a number of batteries (1500 Ah, 24 V) and via an inverter to an RO unit with a nominal power of 1.8 kW. A 500 L/h seawater RO unit driven by a 2.5-kW wind generator (W/G) without batteries was developed and tested by the Centre for Renewable Energy Systems Technology (CREST) UK. The system operates at variable flow, enabling it to make efficient use of the naturally varying wind resource, without need of batteries [49].

Excellent work on wind/RO systems has been done by ITC within several projects such as AERODESA, SDAWES, and AEROGEDESA[50]. Additionally, a wind/RO system without energy storage was developed and tested within the JOULE Program (OPRODES-JORCT98-0274) in 2001 by the University of Las Palmas. The RO unit has a capacity of 43–113 m^3/h, and the W/G has a nominal power of 30 kW [51]. In addition, an excellent job on combining wind/RO was done by ENERCON, the German wind turbine manufacturer. ENERCON provides modular and energy-efficient RO desalination systems driven by wind turbines (grid-connected or standalone systems) for brackish and seawater desalination. Market-available desalination units from ENERCON range from 175 to 1400 m^3/day for seawater desalination and 350 to 2800 m^3/day for brackish water desalination. These units in combination with other system components, such as synchronous machines, flywheels, batteries, and diesel generators, supply and store energy and water precisely according to demand [52]. Table 5 shows several existing wind/RO installations.

Plant Location	Year of Commission	Water Type	Capacity (L/h)	W/T Nominal Power (kW)	Unit Water Cost ($/m³)
Ile de Planier, France	1983	SW/BW	500	4	-
Fuerteventura island, PUNTA JANDIA project	1995	SW	2,333	225	-
Therasia island, Greece	1997	SW	200	15	-
Pozo Izquierdo, Gran Canaria, AEROGEDESA project	2003	SW	800	15	4.4 -7.3
CREST, UK	2004	SW	500	2.5	2.6

Table 5. Installed wind/RO plants

3.4 Geothermal energy

The earth's temperature varies widely, and geothermal energy is usable for a wide range of temperatures from room temperature to well over 300°F. The main advantage of geothermal energy is that thermal storage is unnecessary in such systems. Geothermal reservoirs are generally classified as being either low temperature (<150°C) or high temperature (>150°C). Generally speaking, high-temperature reservoirs are suitable for, and sought out for, commercial production of electricity. Energy from the earth is usually extracted with ground heat exchangers, made of a material that is extraordinarily durable but allows heat to pass through efficiently. The direct use of moderate and high temperatures is for thermal desalination technologies. A high-pressure geothermal source allows the direct use of shaft power on mechanically driven desalination, whereas high-temperature geothermal fluids can be used to generate electricity to drive RO or ED plants.

The first geothermal energy-powered desalination plants were installed in the United States in the 1970s [53–57], testing various potential options for the desalination technology, including MSF and ED. An analysis [58] discussing a technical and economic analysis of an MED plant, with a capacity of 80 m^3/d, powered by a low-temperature geothermal source and installed in Kimolos, Greece showed that high temperature geothermal desalination could be a viable option. A study [59] presented results from an experimental investigation of two polypropylene-made HD plants powered by geothermal energy [60]. Recently, a study [61] discussed the performances of a hybrid system consisting of a solar still in which the feed water is brackish underground geothermal water. Finally, the availability and/or suitability of geothermal energy and other renewable energy resources for desalination is given by [62].

4. General economic assessment of desalination

The cost of desalinated water is usually expressed in US$ per cubic meter of product water. This figure is obtained by dividing the sum of all expenses (capital cost, plus operation and maintenance cost) related to the production of desalinated water by the total amount of desalted water produced. Capital cost includes both direct and indirect costs. Direct capital costs are the land cost, building cost, and all equipment costs. Indirect capital costs include freight, insurance, construction overhead, engineering and legal fees, and contingencies costs. Costs of energy, labor, chemicals, consumables, spare parts, and major replacements or refurbishment required over the lifetime of the plant are included in operational and maintenance costs.

The economies of desalination and the decision as to which approach to select depend on situation-specific parameters. Because energy is the main driver in the cost of operation, economic feasibility of either approach to desalination is highly correlated to the location-specific cost and availability of energy [63]. Table 6 presents a comparative illustration of cost distribution and energy share of total cost for the two widely used conventional systems (RO and MSF) installed in Libya with a capacity of 10 mgbd each.

Type of Plant	Capital Cost (%)	Energy Cost (%)	Maintenance and Repair Cost (%)	Membrane Replacement (%)	Labor (%)	Chemicals (%)
RO	31	26	14	13	9	7
MSF	42	41	8	0	7	2

Table 6. Percentage of cost for conventional systems

In the representative example above, the capital cost is considerably higher for the thermal process than for the membrane process. This reflects the prevailing situation in the desalination industry, in which the construction cost of thermal desalination plants exceeds that of membrane plants. All other main costs related to operating a desalination plant are usually higher for a membrane processes due to the greater complexity of maintenance tasks and operation. Accordingly, cost of chemicals is 7% vs. 2%, maintenance and parts are 14% vs. 7%, and labor cost is 9% vs. 7% of total operating cost for the representative RO and MSF plants, respectively. Membrane replacement, which is listed separately, adds further to the maintenance cost for RO, whereas this cost is obviously absent for thermal processes.

Strong inter-firm competition and advances in technology have resulted in average annual unit cost reductions of close to 6% for MSF processes since 1970. In addition, many MSF desalination plants, which are mostly located in the Middle East, have increasingly taken advantage of economies of scale. RO, which has been used commercially only since 1982, has seen even steeper cost declines since inception. Membrane costs have fallen by 86% between 1990 and 2002 [64]. Steeply declining maintenance cost, in combination with relatively low capital cost, has contributed greatly to the rapidly growing success of membrane technology.

The unit product cost of fresh water differs when it is produced from different plant capacities. Table 7 shows the unit product cost of water produced from plants of different type and capacity. Product unit prices generally take into account all relevant costs originating from direct capital, indirect capital, and annual operating costs.

Type of system and capacity (mgbd)	Product Cost ($/gallon)
MVC (0.03)	1.894
MVC (0.13)	1.220
MVC (1.06)	0.939
MVC (1.20)	0.920
MVC (5.28)	0.174
MSF (7.13-Dual purpose)	0.292
MSF (7.13-Single purpose)	0.621
MSF (Gas turbine, Waste heat boiler)	0.545
MSF (9.99)	0.473
MED (6-Dual purpose)	0.330
MED (6-Single purpose)	0.739
MED (9.99)	0.409
MED (Gas turbine, Waste boiler)	0.496
RO (5.28, Single stage)	0.242
RO (5.28, Two stage)	0.288
RO (0.03)	0.898
RO (1.06)	0.750
RO (1.20)	0.489
RO (9.99)	0.413
RO (30)	0.208
MED- TVC (Single purpose)	0.866
MED- TVC (Dual purpose)	0.496

Table 7. Fresh water cost for different types and capacities

Economic analysis for renewable energy desalination processes

The economics of operating solar desalting units tend to be related to the cost of producing energy with these alternative energy devices. Presently, most of the renewable energy systems have mature technology; but despite the free cost of renewable energy resources, their collecting systems tend to be expensive, although they may be expected to decline as further development of these devices reduces their capital cost. The economic aspects of each renewable energy desalination system will be discussed below.

We first look at the cost distribution of both conventional and renewable energy-operated desalination units. Table 8 shows the comparison of cost distribution for conventional systems (RO and MSF) and plants driven by a renewable energy system [65]. For the renewable systems, the investment costs are the highest and the energy costs are the lowest.

Type of Process	Capital Costs (%)	Operational Costs (%)	Energy Costs (%)
Conventional (RO)	22 - 27	14 - 15	59 - 63
Conventional (MSF)	25 - 30	38 - 40	33 - 35
Renewable	30 - 90	10 - 30	0 -10

Table 8. Distribution of costs for conventional (RO and MF) desalination systems and for systems driven by renewable energy technology

One study has considered the techno-economic viability of solar desalination using PV and low-grade thermal energy using solar ponds [66]. Table 9 presents a comparison of the cost of water produced by a conventional cogeneration system (producing electricity and water) and that of solar-powered MSF and RO systems. The figures in the table are based on a plant capacity of 1 m^3/d and an annual utilization factors of 90% for conventional systems and 75% for solar-based systems.

Parameter	MSF		
	Conventional System	Partial Solar-based System	Complete Solar-based System
Annual Water Production (m^3)	328	274	274
Cost of Water Production ($\$/m^3$)	1.75	1.79	2.84
	RO		
	Conventional System	Partial Solar-based System	Complete Solar-based System
Annual Water Production (m^3)	328	274	274
Cost of Water Production ($\$/m^3$)	1.30	5.70	12.05

Table 9. Cost of desalinated water using conventional and solar-powered MSF and RO systems

The results in Table 9 show that the cost of water produced by a conventional RO system is less than that by a conventional MSF system. However, for solar-based systems, the partial solar-based MSF system gives the lowest cost of water production.

Solar thermal desalination economics

Solar still economic

Because of limited capacity of solar units, the capital costs and operating costs are not as well established as for the other processes. For solar stills, the cost of water production is high due to the low productivity of these stills. However, this type of desalination is only used in remote areas where there is no access to conventional energy resources. Table 10 compares the water costs for simple and multi-effect solar stills [66]. As shown, the water costs for multi-effect solar stills are much lower than for simple stills.

Solar-assisted desalination systems

One study [67] showed that solar-pond desalting systems have considerable potential to be cost effective if favorable site conditions exist. Table 11 presents the cost comparison of solar-pond-powered desalination with conventional seawater RO (SWRO) for two production capacities (20,000 and 200,000 m^3/d). As seen from the table, the unit water-cost difference is relatively small. However, investment costs and specific investment cost for solar-powered systems are still higher compared with the SWRO systems, where the difference decreases as the capacity increases.

Type	Capacity / Productivity	Water Cost ($/m³)	Description	Reference
Solar Stills	4 L/m²d	23.80	20 yrs lifetime, collector cost: $315/m², 5% interest rate	66
Multi-effect Stills	12 L/m²d	9.95	Storage module, 20 year lifetime, 5% interest rate	66
Multi-effect Stills	20 L/m²d	< 9.0*	Non-corroding polymer absorbers, storage, 24-hour operation	66

*Predicted

Table 10. Water costs for simple and multi-effect solar stills

System Type	SWRO		SP-MED		SP-HYB	
	Capacity (m³/d)					
	20,000	200,000	20,000	200,000	20,000	200,000
Investment (mil/$)	20	160	48	380	32	250
Specific Investment ($/m³d)	1000	800	2400	1900	1600	1250
Unit Water Cost ($/m³)	0.77	0.66	0.89	0.71	0.79	0.65

Table 11. Cost comparison of solar pond-powered desalination with conventional SWRO

Using CSP systems with desalination is still in its experimental stage until now but from the several pilot plant projects results, it could be concluded that we need time for this technology to be economically competitive with other desalination technologies.

PV/RO system economics

Cost figures for desalination have always been difficult to obtain. The total cost of water produced includes the investment cost, as well as the operating and maintenance cost. In a comparison between seawater and brackish water desalination, the cost of the first is about 3–5 times the cost of the second for the same plant size. As a general rule, a seawater RO unit has low capital cost and significant maintenance cost due to the high cost of the membrane replacement. The cost of the energy used to drive the plant is also significant. The major energy requirement for RO desalination is for pressurizing the feed water. Energy requirements for SWRO have been reduced to about 5 kWh/m³ for large units with energy recovery systems, whereas for small units (without energy recovery system), this may exceed 15 kWh/m³. For brackish water desalination, the energy requirement is between 1 and 3 kWh/m³. The product water quality ranges between 350 and 500 ppm for both seawater and brackish water units. According to published reports [38–42], the water cost of a PV seawater RO unit ranges from 7.98 to 29 US$/m³ for product-water capacity of 120–12 m³/day, respectively. Also for a PV/RO brackish-water desalination unit, a water cost of about 7.25 US$/m³ for a product-water capacity of 250 m³/day has been reported in the literature [38–42].

PV/ED economics

In general, electrodialysis is an economically attractive process for low-salinity water. EDR has greater capital costs than ED because it requires extra equipment (e.g., timing controllers, automatic valves), but it reduces or almost eliminates the need for chemical pretreatment. In ED applications, the electricity from a PV system can power to electro-mechanical devices such as pumps or to DC devices such as electrodes. The total energy consumption of an ED system under ambient temperature conditions and assuming product water of 500 ppm TDS would be about 1.5 and 4 kWh/m^3 for a feed water of 1500–3500 ppm TDS, respectively. The water cost of a PV-operated ED unit ranges from 16 to 5.8 US$/m^3 [45–46]. The main advantage of PV desalination systems is the ability to develop small-scale desalination plants.

Wind-Renewable Energy economics

Wind energy could be used to drive RO, ED, and VC desalination units. A hybrid system of wind/PV was also used in remote areas. Few applications have been implemented using wind energy to drive a mechanical vapor compression unit, and a number of wind/RO combinations systems have been designed and tested. ENERCON provides modular and energy-efficient RO desalination systems driven by wind turbines for brackish and seawater desalination. The estimated water cost produced from the installed wind/RO unit ranges from 7.2 to 2.6 US$/m^3 of fresh water. According to a published report [68], the water cost of a wind brackish water RO unit (capacity of 250 m^3/day) is of the order of 2 Euro/m^3, whereas for the same feed-water salinity and size, the water cost of a wind/electrodialysis unit is around 1.5 Euro/m^3. For standalone wind-powered MVC units with a capacity range between 5 and12.5 m^3/h, the mean water cost varies between 3.07 and 3.73 Euro/m^3 [69].

5. Conclusion

Desalination technology has been in continuous development during the previous decades, making it possible to include salt water as part of the production of fresh water. However, the current cost of desalinated water is still high because of its extensive use of energy. The selection of a desalination process should be based on a careful study of the specific site conditions and applications. Local circumstances may play a significant role in determining the most appropriate process for an area. The use of renewable energy for desalination is a technically mature option toward emerging energy and water problems. And technological advances will continue to improve system efficiencies and reduce capital costs, making these systems competitive when used in desalination systems. Currently, the cost of fresh-water production from renewable-energy-powered desalinated systems is less than other alternatives in remote areas where access to electricity is not available. Numerous studies on a suitable technical match between renewable energy and desalination process have been reported in the literature. These studies conclude that renewable-energy-powered systems could compete with conventional systems under certain circumstances. Very few solar desalination plants have been reported in the literature. Several studies on a suitable technical match between renewable energy resources and desalination processes propose that solar thermal/MED, solar thermal/MSF, solar PV/RO, solar PV/ED, wind/RO, and geothermal/MED technologies are very promising options. The economic competitiveness of solar thermal/MED and solar thermal/MSF has been shown in a number of theoretical studies. However, this has not been verified experimentally, and therefore, cannot be used as a guide for decision-making regarding technology selection for a particular application. At present, small-scale PV and wind desalination systems appear to be especially suitable in remote regions without access to the electric grid and where water scarcity is a major problem. The large scale of these systems is hindered by non-technical barriers.

6. References

[1] Economic and Social Commission for Western Asia. Energy options for water desalination in selected ESCWA member countries. New York: United Nations; 2001.

[2] CORDIS Database; 2006.

[3] UNEP (United Nations Environment Program) (2003) Key Facts about Water www.unep.org/wed/2003/keyfacts <accessed January 15, 2006>.

[4] http://desaldata.com/.

[5] World Health Organization. Guidelines for drinking water quality, Vol. I, Geneva, 1984.

[6] ARMINES. Technical and economic analysis of the potential for water desalination in the Mediterranean region, RENA-CT94-0063, France; 1996.

[7] O.K.Buros. The desalinating ABC, McGrawhill, New York, 1990.

[8] O.A Hamed (2005). Overview of hybrid desalination systems - current status and future prospects. *Desalination* 186, 207-214.

[9] K. Quteishat and M. Abu-Arabi. Promotion of solar desalination in the MENA region. Middle East Desalination Centre, Muscat, Oman–http://www.menarec.com/docs/Abu-Arabi.pdf [accessed March 28, 2006].

[10] A. Maurel. Desalination by reverse osmosis using renewable energies (Solarwind):cadarache central experiment. In: *Proceedings of the New Technologies for the Use of Renewable Energy Sources in Water Desalination Conference, Session II*, Athens, Greece; 1991. p. 17–26.

[11] B.S. Richards and I.A. Schafer (2002). Design considerations for a solar-powered desalination system for remote communities in Australia. *Desalination* 144: 193–9.

[12] S.A. Avlonitis, K. Kouroumbas, and N. Vlachakis (2003). Energy consumption and membrane replacement cost for seawater RO desalination plants. *Desalination* 157:151–8.

[13] S. Avlonitis, G.P. Sakellaropoulos, W.T. Hanbury (1995). Optimal design of spiral wound modules: an analytical method. *Trans I ChemE* 73(Part A):575–80.

[14] European Commission. Desalination guide using renewable energies, Thermie Programme, Directorate General for Energy (DG XVII), ISBN 960-90557-5-3; 1998.

[15] O. Kuroda, S. Takahashi, S. Kubota, K. Kikuchi, Y. Eguchi, Y. Ikenaga, et al. (1987). An electrodialysis seawater desalination system powered by photovoltaic cells. *Desalination* 167:161–9.

[16] M. Lichtwardt and H. Remmers. Water treatment using solar powered electrodialysis reversal. In: *Proceedings of the Mediterranean Conference on Renewable Energy Sources for Water.*

[17] I. Al-Hayek and O.O. Badran (2004). The effect of using different designs of solar stills on water distillation. *Desalination* 169:121–7.

[18] Economic and Social Commission for Western Asia. ESCWA water development report 3. Role of desalination in addressing water scarcity, New York: United Nations; 2009.
[19] M. Sodha, A. Kumar, G. Tiwari, and R. Tyagi, Simple multiple wick solar still: analysis and performance, *Solar Energy* 26 (1981) 127–131.
[20] J.L. Fernandez and N. Chargoy (1990). Multi-stage, indirectly heated solar still, *Solar energy* 44(4), pp. 215-223.
[21] H. Al-Hinai, M.S. Al-Nassri, and B.A. Jubran (2202). Effect of climatic, design and operational parameters on the yield of a simple solar still. *Energy Conv. Management* 43, 1639–1650.
[22] E. Mathioulakis, V. Belessiotis, and E. Delyannis (2007). Desalination by using alternative energy: Review and state-of-the-art. *Desalination* 203, 346–365.
[23] A. Bohner, Solar desalination with a higher efficiency multi effect process offers new facilities (1989). *Desalination* 73, 197–203.
[24] E. Mathioulakis, V. Belessiotis, and E. Delyannis (2007). Desalination by using alternative energy: Review and state of the art. *Desalination* 203, 346-365.
[25] F.A. Banat, Membrane distillation for desalination and removal of volatile organic compounds from water, Ph.D. thesis , Mcgill University, 1994.
[26] H.M. Qiblawey and F. Banat (2008). Solar thermal desalination technologies, *Desalination* 220, 633–644.
[27] H. Price, et al., (2003). Assessment of Parabolic Trough and Power Tower Solar Technology Cost and Performance Forecasts, Sargent & Lundy LLC Consulting Group, National Renewable Energy Laboratory, Golden, Colorado, 2003 http://www.nrel.gov/solar/parabolic_trough.html
[28] F. Trieb and H. Müller-Steinhagen (2007). Concentrating Solar Power for Seawater Desalination in the Middle East and North Africa, Submitted to *Desalination*.
[29] L. García-Rodríguez (2003). Renewable energy applications in desalination: state of the art, *Solar Energy* 75, 381-393.
[30] J. Zachary and C. Layman, Bechtel Power Corp. Adding Desalination to Solar Hybrid and Fossil Plants, *Electric Power Journal*, May 2010.
[31] M. Al-Shammiri and M. Safar (1999). Multi-effect distillation plants: state of the art. *Desalination* 126:45-59.
[32] K. Quteishat and M. Abu-Arabi, (2006): *Promotion of solar desalination in the MENA region*. Middle East Desalination Centre, Muscat, Oman – http://www.menarec.com/docs/Abu-Arabi.pdf (accessed 28 March 2006).
[33] Collares *et al.* (2003). A new low concentration CPC type collector with convection controlled by a honeycomb TIM Material: A compromise with stagnation temperature control and survival of cheap fabrication materials. In *Proceedings of the ISES Solar World Congress 2003*, 14-19 June, Göteborg, Sweden.
[34] Water corporation, A strategic review of desalination possibilities for western Australia, September 2000.
[35] E. Zarza and M. Blanco. Advanced M.E.D. solar desalination plant: Seven years of experience at the Plataforma Solar de Almería. In: *Proceedings of the Mediterranean Conference on Renewable Energy Sources for Water Production*, Santorini, Greece, 1996.

[36] M. Thomson, J. Gwillim, A. Rowbottom, I. Draisey, and M. Miranda. Batteryless photovoltaic reverse osmosis desalination system, S/P2/00305/REP, ETSU, DT1, UK.

[37] G. Fiorenza, V. Sharma, and G. Braccio (2003). Technoeconomic evaluation of a solar powered water desalination plant. *Energy Convers Manage* 44:2217–40.

[38] S. Kehal. Reverse osmosis unit of 0.85 m^3/h capacity driven by photovoltaic generator in South Algeria. In: *Proceedings of the New Technologies for the Use of Renewable Energy Sources in Water Desalination Conference, Session II,* Athens, Greece; 1991. p. 8–16.

[39] E. Tzen and K. Perrakis. Prefeasibility study for small Aegean islands, RENA-CT94–0063, CRES, Greece; 1996.

[40] M. Thomson and D. Infield (2002). A photovoltaic powered seawater reverse-osmosis system without batteries. *Desalination* 153:1–8.

[41] A. Hafez and S. El-Manharawy (2003). Economics of seawater RO desalination in the Red Sea region, Egypt. Part 1: a case study. *Desalination* 153:335–47.

[42] K. Perrakis, E. Tzen, and P. Baltas. PV-RO desalination plant in Morocco, JOR3-CT95-0066, CRES, Greece.

[43] P. Finken, K. Korupp. Water desalination plants powered by wind generators and photovoltaic systems. In: *Proceedings of the New Technologies for the Use of Renewable Energy Sources in Water Desalination Conference, Session II,* Athens, Greece; 1991. p. 65–98.

[44] M.R. Adiga, S.K. Adhikary, P.K. Narayanan, W.P. Harkare, S.D. Gomkale, and K.P. Govindan (1987). Performance analysis of photovoltaic electrodialysis desalination plant at Tanote in the Thar desert. *Desalination* 67:59–66.

[45] O. Kuroda, S. Takahashi, S. Kubota, K. Kikuchi, Y. Eguchi, Y. Ikenaga, et al. (1987). An electrodialysis seawater desalination system powered by photovoltaic cells. *Desalination* 67:161–9.

[46] M. Lichtwardt and H. Remmers. Water treatment using solar powered electrodialysis reversal. In: *Proceedings of the Mediterranean Conference on Renewable Energy Sources for Water Production*; 1996.

[47] J.A. Carta, J. Gonzalez, and V. Subiela (2004). The SDAWES project: an ambitious R&D prototype for wind—powered desalination. *Desalination* 161: 33–48.

[48] A. Maurel. Desalination by RO using RE (solar & wind): Cadarache Center Experience. *Proceedings of the New Technologies for the Use of RE Sources in Water Desalination,* Greece, 26–28 Sept. 1991, pp. 17–26.

[49] A. Peral, G.A. Contreras, and T. Navarro. IDM—Project: Results of one year's operation. *Proceedings of the New Technologies for the Use of RE Sources in Water Desalination,* Greece, 26–28 Sept. 1991, pp. 56–80.

[50] M. Miranda and D. Infield (2002). A wind powered seawater RO system without batteries. *Desalination* 153:9–16.

[51] De La Nuez Pestana I, Francisco Javier G, Celso Argudo E, Antonio Gomez G. (2004). Optimization of RO desalination systems powered by RE: Part I. Wind energy. *Desalination* 160:293–9.

[52] ENERCON. Desalination units. http://www.enercon.de, accessed 2006.

[53] E. Barbier (2002). Geothermal energy technology and current status: an overview. *Renew Sustain Energy Rev* 6:3–65.

[54] E. Barbier (1997). Nature and technology of geothermal energy. *Renew Sustain Energy Rev* 1(1–2):1–69.

[55] L. Awerbuch, T.E. Lindemuth, S.C. May, and A.N. Rogers (1976). Geothermal energy recovery process, *Desalination* 19, 325–336.

[56] W.J. Boegli, S.H. Suemoto, and K.M. Trompeter (1977). Geothermal desalting at the East Mesa test site. (Experimental results of vertical tube evaporator, MSF and high-temperature ED. Data about fouling, heat transfer coefficients, scaling.) *Desalination* 22, 77–90.

[57] A. Ophir (1982). Desalination plant using low grade geothermal heat, *Desalination* 40, 125–132.

[58] K. Karytsas, V. Alexandrou, and I. Boukis, The Kimolos Geothermal Desalination Project, *Proc. International Workshop on Possibilities of Geothermal Energy Development in the Aegean Islands Region*, Milos Island, Greece, 2002, pp. 206–219.

[59] K. Bourouni, R. Martin, and L. Tadrist (1997). Experimental investigation of evaporation performances of a desalination prototype using the aero-evapocondensation process, *Desalination* 114, 111–128.

[60] B. Bouchekima (2003). Renewable energy for desalination: a solar desalination plant for domestic water needs in arid areas of South Algeria, *Desalination* 153, 65–69.

[61] V. Belessiotis and E. Delyannis, Renewable energy resources, in Encyclopedia of Life Support Systems (EOLSS), Desalination: Desalination with Renewable Energies, available on line:
http:/ /www.desware.net/DeswareLogin/LoginForm.Aspx

[62] V. Belessiotis and E. Delyannis, Renewable energy resources, in *Encyclopedia of Life Support Systems (EOLSS),* Desalination: Desalination with Renewable Energies, available on line: http:/ /www.desware.net/DeswareLogin/LoginForm.Aspx

[63] D. Zejli, O-K Bouhelal, R. Benchrifa, and A. Bennouna (2002). Applications of solar and wind energy sources to sea-water desalination—economical aspects. International Conference on Nuclear Desalination: Challenges and Options. October 2002, Marrakech, Morocco.

[64] Al-Hallaj and J.R. Selman (2002). *A comprehensive study of solar desalination with a humidification - dehumidification cycle.* The Middle East Desalination Research Center - Muscat, Sultanate of Oman.

[65] R.K. Suri, A.M.R. Al-Marafie, A.A Al-Homoud, and G.P. Maheshwari (1989). Cost-effectiveness of solar water production. *Desalination* 71, 165-175.

[66] Templitz-Sembitsky. The use of renewable energy for sea-water desalination; a brief assessment technical information, W16e, Gate information services, Germany, 2000.

[67] P. Glueckstern (1995). Potential uses solar energy for seawater desalination, *Desalination 101,* 11-20.

[68] P. Finken and K. Korupp (1991). Water desalination plants powered by wind generators and photovoltaic systems, *Proceedings of the New Technologies for the Use of Renewable Energy Sources in Water Desalination Conference, Session II*, pp. 65-98, Athens, Greece.

[69] W. Damm, P. Gaiser, and D. Kowalczyk, and U. Plantikow (1996) Wind powered MVC desalination plant - operating results and economical aspects, *Proc. of the Mediterranean Conference on Renewable Energy Sources for Water Production*, pp. 143-146, ISBN 960-905777 0-2, Santorini, Greece.

9

New Trend in the Development of ME-TVC Desalination System

Anwar Bin Amer
Kuwait Foundation for the Advancement of Sciences (KFAS), Research Directorate, Water Resources Program
Kuwait

1. Introduction

Several low temperature Multi-Effect Thermal Vapor Compression (ME-TVC) desalination units have been installed recently in most of the GCC countries. The total installed capacity has increased up to 500 million imperial gallons per day (MIGD) between 2000 and 2010 as shown in (Table 1). The majority of these units were commissioned in the UAE by SIDEM Company. The unit size capacities of these units were increased exponentially from 1 to 8.5 (MIGD) between 1991 and 2008 as shown in Fig.1. The new trend of combining ME-TVC with conventional multi-effect units led to this tremendous increase, more than eight times, during a very short period. Moreover, the unit size capacity of this technology is currently available with 10 MIGD, and it is expected to increase up to 15 MIGD in the near future. Hence, this system has become highly attractive and competitive against Multi Stage Flash (MSF) desalination system and it is predicted to get a considerable increase in the desalination market in future, particularly, in the GCC countries because it includes the following attractive features:

- Operates at lower top brine temperature around 60 to 70 ^oC compared to 90 to 120 ^oC in the MSF, and this reduces the scale formation, corrosion problems, anti-scalant chemicals and maintenance shutdown time (Darwish & Alsairafi, 2004).
- Requires less pumping energy than MSF (1.5~2 kW/m^3 compared to 4~5 kW/m^3), because there is no need to re-circulate large quantities of brine as in MSF system.
- Produces higher gain output ratios (GOR) up to 16 with less number of effects compared to 8 GOR and 21 stages in the MSF (Wade, 2001).
- Uses falling film horizontal tube evaporator (HTE), which gives high heat transfer coefficient and reduces the needed heat transfer area and consequently the capital cost of the desalination plant (Reddy & Ghaffour, 2007).
- Better response to steam supply variation, so, it has more flexibility of operation than MSF (Darwish & Alsairafi, 2004).

Table1 shows that ME-TVC technology is gaining more market shares recently in Bahrain, Saudi Arabia and Qatar with a total installed capacity of 60 MIGD, 176 MIGD and 63 MIGD, respectively.

Year	Location	Country	Unit capacity	No. of units	Total capacity	GOR
1991	Jabal Dhana	UAE	1 MIGD	4	4 MIGD	8
2000	Umm Al-Nar	UAE	3.5 MIGD	2	7 MIGD	8
2001	Layyah	UAE	5 MIGD	2	10 MIGD	8
2002	Al-Taweelah A_1	UAE	3.7 MIGD	14	52 MIGD	8
2005	Sharjah	UAE	8 MIGD	2	16 MIGD	8.4
2006	Al-Hidd	Bahrain	6 MIGD	10	60 MIGD	8.9
2007	Al-Jubail	Saudi Arabia	6.5 MIGD	27	176 MIGD	9.8
2008	Fujairah	UAE	8.5 MIGD	12	100 MIGD	10
2009	Ras Laffan	Qatar	6.3 MIGD	10	63 MIGD	11.1

Table 1. Several projects of ME-TVC commissioned by SIDEM in the GCC countries.

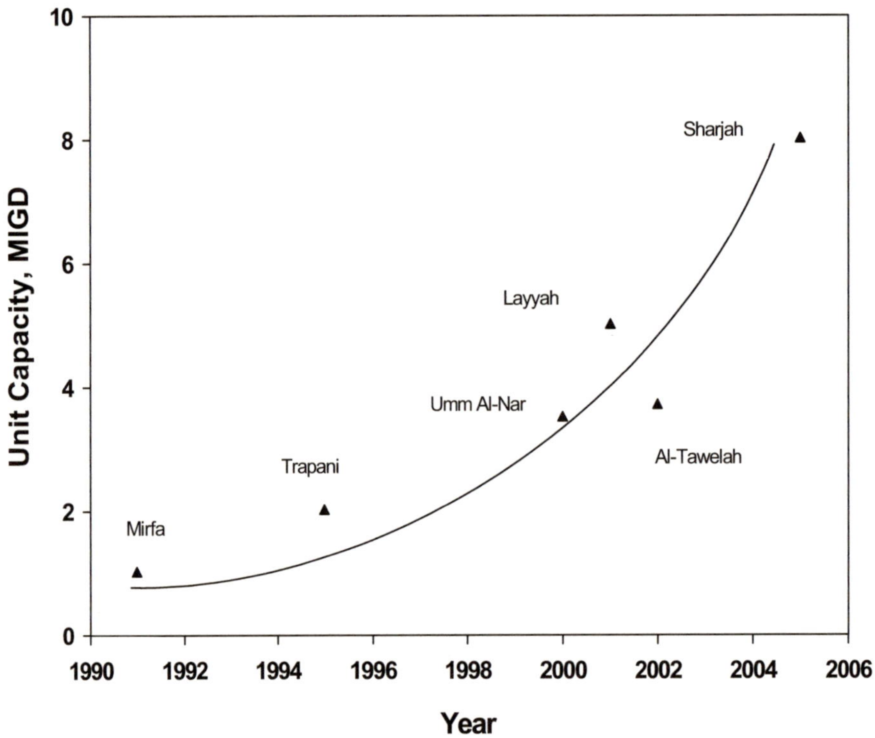

Fig. 1. The increase of unit size capacity of ME-TVC desalination systems.

2. Literature review

Several studies have been published since the early of 1990's concerning ME-TVC desalination system. Some of which include field studies others describe different conceptual designs. Diverse mathematical models have been developed since then, in most of these publications for simulation and economic evaluation purposes. A summary literature review of these studies were reported in (Al-Juwayhel et al, 1997) and (El-Dessouky & Ettouney, 1999). On the other hand, limited studies were published handling ME-TVC desalination system from exergy (Second Law) point of view since the middle of last decade, but it has been carried out in several published works recently.

(Hamed et al., 1996) conducted and evaluated the performance of a ME-TVC desalination system. An exergy analysis was also performed and compared with conventional multi effect boiling (MEB) and mechanical vapor compression (MVC) desalination systems. Results showed that the ME-TVC desalination system is the most exergy-efficient compared to other systems.

(Al-Najem et al., 1997) conducted a parametric analysis using First and Second Laws of Thermodynamics for single and multi effect thermal vapor compression system (ME-TVC). The study revealed that the steam ejector and evaporators are the main source of exergy destruction in the ME-TVC desalination system.

(Alasfour et al., 2005) developed mathematical models for three configurations of a ME-TVC desalination system using energy and exergy analysis. A parametric study was also performed to investigate the impacts of different parameters on the system performance. Results showed that the first effect was responsible for about 50 % of the total effect exergy destructions. The parametric study also showed that the decrease in exergy destructions is more pronounced than the decrease in the gain output ratio at lower values of motive steam pressure. Lowering the temperature difference across the effects, by increasing the surface area, decreases the specific heat consumption. On the other hand, exergy losses are small at low temperature difference and low top brine temperature.

(Choi et al., 2005) presented an exergy analysis for ME-TVC pilot plant units, which was developed by Hyundai Heavy Industries Company. The units have different capacities of 1, 2.2, 3.5 and 4.4 MIGD. Exergy analysis showed that most of the specific exergy losses were in thermal vapor compressor and the effects. The amount of exergy destruction represents more than 70% of the total amount. Results also showed that the increase of entrainment ratio to 120% will decrease the total heat transfer area by 12%.

(Wang & Lior, 2006) presented the performance analysis of a combined humidified gas turbine (HGT) plant with ME-TVC desalination systems using Second Law of thermodynamics. The analysis is performed to improve the understanding of the combined steam injection gas turbine power and water desalination process and ways to improve and optimize it. Results showed that the dual purpose systems have good synergy in fuel utilization, in operation and design flexibility.

(Sayyaadi & Saffari, 2010) developed thermo-economic optimization model of a ME-TVC desalination system. The model is based on energy and exergy analysis. A genetic algorithm is used to minimize the water product cost.

This chapter describes and discusses new developments which have taken place recently in the design, operation and material selection of ME-TVC units. A mathematical model of a ME-TVC desalination system is also developed in this chapter, using Engineering Equation Solver (EES) Software. This model is used to evaluate and improve the performance of some new commercial ME-TVC units having capacities of 2.4, 3.5 and 6.5 MIGD using energy and exergy analysis. The model results were compared against the actual data which showed good agreement. The other aim of this chapter is to develop a mathematical optimization model using MATLAB program. The model is used to determine the optimum operating and design conditions of different numbers of effects to maximize the gain output ratio of the ME-TVC unit, using two optimization approaches: (1) Smart Exhaustive Search Method (SESM) and (2) Sequential Quadratic Programming (SQP).

3. Process description

The arrangement of combining the ME-TVC with conventional Multi-effect consists of two separate rows of effects, each packed into one circular/rectangular vessel along with a thermo-compressor. Both vessels are connected parallel with a third vessel in the middle, which contains a number of effects along with the end condenser.

A schematic diagram of this arrangement is shown in Fig. 2, where two identical ME-TVC units are combined with a single MED unit, where as the vapor produced in the last effect of each ME-TVC unit (D_j) is split into two streams. The first stream D_r is entrained by a thermo-compressor and other part (D_f) is used as a heat source to operate low temperature multi effect distillation unit (LT-MED).

The configuration consists of the following components (1) a number of horizontal falling film evaporators (n effects), (2) two thermo-compressors, (3) a number of feed heaters, (4) five main pumps (distillate, feed, condensate, cooling and brine disposal pumps) to circulate the streams, (5) an end condenser and (6) a number of flashing boxes.

Two streams of motive steam (D_s) are directed at relatively high motive pressure (P_s) into two thermo-compressors. The motive steam is supplied usually either from boiler or steam turbine. Part of the vapor formed in the last effect (D_r) of each ME-TVC unit, is entrained and compressed by the thermo-compressor as mentioned above along with the motive steam (D_s+D_r) into the first effect of each unit where it condenses. The latent heat of condensation is used to heat the feed F_1 from T_{f1} to the boiling temperature T_1 and evaporates part of that feed by boiling equal to D_1. Part of the condensate (D_s) returns to its source and the other part of the condensed vapor (D_r) is introduced to the first flashing box, where a small amount of vapor flashes off due to pressure drop and equal to $D_r\ y$, where $y=C\cdot\Delta T/L_1$. This flashing vapor is passed through the first feed heater along with the vapor formed in the first effect ($D_1+D_r\ y$) heating the feed F_1 from T_{f2} to T_{f1}; then part of it condenses, and the remaining vapor ($D_1+D_r\ y - F_1 y$) flows as a heating source to the second effect and so on up to the last effect n. The brine leaving the first effect (B_1) is directed to the second effect which is at a lower pressure, so that flashing will release additional vapor, which is theoretically equal to $B_1\ C\ \Delta T/L_2$. This process is continued up to the last effect n.

In this configuration, the condensate vapor in each feed heater is assumed to be equal to vapor flashed between the two effects from both accumulated distillate and the brine. This gives an increase in temperature across the feed heaters, which equal to the temperature drop between the effects i.e. ($\Delta T_{fi} = \Delta T$).

The vapor formed in the last effect D_n flows into the end condenser where it condenses by the cooling seawater stream M_c. The latent heat of condensation is used to heat the seawater temperature from T_c to T_f. Part of the cooling stream flows to the effects (F) where it is heated in a series of feed heaters and the remainder (M_c-F) is rejected from the system.

The feed seawater flow (F) rate splits equally into each effect. Each part sprayed over a horizontal tube bundle through nozzles, the spray forms a thin falling film over the tubes of the bundle. The formation of this thin film enhances the heat transfer rate and makes the evaporation process more efficiently. Series feed heaters are also used between the effects to increase the feed seawater temperature and consequently decrease the energy added for evaporation.

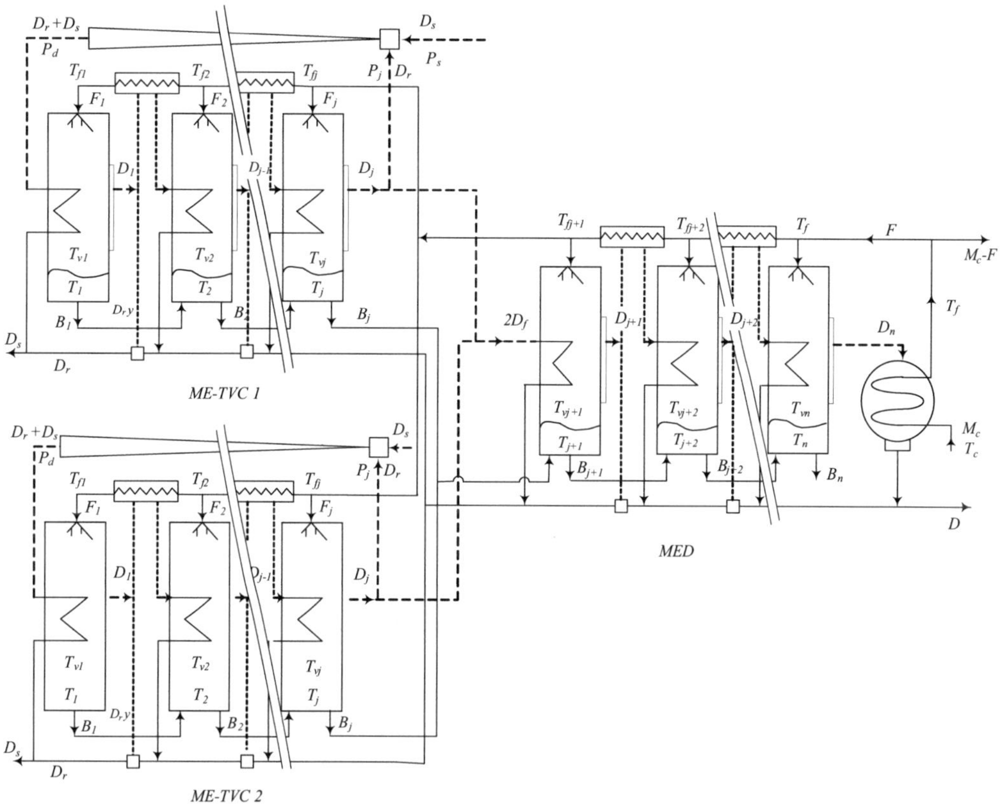

Fig. 2. A schematic diagram of two ME-TVC units combined with a conventional MED unit.

4. Thermal analysis of ME-TVC desalination system

First and Second Laws analysis are used in this section to develop a mathematical model of the ME-TVC desalination system. The model is developed by applying mass and energy conservation laws to the thermo-compressor, evaporators, feed heaters and end condenser. The following assumption were used to simplify the analysis: steady state operation, negligible heat losses to the surrounding, equal temperature difference across feed heaters, salt free distillate from all effects and variations of specific heat as well as boiling point elevation with the temperature and salinity are negligible.

The brine temperature in each effect is less than that of the previous one by ΔT. So, if the brine temperature in the effect i is assumed to be T_i, then the brine temperature in the next effect i+1 and so on up to the last effect n and can be calculated as follow:

$$T_{i+1} = T_i - \Delta T, \qquad i = 1,\ 2, \ldots n \tag{1}$$

The temperature of the vapor generated in the effect i, T_{vi} is lower than the brine temperature by the boiling point elevation plus non equilibrium allowance, where T_{vi} is a saturation temperature corresponding to the pressure in the effect P_i.

$$T_{vi} = T_i - (BPE + NEA), \qquad i = 1,\ 2,\ldots n \tag{2}$$

The temperature difference between the effects is assumed to be the same in this analysis and can be calculated as follows:

$$\Delta T = \frac{T_1 - T_n}{n - 1} \tag{3}$$

The feed seawater temperature flowing into each effect (T_{fi}) can be calculated as follows:

$$T_{fi} = T_f + [n - (i + 1)] \cdot \Delta T \qquad i = 1,\ 2,\ 3\ldots n \tag{4}$$

4.1 Mass and Energy Balance

The feed seawater flow rate F is distributed equally to all effects at a rate equal to F_i which can be calculated as follows:

$$F_i = \frac{F}{n + j} \quad , \quad j = \frac{n}{2} \tag{5}$$

The brine leaving the first effect enters into the second effect and so on up to the last effect n, and the brine from the last effect is rejected. The brine leaving the first, second and last effect can be calculated considering mass balance law as follows:

$$B_i = F_i - D_i \tag{6}$$

$$B_{i+1} = \sum_{i=1}^{j} \left(F_{i+1} - D_{i+1}\right) \tag{7}$$

$$B_n = 2 \cdot \sum_{i=1}^{j} \left(F_i - D_i\right) + \sum_{j+1}^{n} \left(F_{j+1} - D_{j+1}\right) \tag{8}$$

The salt mass conservation law is applied, assuming that the distillate is free of salt, to find brine salinity from the first, second and last effect as follows:

$$X_{bi} = \frac{F_i \cdot X_f}{\left(F_i - D_i\right)} \tag{9}$$

$$X_{bi+1} = \frac{F_{i+1} \cdot X_f}{\sum_{i=1}^{j} \left(F_{i+1} - D_{i+1}\right)} \tag{10}$$

$$X_{bn} = \frac{F_n \cdot X_f}{2 \cdot \sum_{i=1}^{j} (F_i - D_i) + \sum_{j+1}^{n} (F_{j+1} - D_{j+1})} \tag{11}$$

The vapor generated in the first effect by boiling only and can be determined from the energy balance of the first effect as follows:

$$D_1 = \frac{\left[(D_s + D_r) \cdot (h_d - h_{fd}) \right]}{L_1} - F_1 \cdot C \cdot \left(\frac{T_1 - T_{f1}}{L_1} \right) \tag{12}$$

The amount of vapor released from the second up to j can be expressed respectively as follows:

$$D_2 = (D_1 + D_r \cdot y - F_1 \cdot y) \cdot \frac{L_1}{L_2} + B_1 \cdot \frac{C \cdot \Delta T}{L_2} - F_2 \cdot C \cdot \frac{(T_2 - T_{f2})}{L_2} \tag{13}$$

$$D_j = [(D_{j-1} + \sum_{i=1}^{j-2} (D_i + D_r) \cdot y - (j-1) \cdot F_j \cdot y)] \cdot \frac{L_{j-1}}{L_j} + B_{j-1} \cdot \frac{C \cdot \Delta T}{L_j} - F_j \cdot C \cdot \frac{(T_j - T_{fj})}{L_j} \tag{14}$$

The vapor formed in the last effect of each ME-TVC unit D_j is divided into two streams; one is entrained by the thermo-compressor (D_r) and the other is directed to the MED unit.

$$D_j = D_r + D_f \tag{15}$$

The two streams of D_f are used as a heat source to operate low temperature multi effect distillation unit (LT-MED).
So, the vapor formed in first, second and last effect of this unit can be calculated as follows:

$$D_{j+1} = 2 \cdot D_f \cdot \frac{L_j}{L_{j+1}} + 2 \cdot B_j \cdot \frac{C \cdot \Delta T}{L_{j+1}} - F_{j+1} \cdot C \cdot \frac{(T_{j+1} - T_{fj+1})}{L_{j+1}} \tag{16}$$

$$D_{j+2} = \left(D_{j+1} + \sum_{i=1}^{j} (D_i + D_r) \cdot y - (j+n) \cdot F_{j+2} \cdot y \right) \cdot \frac{L_{j+1}}{L_{j+2}} + 2 \cdot B_{j+1} \cdot \frac{C \cdot \Delta T}{L_{j+2}} - F_{j+2} \cdot C \cdot \frac{(T_{j+2} - T_{fj+2})}{L_{j+2}} \tag{17}$$

$$D_n = [(D_{n-1} + \sum_{i=1}^{n-2} (D_i + D_r) \cdot y - (j+n-1) \cdot F_i \cdot y)] \cdot \frac{L_{n-1}}{L_n} + 2 \cdot B_{n-1} \cdot \frac{C \cdot \Delta T}{L_n} - F_i \cdot C \cdot \frac{(T_n - T_f)}{L_n} \tag{18}$$

The total distillate output from all effects is equal to

$$D = 2 \cdot \sum_{i=1}^{j} D_i + \sum_{j+1}^{n} D_{j+1}, \quad i = 1,2,\ldots 3 \tag{19}$$

The energy balance of the thermo-compressor is used to calculate the enthalpy of the discharged steam as shown in equation (20),

$$h_d = \left(\frac{\left(\frac{D_s}{D_r} \right) \cdot h_s + h_{g(j)}}{1 + \left(\frac{D_s}{D_r} \right)} \right) \tag{20}$$

The most essential part in modeling the ME-TVC desalination system is to determine the ratio of motive steam to entrained vapor (D_s/D_r) in such thermo-compressors. An optimal ratio will improve the unit efficiency by reducing the amount of motive steam (Utomo et al., 2008). This ratio is a direct function of discharge pressure (P_d), motive steam pressure (P_s) and entrained vapor pressure (P_j) in terms of compression ratio (CR) and expansion ratio (ER) as follows (El-Dessouky & Ettouney, 2002; Al-Najem et al., 1997):

$$CR = \frac{P_d}{P_j} \tag{21}$$

$$ER = \frac{P_s}{P_j} \tag{22}$$

Several methods are available in the literature to evaluate entrainment ratios; most of these methods need lengthy computation procedures and use many correction factors (El-Dessouky & Ettouney, 2002). Two simple methods are used to evaluate this ratio in this chapter: (1) Power's graphical data method (Al-Najem et al., 1997), (2) El-Dessouky and Ettouney's semi-empirical model (El-Dessouky & Ettouney, 2002). Although Power's method is a straightforward and the entrainment ratio can be extracted directly from Fig. 3

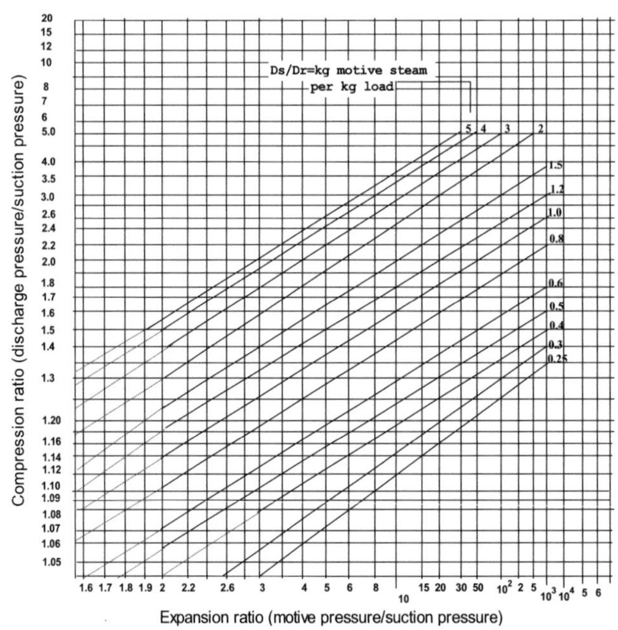

Fig. 3. Entrainment ratio for different compression and expansion ratios (Power, 1994)

in terms of compression ratio and expansion ratio, it is too difficult to use in such optimization and simulation models. The developed semi–empirical model in method 2 is applicable only if the motive fluid is steam and the entrained fluid is water vapor (Al-Juwayhel et al., 1997). The pressure and temperature correction factors were eliminated for simplicity and the model equation is modified as shown in equation (23); results were tested and compared with that obtained by Power's graphical method for validity in the following range of motive pressure $3000 \geq P_s \geq 2000$ (kPa).

$$\left(\frac{D_s}{D_r}\right) = 0.235\frac{(P_d)^{1.19}}{(P_j)^{1.04}}(ER)^{0.015} \tag{23}$$

4.2 Exergy balance

An exergy balance is also conducted to the system to find the exergy destruction (I) in each component; in thermo-compressor, effects, condenser and the leaving streams in kJ/kg according to the following equation:

$$I = T_o \cdot \Delta S = E_{in} - E_{out} \tag{24}$$

Where ΔS is the entropy increase, E_{in} is the input exergy and E_{out} is the output exergy.

4.2.1 Thermo-compressor

The exergy destruction in the thermo-compressor can be expressed as follows:

$$I_{ej} = D_s \cdot \left[(h_s - h_d) - T_o \cdot (S_s - S_d)\right] - D_r \cdot \left[(h_d - h_{gj}) - T_o \cdot (S_d - S_{gj})\right] \tag{25}$$

4.2.2 Effects

The exergy destruction in the first, second and last effect can be expressed respectively as follows;

$$I_{e1} = (D_s + D_r) \cdot \left[(h_d - h_{fd}) - T_o \cdot (S_d - S_{fd})\right] - D_1 \cdot L_1 \cdot \left(1 - \frac{T_o}{T_{v1}}\right) - F_1 \cdot C \cdot \left[(T_1 - T_{f1}) - T_o \cdot In\left(\frac{T_1}{T_{f1}}\right)\right] \tag{26}$$

$$\begin{aligned} I_{e2} = {} & (D_1 + D_r y - F_2 y) \cdot L_1 \cdot \left(1 - \frac{T_o}{T_1}\right) + B_1 \cdot C \cdot \left[\Delta T - T_o \cdot In\left(\frac{T_1}{T_2}\right)\right] - D_2 \cdot L_2 \cdot \left(1 - \frac{T_o}{T_2}\right) \\ & - F_2 \cdot C \cdot \left[(T_2 - T_{f2}) - T_o \cdot In\left(\frac{T_2}{T_{f2}}\right)\right] \end{aligned} \tag{27}$$

$$\begin{aligned} I_{en} = {} & \left[D_{n-1} + \sum_{i=1}^{n-2}(D_i + D_r) \cdot y - (j + n - 1)F_i \cdot y\right] \cdot L_{n-1} \cdot \left(1 - \frac{T_o}{T_{n-1}}\right) + 2 \cdot B_{n-1} \cdot C \cdot \left[\Delta T - T_o \cdot In\left(\frac{T_{n-1}}{T_n}\right)\right] \\ & - D_n \cdot L_n \cdot \left(1 - \frac{T_o}{T_n}\right) - F_i \cdot C \cdot \left[(T_n - T_f) - T_o \cdot \ln\left(\frac{T_n}{T_f}\right)\right] \end{aligned} \tag{28}$$

4.2.3 Condenser and leaving streams

The exergy destruction in the condenser, and in the leaving streams, D_r ,D_f and B_n can be expressed using the following equations respectively:

$$I_c = D_n \cdot L_n \cdot \left(1 - \frac{T_o}{T_n}\right) - M_c \cdot C \cdot \left[\left(T_f - T_c\right) - T_o \cdot \ln\left(\frac{T_f}{T_c}\right)\right] \tag{29}$$

$$I_{Dr} = D_r \cdot C \cdot \left[\left(T_{vj} - T_c\right) - T_o \cdot \ln\left(\frac{T_{vj}}{T_c}\right)\right] \tag{30}$$

$$I_{Df} = D_f \cdot C \cdot \left[\left(T_{vj} - T_c\right) - T_o \cdot \ln\left(\frac{T_{vj}}{T_c}\right)\right] \tag{31}$$

$$I_{Bn} = D_n \cdot C \cdot \left[\left(T_n - T_c\right) - T_o \cdot \ln\left(\frac{T_n}{T_c}\right)\right] \tag{32}$$

4.3 Thermal load

The heat transfer area of an effect can be obtained from the latent heat of condensation (thermal load) of each effect as shown in equation (33), where ΔT_e is the temperature difference across the heat transfer surface.

$$Q = U_e \cdot A_e \cdot \Delta T_e \tag{33}$$

Therefore, the heat transfer area for the first, second and last effect can be obtained as follows:

$$A_{e1} = \frac{(D_s + D_r) \cdot \left[h_d - h_{fd}\right]}{U_{e1} \cdot (T_d - T_1)} \tag{34}$$

$$A_{e2} = \frac{(D_1 + D_r \cdot y - F_1 \cdot y) \cdot L_1}{U_{e2} \cdot (T_{v1} - T_2)} \tag{35}$$

$$A_n = \frac{[(D_{n-1} + \sum_{i=1}^{n-2}(D_i + D_r) \cdot y - (j + n - 1) \cdot F_i \cdot y] \cdot L_{n-1}}{U_{en} \cdot (T_{vn-1} - T_n)} \tag{36}$$

The overall heat transfer coefficient (U_e) depends mainly on the type, design and material of the tubes (El-Dessoukey et al., 2000), and for simplicity it can be calculated as follows (El-Dessouky & Ettouney, 2002):

$$U_{ei} = \frac{\left(1939.4 + 1.40562 \cdot T_i - 0.0207525 \cdot (T_i)^2 + 0.0023186 \cdot (T_i)^3\right)}{1000} \tag{37}$$

The cooling seawater flow rate can be obtained by applying the energy conservation law on the condenser as shown below:

$$M_c = \frac{D_f \cdot L_n}{C \cdot \left(T_f - T_c\right)} \tag{38}$$

The latent heat of condensation of the un-entrained vapor D_f flowing to the condenser is used to increase cooling seawater temperature to feed seawater temperature. The thermal load of the condenser is used to calculate the condenser heat transfer area as follows:

$$A_c = \frac{D_f \cdot L_n}{U_c \cdot (LMTD)_c} \tag{39}$$

The logarithmic mean temperature difference and the overall heat transfer coefficient of the condenser can be obtained from equations (40) and (41) respectively (El-Dessouky & Ettouney, 2002).

$$(LMTD)_c = \frac{\left(T_{vn} - T_f\right) - \left(T_{vn} - T_c\right)}{\ln \frac{\left(T_{vn} - T_f\right)}{\left(T_{vn} - T_c\right)}} \tag{40}$$

$$U_c = 1.7194 + 3.2063 \cdot 10^{-2} \cdot T_{vn} - 1.5971 \cdot 10^{-5} \cdot \left(T_{vn}\right)^2 + 1.9918 \cdot 10^{-7} \cdot \left(T_{vn}\right)^3 \tag{41}$$

Similarly, the heat transfer area of the feed heaters can be expressed as follow assuming that the overall heat transfer coefficient of the feed heaters are equals to that of the condenser.

$$A_{fi} = \frac{(i)F_i \cdot C \cdot \Delta T_f}{U_f \cdot (T_{fi} - T_{fi+1})} \cdot \ln \frac{(T_{vi} - T_{fi+1})}{(T_{vi} - T_{fi})}, \qquad i = 1,\ 2, \ldots n-2 \tag{42}$$

4.4 System performance

The system performance of the ME-TVC model can be evaluated in terms of the following:

4.4.1 Gain output ratio, *GOR*

The gain output ratio is one of the most commonly performance used to evaluate thermal desalination processes. It is defined as the ratio of total distilled water produced (D) to the motive steam supplied (D_s).

$$GOR = \frac{D}{D_s} \tag{43}$$

4.4.2 Specific heat consumption, Q_d

This is one of the most important characteristics of thermal desalination systems. It is defined as the thermal energy consumed by the system to produce 1 kg of distilled water, where L_s is the motive steam latent heat in kJ/kg

$$Q_d = \frac{D_s \cdot L_s}{D} \tag{44}$$

4.4.3 Specific exergy consumption, A_d

The specific exergy consumption is one of the best methods used to evaluate the performance of the ME-TVC based on the Second Law of thermodynamics. It considers the quantity as well as the quality of the supplied motive steam. It is defined as the exergy consumed by the motive steam to produce 1 kg of distillate when the steam is supplied as saturated vapor and leaves as saturated liquid at ambient temperature equal to T_o, according to the following equation (Darwish et al., 2006):

$$A_d = \frac{D_s}{D} \cdot \left[\left(h_s - h_{fd} \right) - T_o \cdot \left(S_s - S_{fd} \right) \right] \tag{45}$$

where h_s, S_s are the inlet motive steam enthalpy and entropy at saturated vapor and h_{fd}, S_{fd} are that of the outlet condensate at saturated liquid.

4.4.4 Specific heat transfer area, A_t

The specific total heat transfer area is equal to the sum of the effect, feed heaters and the condenser heat transfer areas per total distillate product (m²/kg/s).

$$\frac{At_d}{D} = 2 \cdot \sum_{i=1}^{j} \frac{A_{ei}}{D_i} + \sum_{j+1}^{n} \frac{A_{ei}}{D_i} + \sum_{i=1}^{n-2} \frac{A_{fi}}{D_i} + \frac{A_c}{D} \tag{46}$$

4.4.5 Specific exergy destruction, I_t

This term shows the total exergy destruction due to heat transfer and in the thermo-compressor, evaporators, condenser and the leaving streams per unit of distillate water.

$$I_t = \sum \frac{I_i}{D} \tag{47}$$

where I_i is the exergy destruction in each component in kJ/kg.

4.5 Model validity

Engineering Equation Solver (EES) software is used to evaluate the ME-TVC system performance. The validity of the model was tested against some available data of three commercial units having different unit capacities: ALBA in Bahrain (2.4 MIGD), Umm Al-Nar in UAE (3.5 MIGD) and Al-Jubail in KSA (6.5 MIGD). The results showed good agreements as shown in Table 2.

It is also cleared from Table 2 that the available data of Al-Jubail unit is limited in the literature. Hence, the developed mathematical model is used to predict the missing values in order to evaluate the system performance of this plant.

Desalination Plant	**ALBA**		**UMM Al-NAR**		**AL-JUBAIL**	
Number of effects, n	**4**		**6**		**8**	
Operating and Design Parameters	Model	Actual	Model	Actual	Model	Actual
Motive pressure, bar	21	21	2.8	2.8	2.7	2.7
Top brine temperature, °C	63	63	63	62	63	NA
Minimum brine temperature, °C	48	48	44	43	42	NA
Feed sea water temperature, °C	43	43	40	40	40	NA
Motive steam flow rate, kg/s	8.5 × 2	8.3 × 2	11×2	10.65×2	15.5×2	NA
Temperature drop per effect, °C	5	5	3.8	3.8	3	NA
Thermo-compressor Design						
Compression ratio	1.57	NA	1.7	NA	1.75	NA
Expansion ratio	120	NA	18.11	NA	18.7	NA
Motive to entrained vapor ratio	0.58	NA	0.885	NA	0.98	NA
System Performance						
Distillate production, kg/s	123	127	184.2	184.38	340.4	342.22
Gain output ratio	7.23	7.5	8.37	8.6	10.9	9.8
Specific heat consumption, kJ/kg	348.4	NA	292.1	287.5	223	NA
Specific exergy consumption, kJ/kg	127.7	NA	74.6	NA	56.44	NA
Specific heat transfer area, $m^2/kg/s$	244.2	NA	335.6	310	452.2	NA
Specific exergy destruction, kJ/kg/s	94.65	NA	54.24	NA	41.16	NA

Table 2. Mathematical model calculations against some commercial plants.

5. Sensitivity analysis

The new trend of combining ME-TVC desalination system with a conventional Multi effect distillation (MED) unit has been used lately in different large projects and has been also discussed in a few published works (Al-Habshi, 2002), (Darwish & Alsairafi, 2004) and (Bin Amer, 2009). Thus, a sensitivity analysis will be presented in this section to investigate the system performance variations of Al-Jubail ME-TVC unit. This project belongs to Marafiq Company and it is currently considered as the largest ME-TVC desalination plants in the world, it consists of 27 units each of 6.5 MIGD as shown in Fig.4.

The available data of this unit in the literature are: the gain output ratio, number of effect, motive pressure and the unit capacity. These data are used along with the model equations to evaluate the system performance of the plant.

Fig.5. shows the effect of motive steam flow rate on the vapor formed in each effect of this unit, at T_1= 63 °C and ΔT=3 °C. The total distillate production can be controlled by adjusting the motive steam flow rate. The reason is when the motive steam flow rate increases the entrained vapor also increases for constant entrainment ratio (D_s/D_r), this will lead to generate more vapor and consequently more distillate water.

The variation of the gain output ratio and the distillate production as a function of top brine temperature is shown in Fig.6. It is cleared that as the top brine temperature increases the distillate output production decreases and consequently gain output ratio decreases. This is

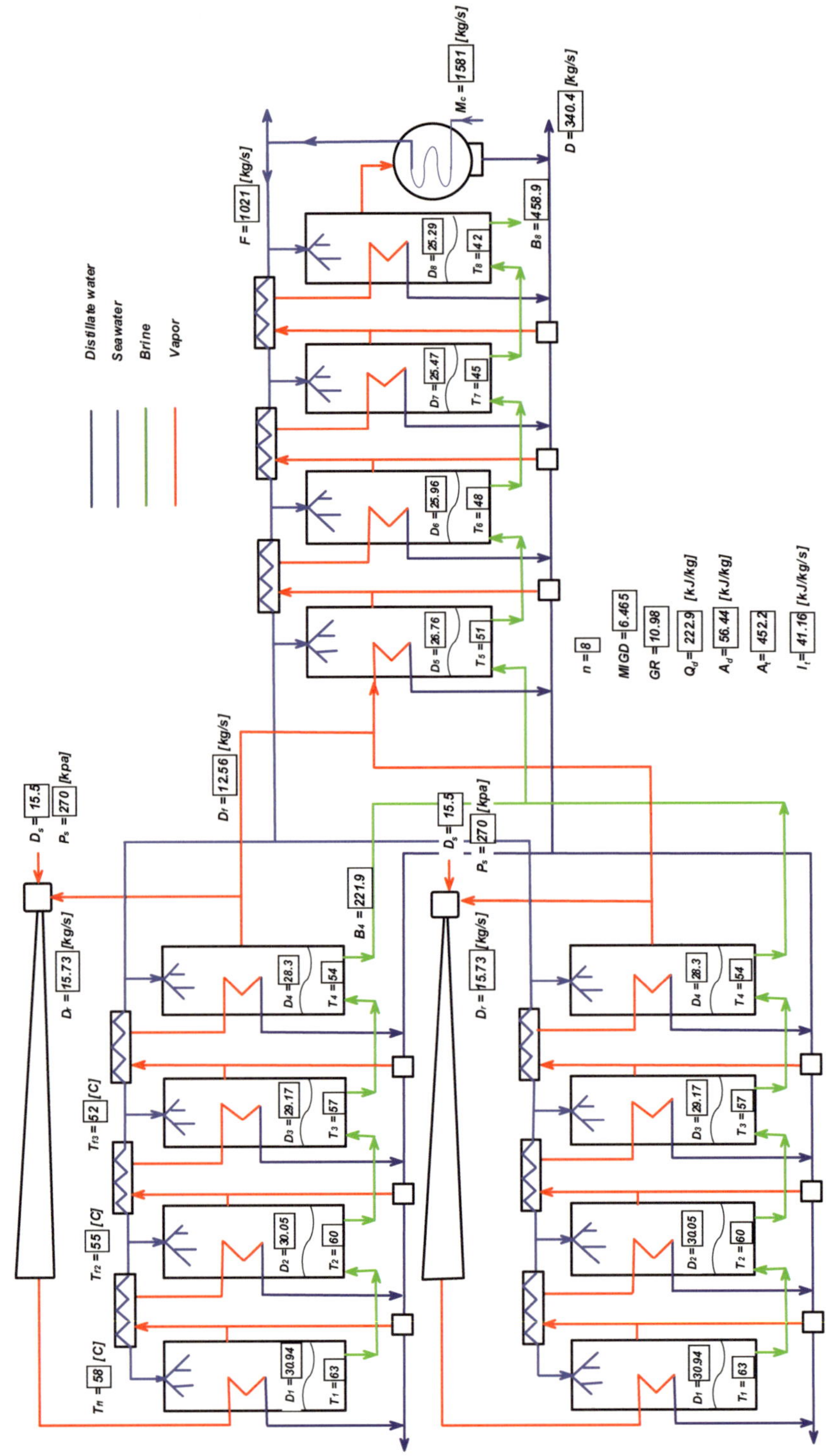

Fig. 4. Schematic diagram similar to Al-Jubail (MARAFIQ) ME-TVC unit, 6.5 MIGD.

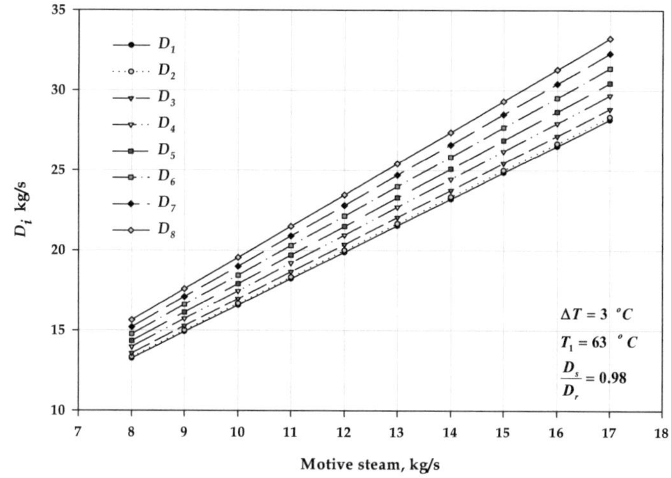

Fig. 5. The effect of motive steam on the distillate production from the effects.

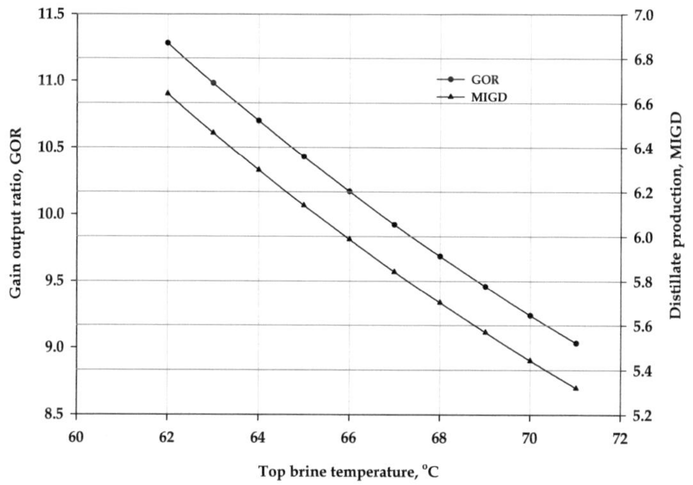

Fig. 6. The effect of top brine temperature on the distillate production and gain output ratio.

because more amount of sensible heating is required to increase the feed seawater temperature to higher boiling temperatures. Additionally, the latent heat of the vapor decreases at higher temperatures.

The direct dependence of the top brine temperature on the specific heat consumption and the specific exergy consumption are shown in Fig. 7. Both of them increase linearly as the top brine temperature increases, because higher top brine temperature leads to higher vapor pressure and consequently larger amount of motive steam is needed to compress the vapor at higher pressures. Fig.8. demonstrates the variations of the specific heat transfer area as a function of temperature difference per effect at different top brine temperatures. The increase in the specific heat transfer area is more pronounced at lower temperature difference per effect than at lower top brine temperatures. So, a high overall heat transfer coefficient is needed to give a small temperature difference at reasonable heat transfer area.

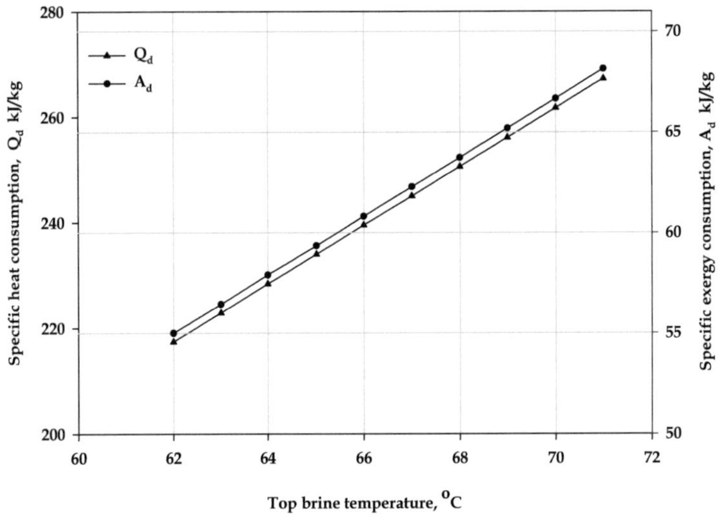

Fig. 7. The effect of top brine temperature on the specific heat consumption and specific exergy consumption.

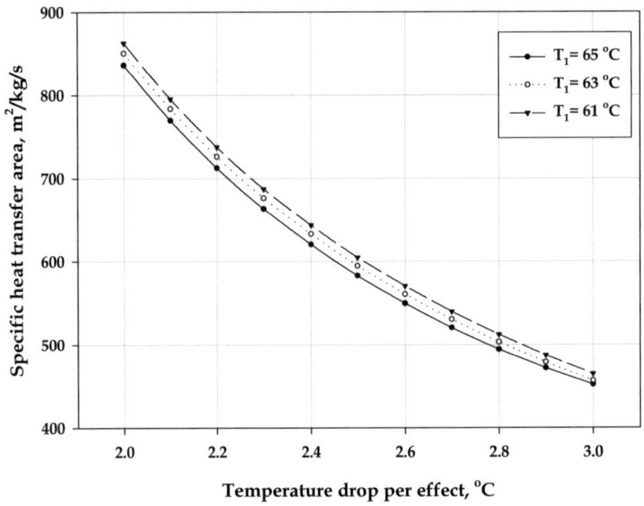

Fig. 8. The effect of temperature drop per effect on the specific heat transfer area.

The exergy analysis is also used to identify the impact of the top brine temperature on the specific exergy destruction for different ME-TVC units as shown in Fig.9. It shows that as the top brine temperature increases, the specific exergy destruction of ALBA, Umm Al-Nar and Al-Jubail plants are increased. It shows also that Al-jubail unit has the lowest values compared to other units. Fig.10 gives detail values of exergy destruction in different components of Al-Jubail units, while Fig.11 pinpoints that thermo-compressor and the effects are the main sources of exergy destruction. On the other hand, the first effect of this unit was found to be responsible for about 31% of the total effects exergy destruction compared to 46% in ALBA and 36% in Umm Al-Nar as shown in Fig.12.

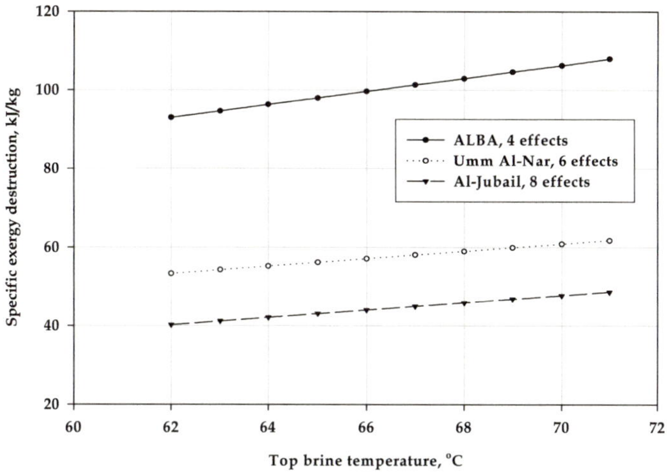

Fig. 9. The effect of top brine temperature on the specific exergy destruction for different units.

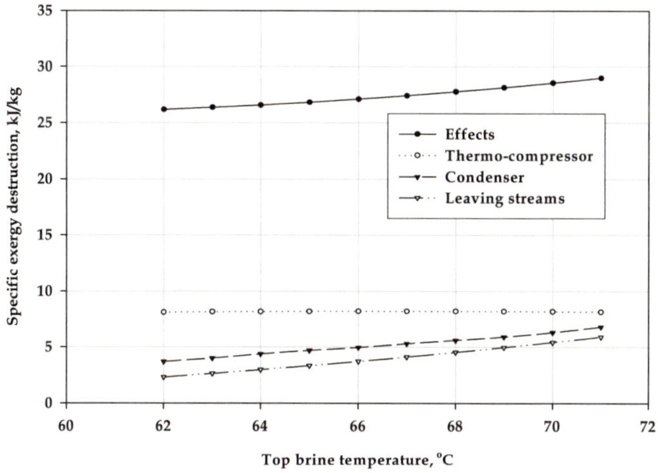

Fig. 10. The effect of top brine temperature on the specific exergy destruction in different components of Al-Jubail ME-TVC unit.

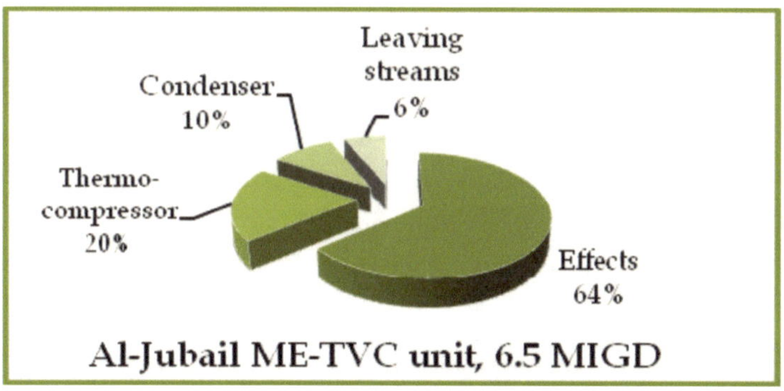

Fig. 11. The exergy destruction in the effects, thermo-compressor, condenser and leaving streams of Al-Jubail unit.

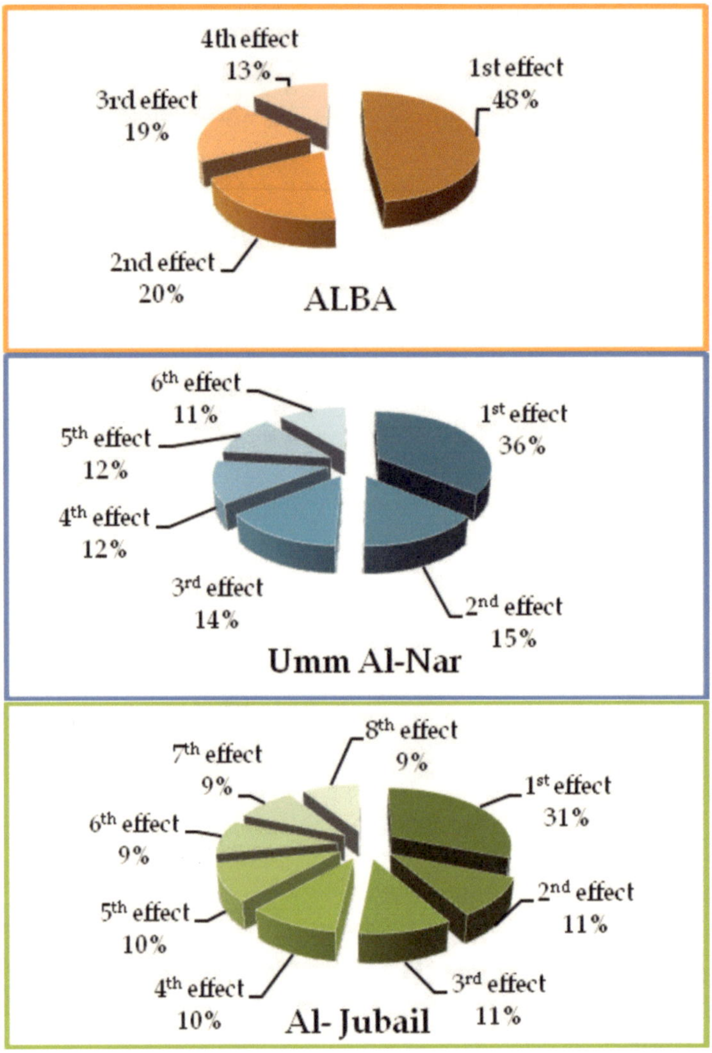

Fig. 12. The exergy destruction in the effects of ALBA, Umm Al-Nar and Al-Jubail units.

6. Development of ME-TVC desalination system.

The first ME-TVC desalination unit of 1 MIGD capacity was commissioned in 1991 in the UAE. It has four effects with a gain output ratio close to 8. A boiler was used to supply steam at high motive pressure of 25 bars (Michels, 1993). The next unit capacity was 2 MIGD which started up in 1995 in Sicily (Italy). It consisted of four identical units; each had 12 effects, with a gain output ratio of 16. The steam was supplied from two boilers at 45 bars to the plant (Temstet, 1996). More units of 1, 1.5 and 2 MIGD were also ordered and commissioned in UAE between 1996 -1999 due to excellent performance of the previous projects (Sommariva, 2001).

The trend of combining ME-TVC desalination system with multi-effect distillation (MED) allowed the unit capacity to increase into a considerable size with less number of effects and at low top brine temperature.

The first desalination project of this type was commissioned in 1999 by SIDEM Company in Aluminum of Bahrain (ALBA). A heat recovery boiler is used to supply high motive steam of 21 bars into four identical units of 2.4 MIGD. Each unit had four effects with a gain output ratio close to 8 (Darwish & Alsairafi, 2004). The next range in size was achieved is 3.5 MIGD in 2000. Two units of this size were installed in Umm Nar; each unit had six effects with a gain output ratio close to 8. The steam was extracted from a steam turbine at 2.8 bars to supply two thermo-compressors in each unit (Al-Habshi, 2002). This project is followed by Al-Taweelah A_1 plant, which was commissioned in 2002 as the largest ME-TVC project in the world at that time. It consists of 14 units; each of 3.8 MIGD. The next unit size that commissioned was in Layyah with a nominal capacity of 5 MIGD (Michels, 2001). The unit size jump to 8 MIGD in 2005 where two units were built in UAE. SIDEM has been also selected to build the largest hybrid plant to date in Fujairah (UAE) which has used two desalination technologies (ME-TVC and SWRO) to produce 130 MIGD as shown in Table 3.

Plant Details	ALBA	Umm Al-NAR	Al-JUBAIL	Al-Fujairah
Country	Bahrain	UAE	KSA	UAE
Year of commission	1999	2000	2007	2008
Source of steam/Arrangement	Boiler	CG-ST/HRSG	CG-ST/HRSG	CG-ST/HRSG
Type of fuel	Diesel oil	Natural gas	Natural gas	Natural gas
Power Capacity, MW	-	1700	2700	2000
Desalination technology	ME-TVC	ME-TVC	ME-TVC	ME-TVC/RO
Unit capacity, MIGD	2.4	3.5	6.5	8.5/RO
Number of units	4	2	27	12/RO
Total capacity, MIGD	9.6	7	176	100+30
Number of effects	4	6	8	10
Water cost, US $/m^3	NA	NA	0.827	0.60

Table 3. Specifications of different ME-TVC desalination units.

6.1 New large projects

This technology is starting to gain more market shares now, in most of the GCC countries for large-scale desalination projects like in Bahrain, Saudi Arabia, and Qatar.

6.1.1 Al-Hidd.

Al-Hidd power and water plant located in northern of Bahrain, consists of three gas fired combined cycle units that produces around 1000 MW. A low motive steam pressure of 2.7 bars is used to feed 10 ME-TVC units, each of 6 MIGD and 9 gain output ratio.

6.1.2 Al-Jubail.

The Independent Water and Power Project (IWPP) MARAFIQ became one of the largest integrated power and desalination plant projects in the world under a BOOT scheme. The project located near Al-Jubail City, north east of Kingdom of Saudi Arabia. It consists of a combined cycle power plant produces 2750 MW along with the world's largest ME-TVC desalination plants of 176 MIGD capacity (27 units × 6.5 MIGD). The units are driven by low motive steam pressure of 2.7 bars. Each unit consisting of 8 effects with gain output ratio around 10.

6.1.3 Ras Laffan.

Ras Laffan is the largest power and water plant in Qatar so far. It will provide the city with 2730 MW electricity and 63 MIGD desalinated water. The power plant consists of eight gas turbines each in conjunction with heat recovery steam generator (HRSG). The high pressure steam enters four condensing steam turbines. A heating steam of 3.2 bars is used to operate 10 ME-TVC units, each of 6.3 MIGD and gain output ratio of 11.1.

6.2 New design and material selection.

Most of the construction materials used in ALBA and Umm Al-Nar desalination plants are almost the same as shown in Table 4. Stainless steel 316L was used for evaporator, condenser and pre-heaters shells, tube-plates, water boxes, spray nozzles and thermo-compressor. Aluminum brass was selected for the tube bundles of the evaporator, except the top rows which were made of titanium in order to prevent erosion corrosion, as water is sprayed from nozzles with high velocities at the upper tubes of the tube bundles (Wangnick, 2004).

Plant	ALBA	Umm Al-Nar	New projects
Evaporator vessel	Cylindrical	Cylindrical	Rectangular
- Shell in contact with seawater	SS 316L	SS 316L	Duplex SS
- Shell in contact with vapor	SS 316L	SS 316L	Duplex SS
- Vapor and distillate boxes	SS 316L	SS 316L	Duplex SS
Heat tube bundles	Titanium	Titanium	Titanium
- Tubes (top rows)	Aluminum	Aluminum	Aluminum
- Tubes (other rows)	brass	brass	brass
- Tube-plates	SS 316L	SS 316L	SS 316L
Demisters	SS 316	SS 316-03	polypropylene
Spray nozzles	SS 316L	SS 316L	SS 316L
Condenser & Pre-heaters	SS 316L	SS 316L	Duplex SS
- Shell & tube-plates	Titanium	Titanium	Titanium
- Tubes	SS 316L	SS 316L	SS 316L
- Water boxes			
Thermo-compressor	NA	SS 316L	Duplex SS

Table 4. Construction materials of the ME-TVC desalination plants.

The new ME-TVC units have rectangular vessel evaporators instead of circular ones as shown in Fig. 13, which gives much more freedom of design (Wangnick, 2004). Additionally, the Duplex stainless steel is also used in these plants instead of 316L Stainless steel as it has better corrosion resistance, higher strength, longer service life as well as lower weight and less market price. (Olsson et al., 2007).

(a) Circular vessel evaporator.

(b) Rectangular vessel evaporator.

Fig. 13. Two types of vessel evaporator used in different ME-TVC units.

In 2005, the first large capacity unit of 8 MIGD was commissioned in UAE, which used the duplex grades stainless steel. It was then used for Al-Hidd plant in Bahrain in 2006 followed by eight units in Libya in 2007, 27 units in Kingdom of Saudi Arabia in 2008 and 12 units in Al-Fujairah in 2009 (Peultier et al., 2009).

6.3 System performance development

The rapid developments in the performance criteria of the ME-TVC during the last ten years can be also observed clearly from Tables 1, 2, 3 and 4 under the following points:

1. This technology is gaining more market shares recently in Bahrain, Saudi Arabia and Qatar in large scale desalination projects with a total installed capacity of 60 MIGD, 176 MIGD and 63 MIGD, respectively.
2. Although the unit size capacities of these desalination projects were almost around six MIGD, their gain output ratios increased gradually to 8.9, 9.8 and 11.1 during 2006, 2007 and 2009 respectively, as shown in Fig. 14.
3. Duplex stainless steels are used in manufacturing the new units instead of 316L stainless steel which have better resistance to corrosion, less costly due to lower contents of nickel and molybdenum, (Olsson et al., 2007).
4. The manufacturer tried to increase the number of effects gradually (4, 6, 8, etc.) in order to increase the size of the units in a compact design.
5. The new generation of large ME-TVC units with high gain output ratio working in conjunction with reverse osmosis as in Al-Fujairah has dramatically decreased the desalinated water production cost as shown in Fig. 15.

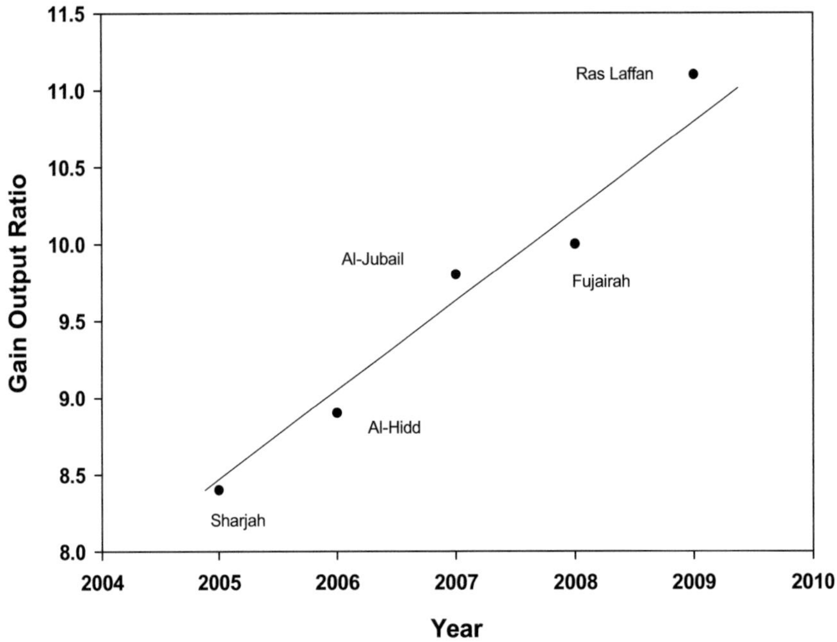

Fig. 14. The increase in the gain output ratio of new ME- TVC projects

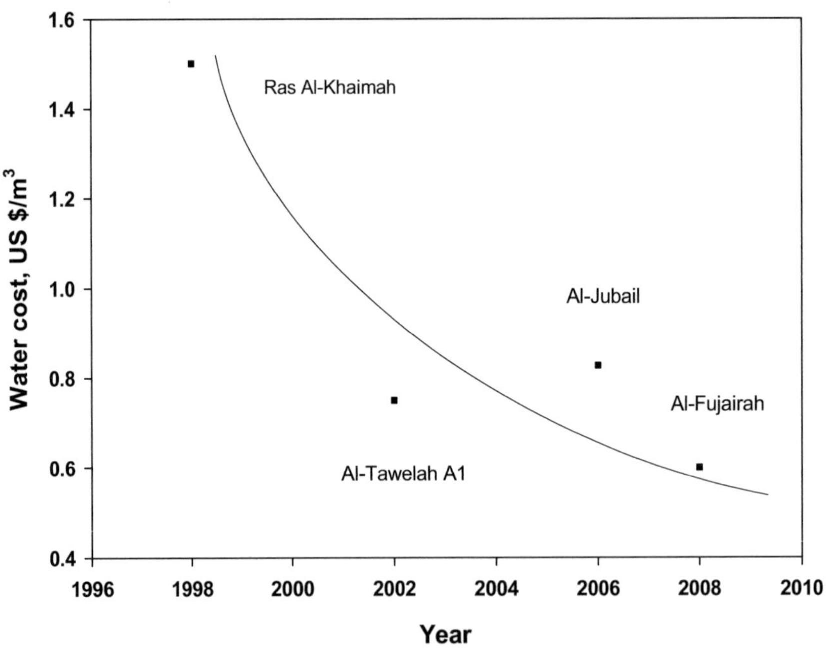

Fig. 15. The drastic decrease in the water cost in the UAE in the last decade.

7. Optimization of ME-TVC desalination system

The schematic diagram consists of *n* number of effects varying from 4 to 16. In any mathematical optimization problem, the objective function, design variables and constrains should be specified in order to formulate the problem properly and to select the appropriate optimization method (Bejan et al., 1996). The general statement of the optimization problem is in the following form:

Find $\chi = \{\chi_1, \chi_2, \dots \chi_N\}$

To Max $f(\chi) = f(\chi_1, \chi_2, \dots \chi_N)$

Subject to $g_j(\chi) \leq 0$, j=1, 2… m

Where N is the number of design variables and m is the number of constraints.

7.1 Optimization approaches

The objective of this optimization work is to find the optimum operating and design conditions of ME-TVC desalination unit for different number of effects to maximize the gain output ratio (*GOR*). MATLAB algorithm solution is used to solve the mathematical model equations by two approaches: (1) Smart Exhaustive Search Method (SESM), which is used for linear and non-linear programming model, based on "for-loops" algorithm, and (2) Sequential Quadratic Programming (SQP), which is a versatile method for solving non-linear constrained optimization problem, based on finding a feasible solution and then start optimization.

The motive steam flow rate is considered to be available at 7 kg/s, directly from a boiler at 25 bars. The cooling and sea seawater temperatures are 30 °C and 40 °C respectively.

The main variables that affect the gain output ratio for a particular number of effects and which can be modified by optimization process are top brine temperature, entrainment ratio and temperature difference per effect (Alasfour et al., 2005).

A set of lower and upper values of those variables were selected as constraints from literatures. Since most ME-TVC plants operate with low top brine temperature (TBT) (not exceeding 75°C) so as to avoid scale formation and corrosion troubles (Al-shammiri & Safar, 1999). The TBT of 76 °C is set here for the upper limit while the lower limit is assumed to be 56 °C (Fisher et al., 1985). The discharged steam temperature T_d is considered to be the hot end temperature of the unit and it is limited by the compression ratio of the steam jet ejector, usually 3 to 5°C above the allowable top brine temperature. In contrast, the last brine temperature, T_n is kept at least 2°C greater than the feed water temperature, T_f (El-Dessouky & Ettouney, 2002), which is assumed to be 10°C greater than the cold end temperature of the model, T_c.

The minimum temperature drop per effect including all thermodynamic losses is close to 1.5 - 2°C (Ophir & Lokiec, 2005) and the maximum temperature drop per effect is set as an upper limit equal to 5°C, and making it higher than this value leads to high top brine temperature and consequently high operating cost (Michels, 2001).

The constraints of entrainment and compression ratios are $\frac{D_s}{D_r} \leq 4$ and $4 \geq CR \geq 1.81$ respectively (El-Dessoukey et al., 2000). The problem can be formulated in a standard design optimization model as shown in Fig. 16.

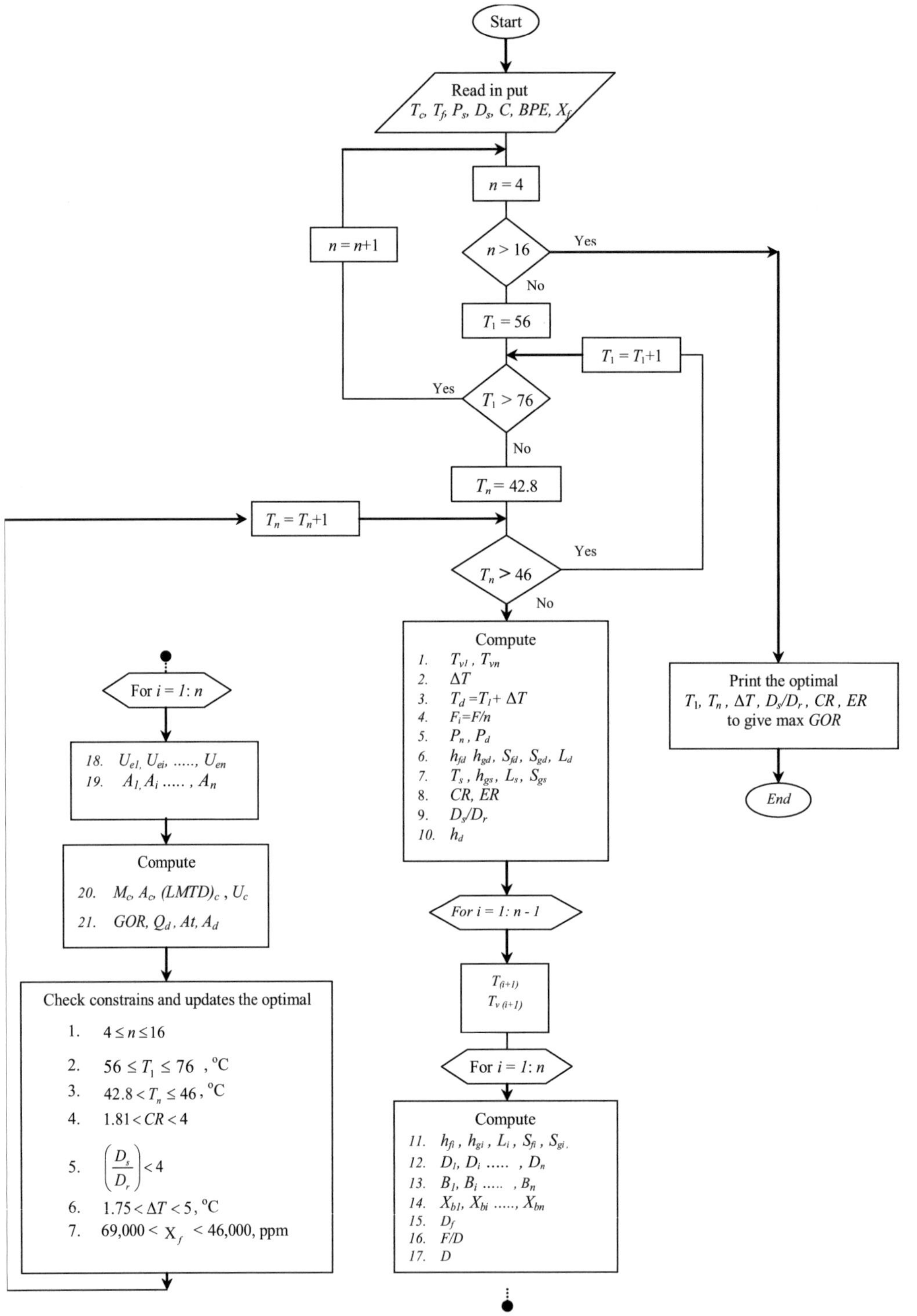

Fig. 16. Solution algorithm of the optimization problem.

7.2 Results and discussion

The optimal computed results of the mathematical optimization problem are displayed below in Table 5.

	Temperature,°C			*Ejector Design*			*System performance*			*GOR*	
n	T_1	T_n	ΔT	*CR*	*ER*	D_s/D_r	Q_d, kJ/kg	A_d kJ/kg	At_d m²/kg/s	*SESM*	*SQP*
4	56	45.8	3.4	1.95	263	0.792	312.8	116.9	450.8	8.18	8.24
5	56	45.8	2.55	1.87	263	0.756	249.6	93.15	699.8	10.27	10.26
6	56	45.8	2.04	1.82	263	0.734	240.1	89.5	947.8	10.67	11.72
7	56	45.3	1.78	1.85	270	0.744	216.3	80.67	1150	11.87	13.28
8	56	43.3	1.81	2	300	0.831	202	75.3	1016.8	12.7	14.57
9	57	42.8	1.77	2.2	307	0.902	187	69.82	982	13.7	15.8
10	59	42.8	1.8	2.43	307	1	174.8	65.42	879	14.61	16.93
11	60.5	42.8	1.77	2.6	307	1.01	161.5	60.5	851.5	15.78	18.1
12	62.5	42.8	1.79	2.85	307	1.22	150	56.36	786.84	16.94	19.41
13	64	42.8	1.76	3	307	1.32	138.3	52	776.5	18.32	20.6
14	66	42.8	1.78	3.33	307	1.47	128.1	48.34	744.6	19.71	21.93
15	67.5	42.8	1.76	3.56	307	1.58	118.2	44.67	752.6	21.31	23.3
16	69.5	42.8	1.78	3.88	307	1.76	109.5	41.47	748.47	22.93	24.74
Time, s										8.89	0.109

Table 5. Optimal operating and design conditions for different number of effects.

In the light of the results shown in Table 5 the following facts can be reported: -

1. The optimal results of GOR obtained by SQP method are close but better than that obtained by SESM and the corresponding total execution time is also less (0.109 sec compared to 8.89 sec, CPU time).
2. The maximum gain output ratio is varied between "8.2 to 24.7" for 4-effects and 16-effects and the optimal top brine temperature varies between 56 to 69.5°C respectively as shown in Fig.17.
3. ME-TVC system can operate at top brine temperature below 60°C with a maximum gain output ratio of 16.9 for 10 effects.

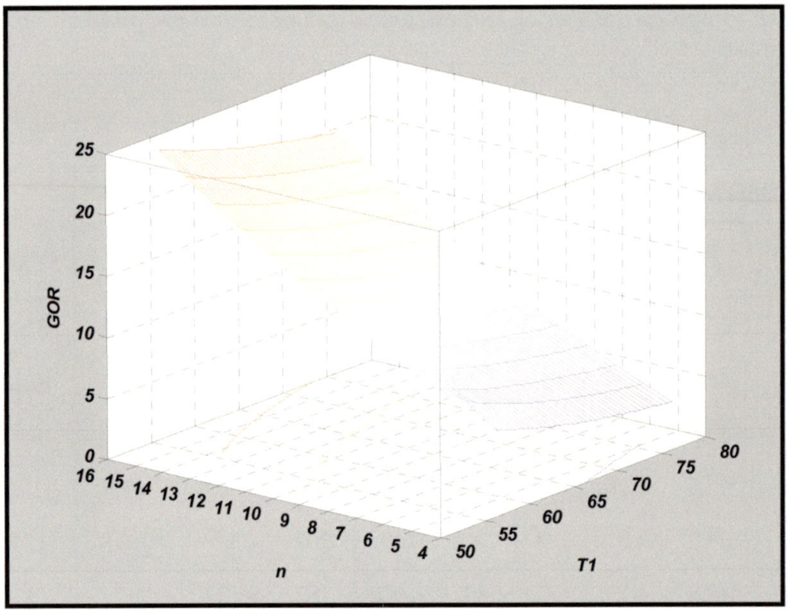

Fig. 17. The impact of top brine temperature and the number of effects on the gain output ratio.

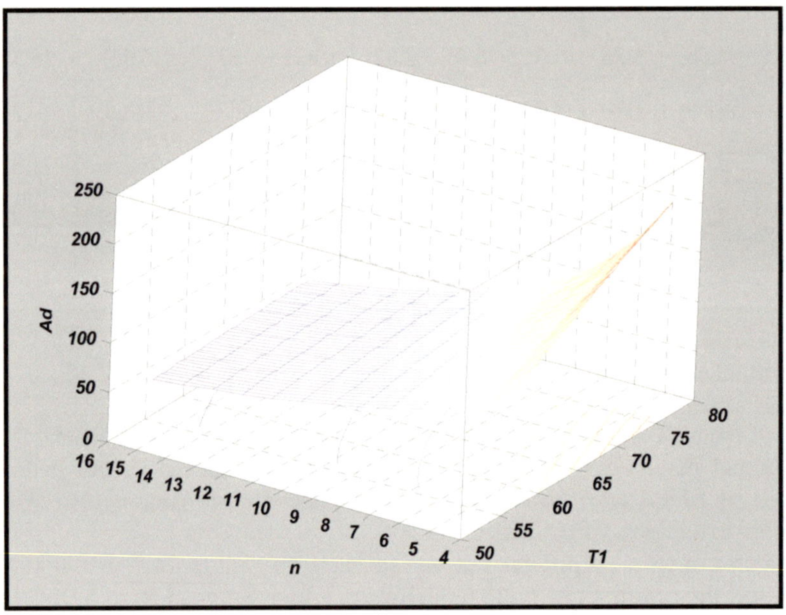

Fig. 18. The impact of top brine temperature and the number of effects on the specific exergy consumption.

4. A maximum gain output ratio of 15.8 can be achieved by ME-TVC, which is close to that of an existing plant (in Sicily), but with low motive pressure (25 bar compared to 45 bar), less number of effects (9 effects compared to 12) and less top brine temperature (57°C compared to 63°C).
5. The optimal entrainment ratios (D_s/D_r) vary from 0.79 for 4 effects to 1.76 for 16 effects.
6. It is clear that as the number of effects increases the gain output ratio, compression ratio and entrainment ratio increases, while the specific exergy consumption decreases as shown in Fig. 18.

8. Conclusion

- This chapter outlines the performance developments in multi-effect thermal vapor compression systems during the last decade in view of some commercial units which were built by SIDEM Company. The new trend of combining ME-TVC desalination system with a conventional Multi effect distillation (MED) unit has been used lately in several large projects. This trend provides an approach to increase the unit capacity with a more compact design.
- Most of the new ME-TVC units are commonly operated with large combined cycle power plants (CC-PP) which are characterized by high efficiency in order to reduce the power and water costs. Al-Fujairah is an ideal example of a large hybrid desalination project which led to considerable reduction in the desalinated water cost.
- Greater understanding of the behavior of the material at different operating conditions led the manufacturer to use Duplex grades of stainless steel in different parts of the new units instead of conventional material (316L). Titanium is being selected also for the tube bundles instead of aluminum brass.
- Exergy analysis shows that the specific exergy destruction in ALBA unit (94.65 kJ/kg) is almost twice that in Umm Al-Nar and Al-Jubail units (54.24 kJ/kg and 41.16 kJ/kg respectively) because high motive pressure of 21 bars is used in ALBA compared to low motive pressure of 2.8 bars in other units. The analysis indicates that thermo-compressor and the effects are the main sources of exergy destruction in these units. On the other hand, the first effect of this unit was found to be responsible for about 31% of the total effects exergy destruction compared to 46% in ALBA and 36% in Umm Al-Nar. The specific exergy destruction can be reduced by increasing the number of effects as well as working at lower top brine temperatures.
- The manufacturer has tried to improve the new ME-TVC desalination system projects based on their experience in the previous projects. Further developments can be achieved by technical optimization in order to reduce the desalinated water cost.
- A MATLAB algorithm was developed and used to solve a mathematical model optimization problem, where different numbers of effects were tested to maximize the gain output ratio using: (1) Smart Exhaustive Search Method and (2) Sequential Quadratic Programming. The maximum gain output ratio varied between 8.24 to 24.74 for 4 and 16 effects with an optimal top brine temperature ranging between 56 to 69.5°C and reasonable specific heat transfer area. The optimal ranges of compression and entrainment ratios were between 1.82 to 3.88 and 0.734 to 1.76, respectively. The

optimal results of GOR obtained by SQP method are close but better than that obtained by SESM and the corresponding total execution time is also less (0.109 sec compared to 8.89 sec, CPU time).

- To conduct a complete and successful optimization in a multi effect thermal vapor compression desalination system, exergo-economic analysis must be understood to know the behavior of the quality of the energy from a cost point of view and this chapter can be an introduction to exergo-economic optimization design in future work.

9. References

Alasfour, F.; Darwish, M. & Bin Amer, A. (2005). Thermal analysis of ME- TVC + MEE desalination system. *Desalination*, Vol. 174 (2005) 39-61.

AL-Habshi, S. (2002). Simulation and economic study of the MED-TVC units at Umm Al-Nar desalination plant. *Thesis report*, (2002) UAE.

Al-Juwayhel, F.; El-Dessouky, H. & Ettouney, H. (1997). Analysis of single-effect evaporator desalination systems combined with vapor compression heat pumps. *Desalination*, Vol. 114 (1997) 253-275.

Al-Najem, N.; Darwish, M. & Youssef, F. (1997). Thermo-vapor compression desalination: energy and availability analysis of single and multi-effect systems. *Desalination*, Vol. 110 (1997) 223 – 238.

Al-Shammiri, M. & Safar, M. (1999). Multi-effect desalination plants: state of the art. *Desalination*, Vol. 126 (1999) 45-59.

Ashour, M. (2002) Steady state analysis of the Tripoli West LT-HT-MED plant. *Desalination*, Vol. 152 (2002) 191-194.

Bejan, A.; Michael, J. & George, T. (1996). *Thermal design and optimization*, John Wiley & Sons, 0471584673, New York.

Bin Amer, A. (2009). Development and optimization of ME-TVC desalination system. *Desalination*, Vol. 249 (2009) 1315-1331.

Choi, H.; Lee, T.; Kim, Y. & Song, S. (2005). Performance improvement of multiple-effect distiller with thermal vapor compression system by exergy analysis. *Desalination*, Vol. 182 (2005) 239-249.

Cipollina, A.; Micale, G. & Rizzuit, L. (2005). A critical assessment of desalination operation in Sicily. *Desalination*, Vol. 182 (2005) 1-12.

Darwish, M. & Alsairafi, A. (2004). Technical comparison between TVC/MEB and MSF. *Desalination*, Vol. 170 (2004) 223-239.

Darwish, M.; Alasfour, F. & Al-Najem, N. (2002). Energy consumption in equivalent work by different desalting methods case study for Kuwait. *Desalination*, Vol. 152 (2000) 83-92.

Darwish, M.; Al-Juwayhel, F. & Abdulraheim, H. (2006). Multi–effect boiling systems from an energy viewpoint. *Desalination*, Vol. 194 (2006) 22-39.

El-Dessouky, H. & Ettouney, H. (1999). Multiple-effect evaporation desalination systems: thermal analysis. *Desalination*, Vol. 125 (1999) 259-276.

El-Dessouky, H.; Ettouney, H. & Al-Juwayhel, F. (2000). Multiple effect evaporation- vapor compression desalination processes. *Trans IChemE*, Vol. 78, Part a, (May 2000) 662-676.

El-Dessouky, H. & Ettouney, H. (2002). *Fundamental of salt-water desalination*, Elsevier Science, 0444508104, Amsterdam.

Fisher, U.; Aviram, A. & Gendel, A. (1985). Ashdod multi-effect low temperature desalination plant report on year of operation. *Desalination*, Vol. 55 (1985) 13-32.

Hamed, O.; Zamamiri, A.; Aly, S. & lior, N. (1996). Thermal performance and exergy analysis of a thermal vapor compression desalination system. *Energy Conversion Mgmt*, Vol. 37 (1996) 379-387.

Michels, T. (1993). Recent achievements of low temperature multiple effect desalination in the western areas of Abu Dhabi, UAE. *Desalination*, Vol. 93 (1993) 111-118.

Michels, T. The modern MED pursuing its way in the Gulf Region. *European Desalination Society, Newsletter*, 11 January 2001.

Olsson, J. & Snis, M. (2006). Duplex- A new generation of stainless steel for desalination plants. *Proceeding of EDS congress on desalination strategies in South Mediterranean Countries*, Montpelier, 2006.

Ophir, A. & Lokiec, F. (2005). Advanced MED process for most economical sea water desalination. *Desalination*, Vol. 182 (2005) 187-198.

Peultier, J.; Baudu, V.; Boillot, P. & Gagnepain, J. (2009). New trends in selection of metallic material for desalination industry. *Proceeding of IDA world congress on desalination and water reuse*. Dubai, November, 2009, UAE.

Power, R. (1994). *Steam Jet Ejectors for the Process Industries*, McGraw-Hill, 0070506183, NewYork.

Reddy, K. & Ghaffour, N. (2007). Overview of the cost of desalinated water and costing methodologies. *Desalination*, 205 (2007) 340-353.

Sayyaadi, H & Saffaria, A. (2010). Thermo-economic optimization of multi effect distillation desalination systems. *Applied Energy*, Vol. 87 (2010) 1122-1133.

Sommariva, C.; Hogg, H. & Callister, K. (2001). Forty-year design life: the next target material selection and operation conditions in thermal desalination plants. *Desalination*, Vol. 136 (2001) 169-176.

Sommariva, C. & Syambabu, V. (2001). Increase in water production in UAE. *Desalination*, Vol. 138 (2001)173-179.

Temstet, C.; Canon, G., Laborie, J. & Durante, A. (1996). A large high-performance MED plant in Sicily. *Desalination*, Vol. 105 (1996) 109-114.

Utomo, T.; Ji, M.; Kim, P.; Jeong, H. & Chung, H. (2008). CFD analysis of influence of converging duct angle on the steam ejector performance. *Proceeding of the International Conference on Engineering Optimization*, Rio de Janeiro, June 2008, Brazil.

Wade, N. (2001). Distillation plant development and cost update. *Desalination*, 136 (2001) 3-12.

Wangnick, K. (2004). Present status of thermal seawater desalination techniques. http://www.idswater.com/Common/Paper/Paper_51/Klaus.pdf.

Wang, Y. & Lior, N. (2006). Performance analysis of combined humidified gas turbine power generation and multi-effect thermal vapor compression desalination systems – Part 1: The desalination unit and its combination with a steam-injected gas turbine power system. *Desalination*, Vol. 196 (2006) 84-104.

Part 3

Environmental and Economical Aspects

10

Solar Desalination

Bechir Chaouachi
Unit of research: Environment, Catalysis & Processes Analysis
National School of Engineers of Gabes, University of Gabes,
Omar Ibn ElKhattab Street - 6029 Gabes,
Tunisia

1. Introduction

Water is life in all its forms. All living organisms contain water: the body of a human being is composed of approximately 60% of water, a fish of 80%, plants between 80 and 90%. Water is necessary for the chemical reactions that occur in living cells and is also in the middle of this water that these cells are formed. Water is essential to sustainable food production as well as all living ecosystems; human development is based entirely on the hydrological cycle.
Water covers about 70% of the globe area. Furthermore, 97% of this water (salty, non-potable and unsuitable for irrigation) is located in the oceans. Freshwater is only 3% of total water on our planet. In this low percentage, rivers and lakes are 0.3%, while the rest is stored in the polar caps and glaciers.
Freshwater tanks are very unevenly distributed on the surface of the globe. While Western countries for example have the chance to have huge reserves which will renew each year to feed a population that acknowledges a low population growth for most. Many tropical and island countries lack sufficient water, however suffer rampant demographic growth and know an extremely bad supply difficulties. Arid regions are in a situation of severe water stress and simply a drought to decimate the weaker populations and livestock.
We fought for the strategic islands or for black gold, we will fight soon for «blue gold" if everyone does not share its resources, and does not reduce consumption and losses.
Drinking water demand is also growing more and more, and the inadequacy of this water can be considered to be a danger that continued to disturb the humanity until our days (and in the future), causing thus disruption or even a braking of economic activities and a deterioration of living standards [1, 2]. Similarly this lack can be linked directly to 80% of diseases affecting the world's population and 50% of cases of infant mortality [3]. All these data so eloquent drew our attention on the need to search other sources of drinking water.
On the other hand, and worldwide distribution of drinking water is not commensurate with the needs of each region. This is manifested by finding a surplus of water in regions, while others have chronic shortages. For the latter, the desalination of brackish or sea water is becoming the inevitable solution.
Furthermore, in addition to the vital need for water, human beings live also have a crucial need for energy. This is particularly true for human beings who consume increasing energy not only for food, dress, heat, move, entertainment and treat, but also for product all manufactured objects quantities.

The quality of life of the world population largely depends on energy at its disposal, not only in quantity but also in quality. It is determined by the choice of modes of production, distribution and consumption. Resolve the crucial energy in the world by providing men energy that they need on their housing and production sites is certainly a factor of peace.
In these circumstances, made to find a source of energy other than those of fossil energies and responding to environmental requirements, seems crucial. In this context, renewable energy have a certain interest and, in particular, solar energy.
Desalination processes fall into two categories; a distillation processes (requiring a phase change, vaporization/condensation) and in the other hand the membrane processes (membrane separation).
For its operation, the distillation process requires, for much part, the thermal energy for heating salty water. For seawater, for example, 100-50. 10^3 kcal per m^3 of water's produced following the performance of the unit. In addition, this thermal energy must be provided at a relatively low temperature, between 120 and 60 ° C according to the technology adopted. The heat source can be provided, in the case of a coupling solar, by solar flat plate or concentrator collectors.
The usually used processes which are likely to be coupled to solar energy are:

- Direct solar distillation greenhouse is a strictly a solar process.
- Classical distillation processes such as Multi-stage flash, multiple-effects, vapour compression process.

Solar energy can be converted from appropriate converters to other forms of energy such as electrical, mechanical, thermal, etc. In the thermal energy conversion there are two modes of conversion: at low temperature, where heating fluid temperature remains below 100 ° C [4-7] and at average and high temperature when it exceeds 100 ° C. For the first case, this level of temperature is reached by means of a flat plate collector, while in the second case, a concentrator collector is required [7, 8]. Several types of collectors were made until today we quote:

- conical concentrator,
- spherical concentrator,
- cylindro-parabolic concentrator and
- parabolic concentrator.

The process of solar distillation is used to distill brackish/saline water by using solar energy. The systems involved in solar distillation operate under two modes: passive and active.
Many prototypes of solar stills have been constructed and experimented by various researchers. A solar distillation system may consist of two separated devices - the solar collector and the distiller - or of one integrated system. The first case is an indirect solar desalination process, and the second one is a direct solar desalination process. Many small-size systems for direct solar desalination and several pilot plants of indirect solar desalination have been designed and implemented [9-11].

2. Desalination processes

These are separation processes that rely on a technique or technology for transforming a mixture of substances into two or more distinct components. The purpose of this type of process is to purify the saline water of its impurities.

The principle of a separation process is to use a difference of properties between the interest compound and the remaining mixture. When the difference property will be greater, the separation is easy. So the choice of the separation process starts with a good knowledge of the mixture composition and properties of different components. The desalination processes are divided into two main categories: on the one hand, the distillation process (which requires a phase change, evaporation / condensation) and on the other hand the membrane processes (filtration).
The most current techniques of desalination are thermal distillation - for the treatment of great volumes of water (55 000 m^3/jour) - and the membranes technology: electrodialysis and reverse osmosis. The ability of treatment with membrane technology can be adapted according to the intended use (the great plants have a capacity of more than 5000 m^3/day, the averages plant between 500 and 5000 m^3/day, while that small installations have a maximum capacity of 500 m^3/day).
It is noticed that these processes use thermal energy and / or electrical energy and consequently are consumer's energy and pollutants. The energy, conventional methods commonly used, can be of solar origin either a partial or total depending on production capacity and in this way we minimize significantly the consumption of energy while protecting the environment. Future research in this area is oriented toward the maximum utilization of solar energy, which is free and clean, or through technological innovation and/or improvements on conventional methods.

2.1 Solar thermal distillation

For their operation, the distillation processes require for much of the thermal energy for heating salt water. Furthermore, this thermal energy must be supplied at a relatively low temperature, between 60 and 120 ° C. Heat can be provided in the case of the use of solar energy by solar flat plate or concentrator collector according to working conditions.
The processes most commonly used and which are likely to be coupled to a source of solar energy are:

- The direct solar greenhouse distillation is a properly solar process.
- The conventional distillation processes such as multi-stage flash, multi-effects, vapor compression

2.1.1 Direct solar greenhouse distillation

This process consists in heating water directly by the solar radiation in a closed enclosure covered with glazing. The produced vapor, which condenses on the colder glazing and slightly inclined, east collects in the form of condensed in gutters. The principle is very simple, reliable and does not require any maintenance. But its output is relatively weak, 4 to 6 liters/day.m^2 [12, 13]. They are however two types of manufacturing distillers, they can be built either:

- In the form of modular product, it is usually a tray (plastic, metal, wood ...) isolated from below and covered with a glass top. Several distillers can be fed simultaneously to form a distillation unit. The number of distillers depends on the desired produced water capacity. This model is used only for very small product capacities, a few liters per d. It is practical when the need for distilled water is not very important (laboratory analysis, auto park ...).

They are however several variants include plat distillers, cascading wick, with multiple effects, spherical ... etc.

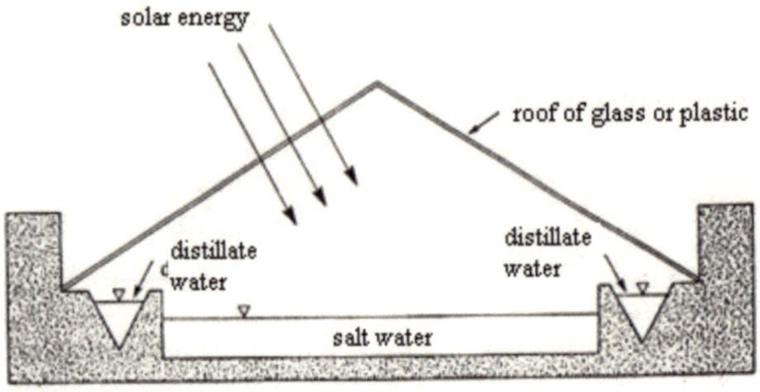

Fig. 1. Solar distillation by greenhouse effect.

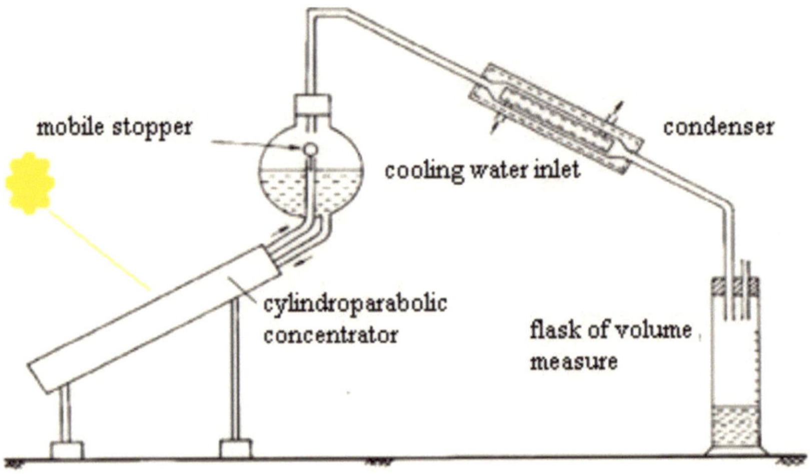

Fig. 2. Solar distiller with cylindroparabolic concentrator

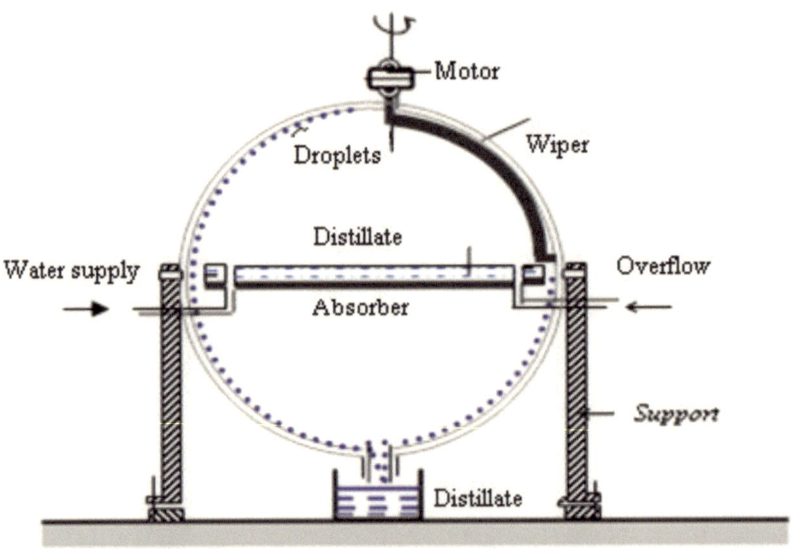

Fig. 3. Spherical solar distiller with sweeping

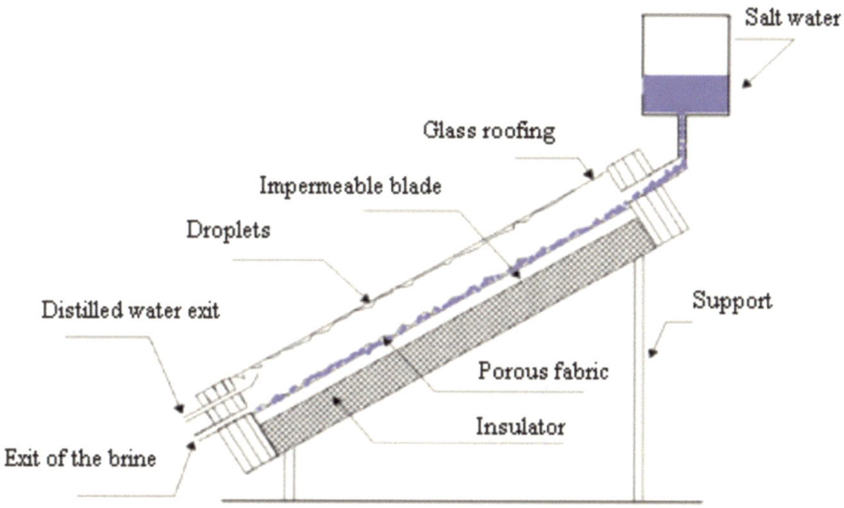

Fig. 4. Solar distiller with wick

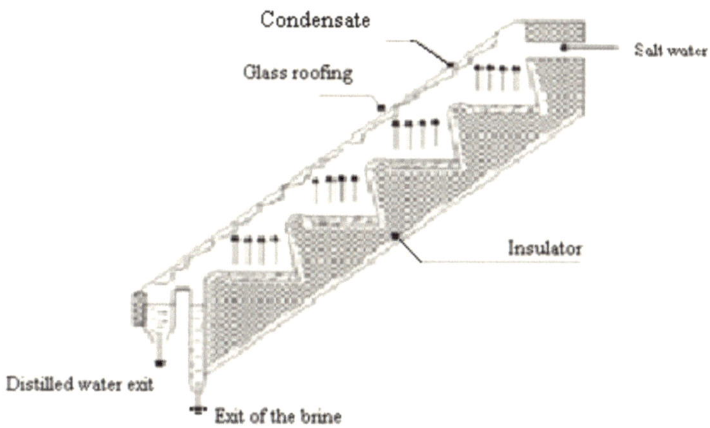

Fig. 5. Solar distiller with cascade

- When the needs are greater and to increase the production of fresh water, we can juxtapose several distillers or build a distiller of large surface. The first construction of this type of distillers was held in 1872 at Las Salinas (Chile) with an area of 4700 square meters and a production of 23 m^3 / d of fresh water [14]. In Tunisia, a desalination plant was built in 1929 near Ben Gardanne to support French military troops [15]. The first large pools (439 and 1300 m^2) were built during the 60s in the regions of Chakmou and Mahdia. Their daily production is respectively 0.57 and 4.48 m^3 [14, 16, 17].

The theoretical analysis is based on the heat balance of the distiller who allows to determine its output according to the various parameters.

2.1.2 Distillation with multi-stage flash (MSF)

This process usually profitable only for large capacity (several hundreds of thousands of m^3), is not flexible and presents difficulties of setting in mode for a solar application. The number of effects depends on the pressure difference that exists between the first and last stage. It is noticed that the contribution of thermal energy can be completely or partially solar and this is function of the desired production.

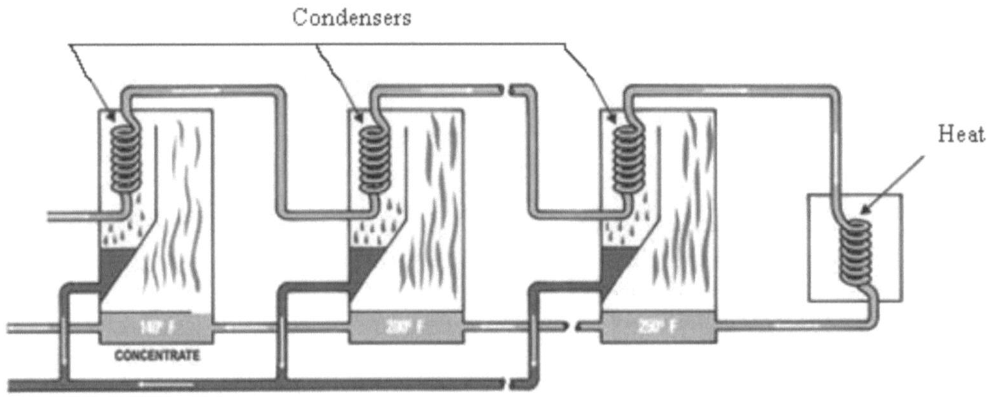

Fig. 6. Multi stage flash distiller

2.1.3 Distillation by vapour compression

It is a process involving a series of evaporators; however, its performance is improved by recycling vapor from the last effect (at the lowest heat) by compressing and then used as heating steam to the first effect. This method can use solar energy as heat source, but requires more energy to compress vapor. This is done either with a supercharger (mechanical compression) or a steam ejector (thermal compression)

2.1.4 Distillation by multiple effects

In this category, there are two processes: some use vertical tubes, the other horizontal tubes. The advantage goes to the horizontal tubes for low pumping power used and a global coefficient of heat exchange important.

An example of multiple effect distillers is shown in Fig. 7 [14].

This distiller is composed of a series of vertical and parallel plates, a storage tank for hot water and a solar panel. The first plate is heated by hot water circulating in the pipe welded at its left part. The last plate is cooled by circulating salt water in a tube in contact with it. After that, the heated salt water supplies distributors at the top and right side plates. These distributors provide a falling film flow along them. The contribution of energy provided by hot water at the first stage, will give rise to the formation of a quantity of steam in the right side of this plate. The steam is condensed in the left side of the plate after evaporating a quantity of water falling film flowing on the right side of this plate and so on. The condensate is collected at the bottom of the plates.

The storage tank allows the multiple effect process to operate during periods of absence or insufficient solar radiation. Hence the advantage of this system compared to those using solar energy directly.

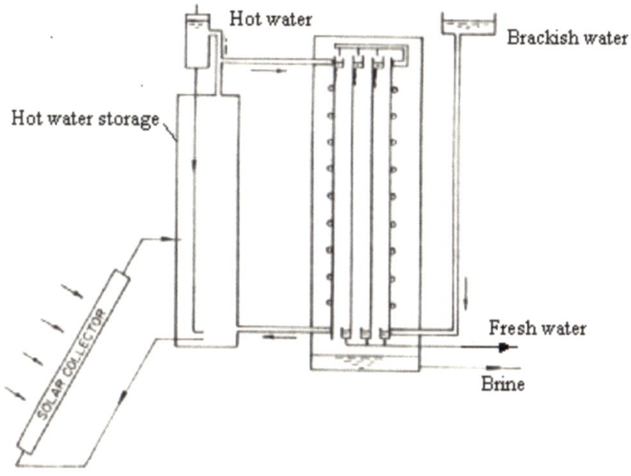

Fig. 7. Multiple effects solar distillation.

It should be noted that the multiple effect solar distillation at atmospheric pressure cannot always compete with one single effect. Thus, several studies have been conducted to improve the performance of these distillers. Among these works, there are those that replace the flat-plate by parabolic concentrator in order to produce steam for the initiation of multiple effect distillation [14].

2.2 Solar membrane processes

The main membrane processes used in the field of desalination are electrodialysis and reverse osmosis.

2.3 Electrodialysis

This process requires, for its operation, the application of an electric field between a cathode and an anode to allow the migration of the ions (positive and negative) through the membranes. It is a large consumer of energy, which makes its solar application possible, only for brackish water of very low salinity.

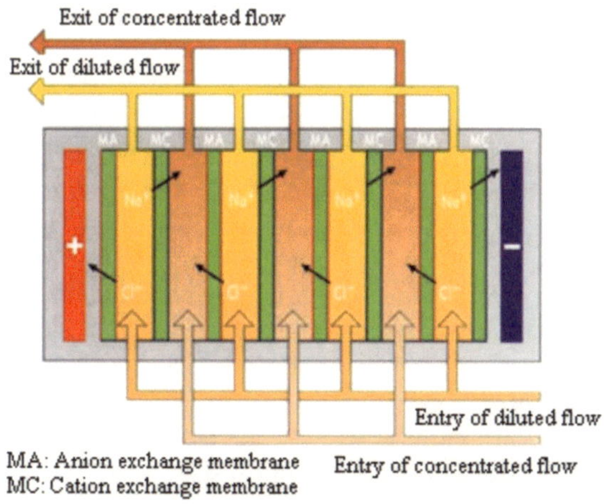

Fig. 8. Electrodialysis desalination process

2.4 Reverse osmosis

The principle of this process is to move under the influence of pressure, pure water through a semi permeable membrane which has the characteristic of retaining the salts dissolved in water. So we need the necessary energy to supply a high pressure pump, which can be provided economically (for small plants) by a photovoltaic generator or an aerogenerator.

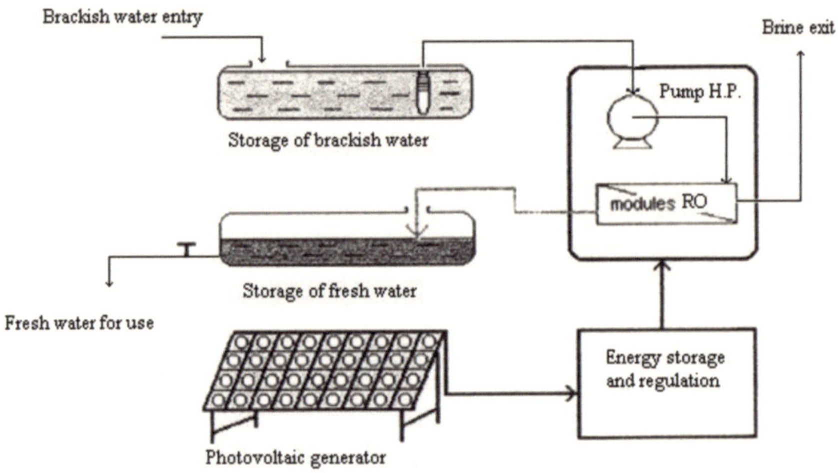

Fig. 9. Reverse osmosis desalination process

A certain number of units are in operation worldwide, and the results obtained through various studies and experiments make its application very encouraging.

2.5 Vacuum membrane distillation

Membrane distillation is a relatively recent process. This process uses hydrophobic porous membranes to separate a solution physically. The process driving force corresponds to the pressure and temperature variation between the two sides of membrane. The principle of separation by the membranes distillation is based on the balance liquid/vapour which controls the selectivity of the process [18].

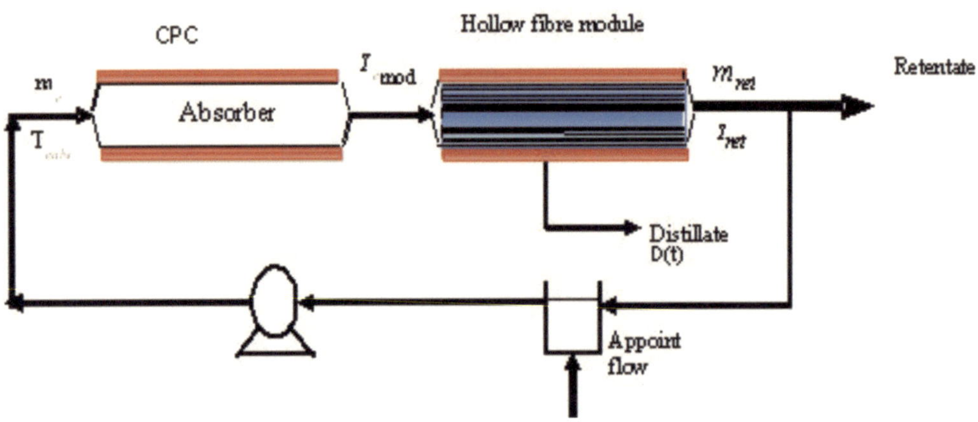

Fig. 10. Diagram of the desalination unit

The principal interests of membrane distillation compared to the conventional distillation process are the great contact specific area due to the installations compactness's, the modularity and the possibility of automating the process easily.
Among the membrane distillation techniques we are interested to the vacuum membrane distillation (VMD). It is an evaporative process which uses a hydrophobic porous membrane, whose function is to separate and put in contact a liquid and a gas phase [19]. This process is interesting for the seawater desalination. The analysis of the operating conditions shows that the parameters keys are a relatively low temperature and pressure. Moreover, the process coupling vacuum membrane distillation with a source of energy (solar or geothermal) could compete with reverse osmosis [20].
An example of a solar vacuum membrane distillation (SVMD) is represented on the figure 1. It is composed by a solar desalination system in series with a tank which receives seawater. This one feed by retentate flow and a auxiliary flow

3. Solar energy

The solar energy received at the ground level is very abundant and far exceeds the energy requirements of the current world population. It is free and accessible to all by its decentralized appearance. It reaches the user without hazardous waste or pollution risk.

3.1 Different forms of solar radiation

During the passage of solar radiation by the atmosphere, some is absorbed (UV and X rays), another part is dispersed by air molecules or suspended particles (dust). This gives rise to diffuse solar radiation. The remaining part arrives directly at the surface of the ground and constitutes the direct radiation. The total solar radiation is made of direct and diffuse radiation.

3.2 Solar constant

It is the total energy sent by the sun to the limit of the terrestrial atmosphere on a perpendicular surface to the radiation. The average value selected is of 1353 W/m^2 [21].

3.3 Evaluation of solar radiation reaching the collector

Several mathematical models have been developed to measure the solar irradiance at the collector. Among these models are cited that of EUFRATE which is based on the synthesis work of PERRIN DE BRICHAMBAUT, KASTEN and HAY. This model uses the factor of disturbance of LINKE (TL), latitude and declination [22]. Because the concentrators do not collect the diffuse radiation [21], we will be interested in this part only to the direct radiation. The various equations describing the EUFRAT model are:

3.3.1 Correction of the earth-sun distance (r)

$$r = 1 + 0{,}034 \cos(0{,}986j - 2) \tag{1}$$

3.3.2 Estimation of the disturbance factor of LINKE (TL)

$$TL = 1{,}6 + 16\beta A + 0{,}5 \ln(P/100) \tag{2}$$

The disturbance coefficient of Angström (βA) varies according to the sky type. The following table illustrates this variation:

Type of the sky	βA
Major blue	0,02
Pure blue	0,05
Light blue	0,10
Milky blue	0,20
Whitish	0,50

Table 1. Angström disturbance coefficient βA.

In the same way, the factor TL can be estimated by geographical area with the help of the equation:

$$TL = TO + u\cos(0{,}986j) + v\sin(0{,}986j) \tag{3}$$

The TL value must be corrected according to altitude (-0,35 par 1000 m).

Zone		TO	u	v
Mediterranean	coast	3,25	-1,1	-0,15
	inside	3,15	-0,5	-0,05
Atlantic		3,25	-0,7	-0,15
Continental		3,75	-0,2	-0,05
Great agglomerations		4,05	-0,5	-0,10

Table 2. Values of the coefficients TO, u et v.

3.3.3 Equation of time (Et) [21]

$$Et = 0{,}123\,(t + 87) - \frac{1}{6}\sin(2(t + 10)) \tag{4}$$

$$t = 0{,}988\,[\,D + 30{,}3\,(\,M - 1\,)\,] \tag{5}$$

3.3.4 Real solar time (TSV)

$$TSV = TU + \frac{Lo}{15} + Et \tag{6}$$

3.3.5 Hour Angle (ω)

$$\omega = 15\,(TSV - 12) \tag{7}$$

3.3.6 Declination (δ)

$$\delta = 23{,}45\sin(0{,}986j - 80) \tag{8}$$

3.3.7 Height or the angle between the direction of the sun with its projection (γ) [23]

$$\sin(\gamma) = \sin(\delta)\sin(La) + \cos(\delta)\cos(La)\cos(\omega) \tag{9}$$

3.3.8 Optical way (AM)

$$AM = \frac{(1 - 0{,}1\ .z))}{\sin(\gamma)} \tag{10}$$

3.3.9 Direct solar irradiance received on a normal surface with the rays (E)

$$E = Bo\ r\ \exp\left[\frac{-\ AM\ TL}{0{,}9\ AM + 9{,}4}\right] \tag{11}$$

Bo = 1367 W/m^2.

3.4 Conversion of solar energy

Solar energy can be converted using appropriate converters in other forms of energy such as electrical, mechanical, thermal, etc.. In this case, we are interested to the thermal conversion. We distinguish then two conversion modes: at low temperature, where the fluid temperature to heat remains below 100 ° C and at medium and high temperature when it exceeds 100 ° C. For the first case, the temperature level is achieved by means of a flat plate collector, whereas in the second case, a concentrator collector is required.

Several types of concentrators have been made to this day which we quote:

- the conical concentrator,
- the spherical concentrator,
- the cylindro-parabolic concentrator and
- the parabolic concentrator.

This last which will be the subject of this study, was used for a long time, because it ensures an high level of temperature and power [24]. A parabola is the whole of the points located at equal distance from a line fixes D, called director of the parabola, perpendicular to axis (OZ), passing by the X-coordinate (- F) and the focal point F (fig.11). Any luminous ray parallel with axis (OZ) will be deviated by the reflective surface of the paraboloid towards the point F of X-coordinate (F) representing the focal distance. The equation giving the focal distance (F) is [25]:

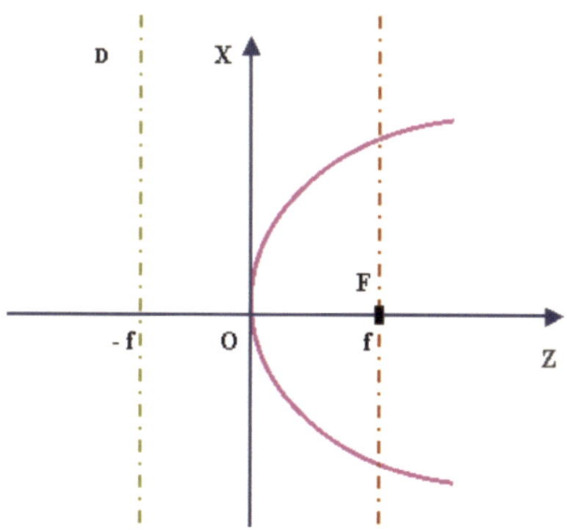

Fig. 11. Status of the paraboloid from the director D and demonstration of the focus F.

$$f = \frac{a}{4tg(\frac{\Phi o}{2})} \quad (12)$$

The depth of the paraboloid (Po) is given by:

$$Po = \frac{a^2}{16f} \quad (13)$$

3.4.1 Concentration of the solar radiation

a. Geometrical concentration (Cg)

The geometric concentration is defined as the ratio between collector area (S) of the concentrator obtained by orthogonal projection of the pupil surface on a perpendicular plane to the incident radiation and surface receptor (s) of the converter [26]. To determine the equation of the geometric concentration factor, we proceed as follows: Consider a paraboloid which approaches the stigmatism (small diameter of the paraboloid and low angle of incidence [27]), in I point, it receives a conical light beam of angle ε (fig.12).

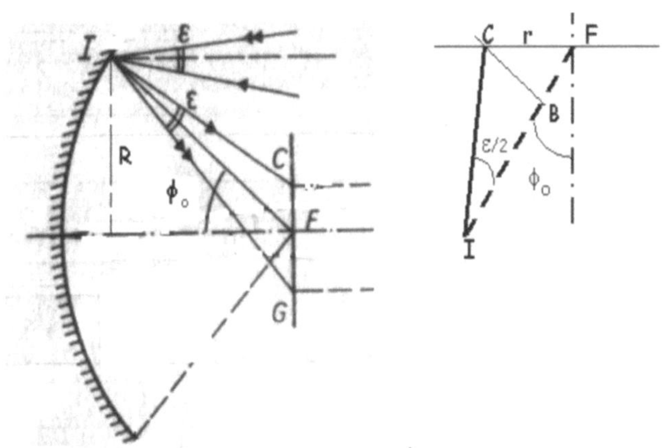

Fig. 12. Image of a solar beam on the focal plane[21].

As defined, the expression of the geometric concentration factor is presented in the following form:

$$C_g = \frac{S}{s} = \frac{\pi R^2}{\pi r^2} \quad (14)$$

When we replace (r) and (R) by their expression in equation (14), we find:

$$C_g = \frac{\sin^2(2\varphi_o)}{4tg^2(\varepsilon/2)}(1+tg(\varphi_0)tg(\frac{\varepsilon}{2}))^2 \quad (15)$$

b. Effective geometrical concentration (Cge)

The effective geometrical concentration (Cge) is the ratio between the surface of pupil (So) and the absorbing surface (so) [5]. In the case of a pointed concentrator with a flat absorber, the effective geometric concentration is equal to the geometric (Cg = Cge).

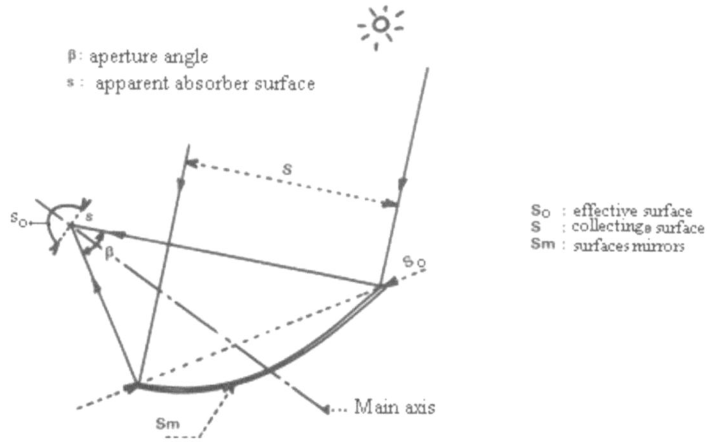

Fig. 12′. Diagram of a mirror concentrator

3.4.2 Heat balance of a concentrator

The heat balance of a solar concentrator in permanent pattern is the following:

$$Pa = Pu + Pe \tag{16}$$

The power of absorbed radiation can be calculated by:

$$Pa = ECgs\rho\gamma\tau\alpha \tag{17}$$

The coefficient of interception is given by the following equation [28,29]:

$$\gamma = 1\text{-}\exp\left[-820\left(0{,}7\frac{r}{f}\right)^2(1+\cos(\varphi_o))\right] \tag{18}$$

If we consider that on the level of the absorber we have only one transfer of energy without phase change, the useful power [30] will be:

$$Pu = qc(T_S - T_e) \tag{19}$$

The heat losses of the converter are given by:

$$Pe = Cps(T_{moy} - T_a) \tag{20}$$

The average temperature T_{moy} of the absorber is expressed according to the exit and inlet temperatures of the coolant and the temperature gradient heat transfer between the absorbing surface and the coolant ΔT. This is given by the following expression:

$$T_{moy} = \frac{T_s + \frac{T_s + T_e}{2}}{2} + \Delta T \tag{21}$$

Substituting Pa, Pu and Pe in equation (22) by their expressions, we find:

$$ECgs\rho\tau\alpha\gamma = q_c c_c(T_S - T_e) + Cps(0{,}25(3Ts+Te) + \Delta T - T_a) \tag{22}$$

The instantaneous global efficiency of the concentrator η_g is determined by:

$$\eta_g = \frac{P_u}{C_g s E} \tag{23}$$

By replacing P_u by $EC_g s\rho\tau\alpha\gamma$ - $Cp(T_{moy}-T_a)s$ in equation (23), we obtained:

$$\eta_g = \alpha\rho\tau\gamma - Cp\,\frac{(T_{moy} - T_a)}{C_g\,E} \tag{24}$$

We note here, that when the geometric concentration is higher, the instantaneous global efficiency is better. The instantaneous internal efficiency η_i of the concentrator is:

$$\eta_i = \frac{P_u}{P_a} \tag{25}$$

After replacement of Pu and Pa by their expression we obtained:

$$\eta_i = 1 - Cp\frac{(T_{moy} - T_a)}{EC_g\rho\alpha\tau\gamma} \tag{26}$$

a. Conductance of the thermal losses [23,30,31]

The thermal losses are located for a concentrator on the level of the absorber. These losses are by convection and radiation on the illuminated face and by conduction on the no enlightened face. The conductance of the thermal losses is given by the following equation:

$$Cp = hr + hcv + har \tag{27}$$

b. Exchange coefficient by conduction

The coefficient of exchange by conduction is given by:

Absorber cavity shape

$$har = \frac{so\,\lambda ar\,(Tmoy - Tar)}{s\,e\,(Tmoy - Ta)} \tag{28}$$

Absorber plane shape

$$har = \frac{\lambda ar\,(T_{moy} - T_{ar})}{e\,(T_{moy} - T_a)} \tag{29}$$

c. Exchange coefficients by radiation and convection

The exchange coefficients by radiation and convection are given following the case where the absorbing surface is protected or not.

Unprotected absorbent surface on the enlightened face. (fig. 13)

The following diagram shows the case of an unprotected absorber.
In this case, the coefficient of exchange by radiation is given by the following expression for the two shapes of the absorber:

$$h_r'' = \varepsilon_a\,\sigma\,\frac{(T_{moy}^4 - T_v^4)}{(T_{moy} - T_a)} \tag{30}$$

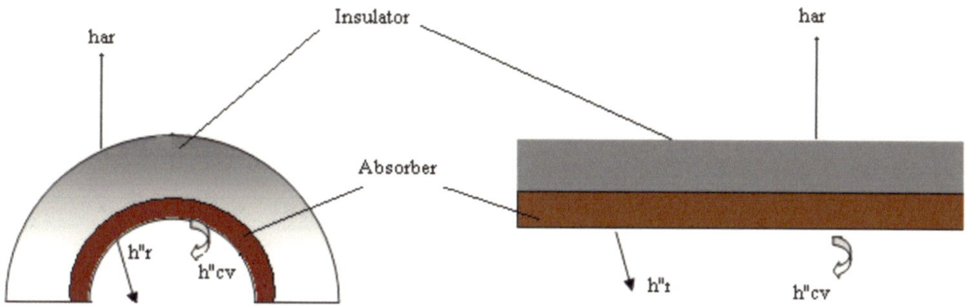

a / cavity shape b / plane shape

Fig. 13. Unprotected absorber.

For an absorbing surface with $100 < T_{moy} < 500°C$, the exchange coefficient by convection is given by:

Absorber cavity shape

for $0 < V < 4$ m/s

$$h''_{cv} = 7{,}5 + 4V \tag{31}$$

for $4 \leq V < 40$ m/s

$$h''_{cv} = 7{,}3V^{0{,}8} \tag{32}$$

Absorber plane shape

for $0 < V < 4$ m/s

$$h''_{cv} = 2{,}2(T_{moy} - T_a)^{0{,}25} + 4V \tag{33}$$

for $4 \leq V < 40$ m/s

$$h''_{cv} = 7{,}5V^{0{,}8} \tag{34}$$

Protected absorbing surface : greenhouse effect- (figure 14)

In this case, we have partial conductances presented in the following diagram:

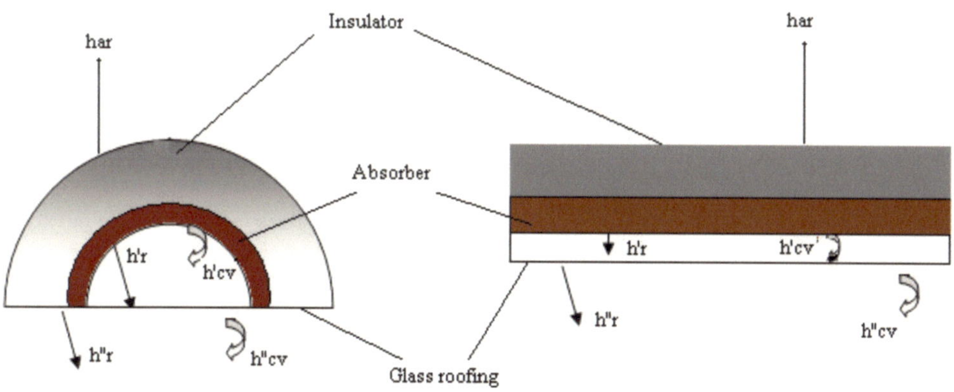

a / cavity shape b / plane shape

Fig. 14. Protected absorber (greenhouse effect).

The expressions which give these conductances are the following ones:

Absorber cavity shape

$$h_r^{'} = \varepsilon_{ac}\,\sigma\,\frac{(T_{moy}^4 - T_{cmoy}^4)}{(T_{moy} - T_{cmoy})} \tag{35}$$

With

$$\varepsilon_{ac} = \frac{1}{\frac{1}{\varepsilon_a} + \frac{1}{\varepsilon_c} - 1}$$

$$h'cv = 3{,}30\ W/m^2°C \tag{36}$$

$$h_r^{''} = \frac{\varepsilon_a\,\sigma\,(T_{cmoy}^4 - T_v^4)}{(T_{cmoy} - T_a)} \tag{37}$$

$$h''cv = 7{,}5 + 4V \tag{38}$$

for $0 < V < 4$ m/s

or

$$h''cv = 7{,}3V^{0{,}8} \tag{39}$$

for $4 \leq V < 40$ m/s

Absorber plane shape

$$h_r^{'} = \varepsilon_{ac}\sigma\frac{(T_{moy}^4 - T_{cmoy}^4)}{T_{moy} - T_{cmoy}} \tag{40}$$

$$h'cv = 1{,}1\ (T_{moy} - T_{cmoy}) \tag{41}$$

$$h_r^{''} = \varepsilon_a\,\sigma\frac{(T_{cmoy}^4 - T_v^4)}{(T_{cmoy} - T_a)} \tag{42}$$

$$h_{cv}^{''} = 2{,}2\ (T_{moy} - T_a)^{0{,}25} + 4V \tag{43}$$

For $0 < V < 4$ m/s

or

$$h_{cv}^{''} = 7{,}5\ V^{0{,}8} \tag{44}$$

For $4 \leq V < 40$ m/s

The expression of the resulting coefficient of exchange for the front face is given by:

$$hr + hcv = \frac{1}{\frac{1}{h'r + h'cv} + \frac{1}{h''r + h''cv}} \tag{45}$$

3.4.3 Example

The thermal conversion of solar energy by means of solar concentrators makes it possible to reach high temperatures able to boil the salted water with pressures higher or equal to the atmospheric one. In order to test these concentrators in the brackish water desalination field, Chaouachi and al. have designed, dimensioned and built in their laboratory a small solar desalination unit equipped with a paraboloid concentrator.

The experimental device shown in Fig. 1, is composed of a solar parabolic collector type. The support of the parabolic concentrator is made of molded fiberglass with 1.8 m in dish aperture diameter. This dish surface was covered with rectangular stainless steel sheet segments with a thickness of 1 mm [32,33]. The obtained reflecting surface presents some imperfections due to the imperfect attachment of the sheets to the mother dish. The absorber is mounted at its focus, which is shaped like a cylindrical vase, with a receiving surface of 0.013 m2 and a geometric concentration of 195. This absorber is completely insulated except the part lit by the solar rays reflected by the parabolic surface. The sun tracking mechanism for this solar distiller has two axes according to previous research [8,9] and it is a manual system. The brackish water supply to the absorber is kept continuous in order to keep a constant volume of water in the absorber. The steam produced passes in a coil condenser where it is condensed. The cooling water circulates with counter flow in the shell and with a salt water flow of 40 L/h. Condensation is made inside the horizontal copper tubes and it is out of film. In the same way, it works at the atmospheric pressure and it is followed by a stage of under-cooling.

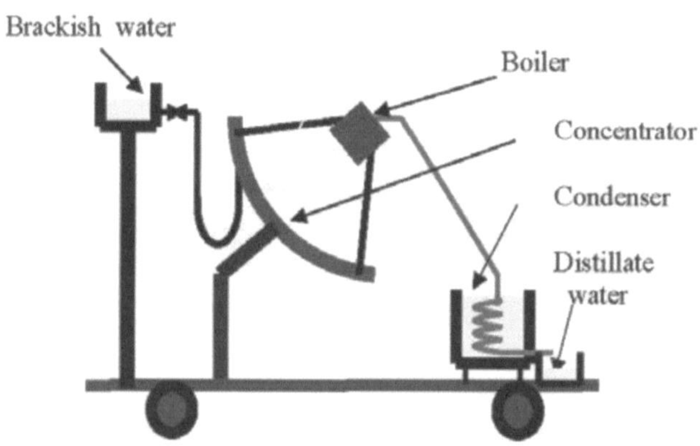

Fig. 15. Experimental device.

The capacity of production can reach 4,5 l/m^2.day.

4. Conclusion

The lack of drinking water, exhaustion and the high cost of energy and pollution problem, encourage us to find appropriate solutions to meet these challenges. The desalination of brackish or sea water using renewable energy such as solar energy represents a promising way. Future research in this area is oriented toward the maximum utilization of solar energy, which is free and clean, or through technological innovation and/or improvements on conventional desalination processes.

Nomenclature

Capital letters :

Cp	Apparent conductance of heat loss ($W/m^2°C$).
D	Day of the month
La	Latitude (degree).
Lo	Longitude (degree).
M	Number of months
P	Vapour pressure (Pa)
Pa	Incident power of absorbed radiation (W/m^2)
Pe	Power of heat loss (W/m^2)
Pu	Useful power (W)
R	Radius of the pupil surface (m)
S	Collecting area (m^2)
T	Temperature (K)
TU	Universal Time (h)
V	Wind speed (m/s)

Small letters:

a	Aperture diameter of the paraboloid (m).
c	The specific heat (J/kg°C).
e	Thickness of the insulation on the back of the absorbers (m).
f	Focal or friction factor.
h	Exchange coefficient ($W/m^2°C$).
h'	Internal heat transfer coefficient ($W/m^2°C$).
h''	External heat transfer coefficient ($W/m^2°C$).
qc	Mass flow of coolant (kg/s).
R	Radius of the absorber or correction of the earth-sun distance (m).
so	Surface receptor (m^2).
s	Collecting surface (m^2)
z	Altitude (km).

Greek letters

ϕo	Aperture Half angle of the paraboloid (degree).
α	Absorption coefficient of the absorber (%).
ε	The angle of a conical light beam (degree).
εa	Emissivity of the absorber (%).
εc	Emissivity of the cover (%).
εac	Apparent emissivity of the system (%).
λ	Thermal Conductivity (W/m°C).
ρ	Reflection coefficient of the paraboloid (%).
σ	STEFAN-BOLTZMANN constant.
τ	Transmittivity of the cover (%).
ω	Hour Angle (degree).

Indices :

a	Absorber or ambient.
ar	Rear wall insulation.

cmoy	Average cover.
moy	Average absorber
cv	Convection
r	Radiation.
s	Fluid outlet of the concentrator.
v	Steam or vault surrounding.

5. References

[1] G.N.Tiwari, H.N.Singh, R.Tripathi (2003) .Present status of solar distillation. *Solar Energy,* 75, pp. 367-373.

[2] L.Zhang, H.Zheng, Y.Wu (2003). Experimental study on a horizontal tube falling film evaporation and closed circulation solar desalination system. *Renewable Energy,* 28, pp. 1187-1199.

[3] R.DESJARDINS (1988). *Traitements des eaux.* 2ème édition, Editions de l'école polytechnique de Montréal.

[4] A.Al-kharabshesh, Y. Goswami (2003). Experimental study of an innovative solar water desalination system utilizing a passive vacuum technique. *Solar Energy,* 75, pp. 395-401.

[5] S.K. Shukla, V.P.S. Sorayan (2005). Thermal modeling of solar stills: an experimental validation, *Renewable Energy,* 30, pp. 683-699.

[6] H.D. Ammari, Y.L. Nimir (2003). Experimental and theoretical evaluation of the performance of a tar solar water heater. *Energy Conversion and Management,* 44, pp. 3037-3055.

[7] S.A. Kalogirou (2004). Solar thermal collectors and applications. *Progress in Energy and Combustion Science,* 30, pp. 231-295.

[8] R.Y. Nuwayhid, F. Mrad, R. Abu-Said (2001). The realization of a simple solar tracking concentrator for university research applications. *Renewable Energy,* 24, pp. 207-222.

[9] H.E.S. Fath(1998). *Desalination,* 116, 45.

[10] E. Delyannis, and V. Belessiotis, Mediterranean. Conference on Renewable Energy Sources for Water Production. European Commission, EURORED Network, CRES, EDS, Santorini, Greece, 1996, pp. 3-1 9.

[11] E.E. Delyannis(1987). *Desalination,* 67, pp. 3.

[12] J. GIRI, B. MEUNIER, (1980). *Evaluation des énergies renouvelables pour les pays en développement.* Volume 2, Commissariat à l'énergie solaire, France, pp. 194, 199-201.

[13] J. R. VAILLANT, (1978). *Utilisation et promesses de l'énergie solaire.* EYROLLES, Paris, pp. 178,183.

[14] A. A. M. SAYIGH, (1977). *Solar energy engineering.* Academic press, New York, pp. 434, 437,449,455,459.

[15] F. BEN JEMAA et al, (1998). Desalting in Tunisia : Past experience and future prospects. *Desalination* 116, pp. 124.

[16] F. BEN JEMAA et al, (1998). Potential of renewable energy development for water desalination in Tunisia. *Renewable energy,* December, pp. 6.

[17] I. HOUCINE et al, (1999). Renewable energy sources for water desalting in Tunisia. *Desalination,* 125, p p. 126.

[18] N. COUFFIN, C. CABASSUD et V. LAHOUSSINE-TURCAUD, (1998). A new process to remove halogenated VOCs for drinking water production: vacuum membrane distillation. *Desalination*, 117 pp. 233-245

[19] D. WIRTH, (2002). *Etude de la distillation pour le dessalement de l'eau de mer*, Thèse de Doctorat, Institut National des Sciences Appliquées de Toulouse.

[20] D. WIRTH, C. CABASSUD, (2002). Water desalination using membrane distillation: comparison between inside/out and outside/in permeation. *Desalination*, 147, pp. 139-145.

[21] R. BERNARD, G. MENGUY, M. SCHARTZ, (1980). *Le rayonnement solaire conversion thermique et applications*. 2ème édition, Technique et documentation, Paris, pp. 30,39,149,197.

[22] B. BOURGES, L. BERTOLO, (1992). *Données climatiques utilisées dans le bâtiment*. Technique de l'ingénieur, B 2015, Paris, pp. 22.

[23] A. A. SFEIR, G. GUARRACINO, (1981). *Ingénierie des systèmes solaires applications à l'habitat*. Technique et documentation, Paris, pp. 55.

[24] J. GLEN, K. LOVEGROVE, A. LUZZI, (2003). Optical performance of spherical reflecting elements for use with paraboloidal dish concentrators. *Solar energy*, 74, pp. 133.

[25] R. HOUZE, (1989).*Les antennes du fil rayonnant à la parabole*, Tome 2, EYROLLES, Paris, pp. 150,154.

[26] R. PASQUETTI, (1987). Chauffage des fluides par capteurs solaires à concentration. *Technique de l'ingénieur*, B 2420, Paris, pp. 4-7,13,16.

[27] M. HENRY, (1981). Optique géométrique. *Technique de l'ingénieur*, A 190, Paris, pp. 5.

[28] P. GALLET, F. PAPINI, G. PERI, (1980). *Physique des convertisseurs héliothermiques*. EDISUD, Aix en Provence, pp. 131,135,136,144,145.

[29] J. DUFFIC, B. WILLIAM, (1974). *Solar energy thermal processes*. John Wiley & Sons inc, New York, pp. 191,194,196.

[30] J. DESAUTEL, (1979). *Les capteurs héliothermiques*. EDISUD , Paris, pp.16-19, 80- 83.

[31] A. S. KENKARE, J. P. YIAMMOULLOU, (1983). The performance of a concentrating solar collector in UK weather conditions. *Solar world congress*. 2, Pergamon press,U.K., pp. 1043.

[32] R.Y. Nuwayhid, F. Mrad and R. Abu-Said, (2001) .The realization of a simple solar tracking concentrator for university research applications. Renewable Energy, 24, PP. 207–222.

[33] V.V. Pasichny and B.A. Uryukov, (2002). Theoretical aspects for optimization of solar radiation concentrators with plane facets. Solar Energy, 73, pp.239.

[34] B. Chaouchi, A. Zrelli, S. Gabsi (2007). Desalination of brackish water by means of a parabolic solar concentrator. Desalination 217, pp. 118–126.

11

Reject Brine Management

Muftah H. El-Naas
United Arab Emirates University
UAE

1. Introduction

Desalination has been growing rapidly as an industry and as a field of research that combines engineering and science to develop innovative and economical means for water desalting. Many countries in the world, especially in the Middle East, depend heavily on seawater desalination as a major source of drinking water and have invested considerable efforts and financial resources in desalination research and training. Desalination plants have seen considerable expansion during the past decade as the need for potable water increases with population growth. It is estimated that the world production of desalination water exceeds 30 million cubic meters per day and the desalination market worldwide is expected to reach $ 30 billion by 2015.

One of the major economical and environmental challenges to the desalination industry, especially in those countries that depend on desalination for potable water, is the handling of reject brine, which is the highly concentrated waste by-product of the desalination process. It is estimated that for every 1 m^3 of desalinated water, an equivalent amount is generated as reject brine. The common practice in dealing with these huge amounts of brine is to discharge it back into the sea, where it could result, in the long run, in detrimental effects on the aquatic life as well as the quality of the seawater available for desalination in the area.

Although technological advances have resulted in the development of new and highly efficient desalination processes, little improvements have been reported in the management and handling of the major by-product waste of most desalination plants, namely reject brine. The disposal or management of desalination brine (concentrate) represents major environmental challenges to most plants, and it is becoming more costly. In spite of the scale of this economical and environmental problem, the options for brine management for inland plants have been rather limited. These options include: discharge to surface water or wastewater treatment plants; deep well injection; land disposal; evaporation ponds; and mechanical/thermal evaporation. Reject brine contains variable concentrations of different chemicals such as anti-scale additives and inorganic salts that could have negative impacts on soil and groundwater.

This chapter highlights the main concerns as well as the environmental and economical challenges associated with the generation of large amounts of reject brine as a by-product of the desalination process. The chapter also outlines and compares the most common options for the treatment or disposal of reject brine. The chapter focuses on a novel approach to the management of reject brine that involves chemical reactions with carbon dioxide in the presence of ammonia, based on a modified Solvay process. Reject brine is mixed with ammonia and then exposed to carbon dioxide using different contact techniques. The end result is the conversion of NaCl and CO_2 into a useful solid product, namely sodium bicarbonate, and the reduction of the salinity of the treated brine, which may then be used for irrigation. Besides brine management, the new approach will reduce the emissions of CO_2 as a major contributor to global warming. Carbon dioxide can be used as a pure gas from gas sweetening units or in the form of flue or exhaust gas from chemical or power plants.

2. Current brine disposal options

Since desalination processes generate considerable amounts of reject brine, the industry has adopted numerous disposal options that usually depend on the location of the desalination plant and type of process used. These options include: discharge to surface water or wastewater treatment plants; deep well injection; land disposal; evaporation ponds; and mechanical/thermal evaporation. Management of reject brine has recently become an increasingly difficult challenge due to many factors that include: growing number and size of desalination plants which limits disposal options; increased regulations of discharges that make disposal more difficult; increased public concern with environmental issues; increased number of desalination plants in semi-arid regions where conventional disposal options are limited (Mickley, 2006). Cost plays an important role in the selection of a brine disposal method and it is believed to range from 5% to 33% of the total cost of desalination (Ahmed et al, 2001). Mickley et al. (1993) identified the factors that influence the selection of a disposal method. These include the quantity and quality of the brine; composition of the concentrate; physical or geographical location of the discharge point of the concentrate; availability of receiving site, permissibility of the option, public acceptance, capital and operating costs, and ability for the facility to be expanded. The cost of disposal depends on the characteristics of reject brine, the level of treatment before disposal, means of disposal, volume of brine to be disposed of, and the nature of the disposal environment (Ahmed et al, 2001). A detailed review of the different brine disposal methods can be found in a report by Mickley (2001). The following sections will present a brief summary of the main brine disposal options and highlight the main drawbacks of each option.

2.1 Discharge into surface water

It has been a common practice for coastal desalination plants to dispose reject brine into the close-by surface water body, namely sea or ocean. For these plants, such disposal operation has always been deemed the most practical and least expensive. Costs for disposal are typically low provided that pipeline conveyance distances are not excessively long and the concentrate is compatible with the environment of the receiving water body. An assessment of salinity or TDS impact as well as those of specific constituents on the receiving stream must always be considered (Mickely et al, 2006). The main factors that determine the costs of reject brine discharge to surface water include: costs to transport the brine from the desalination plant to the surface water discharge outfall; costs for outfall construction and operation; and costs associated with monitoring the environmental effects of the brine discharge on the surface waters (Mickely et al, 2006).

The impact of brine disposal operations on coastal and marine environment is still largely unknown, but the high temperature and salinity associated with reject brine may have detrimental effects on marine life. Moreover, the high level of chemicals could reduce the amount of dissolved oxygen available for the marine organisms. Other harmful chemicals that may be present in the reject brine such as hydrogen sulfide and chloride may have negative effect if the brine is not treated before disposal. In addition, the continuous disposal of reject brine into water body near the desalination plants could, in the long run, affect the suitably of the feed water. This is especially true for small and rather closed water bodies such as the Arabian Gulf, where most of the desalination activities in the world take place.

2.2 Deep well injection

Deep well injection is often considered for the disposal of industrial, municipal and liquid hazardous wastes (Saripalli et al, 2000). In recent years, this approach has been given serious consideration as an option for brine disposal from inland desalination plants, where surface water discharge is not viable or very costly. Deep wells can offer a feasible and reliable solution to disposing reject brine. However, deep wells are not feasible in areas subject to earthquakes or where faults are present that can provide a direct hydraulic connection between the receiving aquifer and an overlying potable aquifer (Mickely et al, 2006). Therefore, prior to drilling any injection well, a careful assessment of geological conditions must be conducted in order to determine the depth and location of suitable porous aquifer reservoirs (Glator and Cohen, 2003). The capital cost for deep well injection is usually higher than surface water disposal, where the latter method does not require long brine transport pipelines. Although deep well injection may be a feasible option for reject brine disposal, it still suffers from many drawbacks such as the need for selecting a suitable well site; the extra costs involved in conditioning the reject brine; corrosion and subsequent leakage in the well casing; and seismic activity which could cause damage to the well and subsequently contamination of groundwater (Glator and Cohen, 2003). Performance, design consideration and modeling of deep well injection have been addressed by many researchers (Rhee and Reible, 1993; Saripalli et al, 2000; Skehan and Kwiatkowski, 2000).

2.3 Evaporation ponds

This option has always been considered the most effective and economical method for brine disposal for inland desalination plants, especially for dry, arid regions similar to those in North Africa and Middle East. Inland plants in these regions are usually located in areas known to have high dry weather, relatively high temperature and, consequently, high evaporation rates. Ahmed et al. (2000) reviewed the relevant literature and presented the design aspects of evaporation ponds, highlighting the importance of selecting the main design parameters, namely surface area and pond depth. In another study (Ahmed et al, 2001), the authors surveyed the application of evaporation ponds in Arabian Gulf countries, namely United Arab Emirates and Oman. The authors reported that the newer plants have lined evaporation ponds, whereas the older ones have unlined disposal pits. The primary environmental concern associated with evaporation pond disposal is pond leakage, which may result in subsequent contamination of groundwater in the region. Recent evaporation ponds are always lined with polyethylene or other polymeric materials to prevent leakage and seepage of contaminants into the nearby groundwater.

A key factor in the effectiveness of evaporation ponds is the evaporation rate, which depends heavily on the weather conditions, mainly humidity and surrounding temperature. Attempts have been made, with limited success, to improve evaporation through the use of wind-aided intensified evaporation (Gilron et al, 2003). This technique claims to increase the evaporation rate by 50% for dry climate, but still depends on weather conditions. Improving the evaporation rate could in principal reduce the size of the evaporation ponds and enhance their efficiency and potential of application in many parts of the world. Although high temperature and, consequently, high evaporation rates may speedup water reduction, evaporation ponds still suffer from many drawbacks including the need for huge areas and the possibility of contaminants dissipation into soil and groundwater.

3. Characteristics of reject brine

By definition, brine is any water stream in a desalination process that has higher salinity than the feed. Reject brine is the highly concentrated water in the last stage of the desalination process that is usually discharged as wastewater. Several types of chemicals are used in the desalination process for pre- and post-treatment operations. These include: Sodium hypochlorite (NaOCl) which is used for chlorination to prevent bacterial growth in the desalination facility; Ferric chloride ($FeCl_3$) or aluminum chloride ($AlCl_3$), which are used as flocculants for the removal of suspended matter from the water; anti-scale additives such as Sodium hexameta phosphate $(NaPO_3)_6$ are used to prevent scale formation on the pipes and on the membranes; and acids such as sulfuric acid (H_2SO_4) or hydrochloric acid (HCl) are also used to adjust the pH of the seawater. Due to the presence of these different chemicals at variable concentrations, reject brine discharged to the sea has the ability to change the salinity, alkalinity and the temperature averages of the seawater and can cause change to marine environment. The characteristics of reject brine depend on the type of feed water and type of desalination process. They also depend on the percent recovery as well as the chemical additives used (Ahmed et al., 2000). Typical analyses of reject brine for different desalination plants with different types of feed water are presented in Table 2.1.

Parameters	Abu-fintas Doha/Qatar Seawater	Ajman BWRO	Um Quwain BWRO	Qidfa I Fujairah Seawater	Qidfa II Fujairah Seawater
Temperature, °C	40-44	30.6	32.4	32.2	29.1
pH	8.2	7.46	6.7	6.97	7.99
Electrical conductivity	NR	16.49	11.33	77.0	79.6
Ca, ppm	1,300-1,400	312	173	631	631
Mg, ppm	7,600-7,700	413	282	2,025	2,096
Na, ppm	NR	2,759	2,315	17,294	18,293
HCO_3, ppm	3,900	561	570	159	149.5
SO_4, ppm	3,900	1,500	2,175	4,200	4,800
Cl, ppm	29,000	4,572	2,762	30,487	31,905
TDS, ppm	52,000	10,114	8,276	54,795	57,935
Total hardness, ppm	NR	NR	32	198	207
Free Cl_2, ppm	Trace	NR	0.01	NR	NR
SiO_2, ppm	NR	23.7	145	1.02	17.6
Langlier SI	NR	0.61	0.33	NR	NR

Table 2.1. Characteristics of reject brine from desalination plants in the Gulf region (adapted from Khordagui, 1997). NR: Not reported; BWRO: brackish water reverse osmosis.

More data about the characteristics of reject brine and feed water for several desalination plants in Gulf counties such as Oman, UAE and Saudi Arabia can be found elsewhere (Ahmed et al, 2001; Mohamed et al, 2005).

4. Environmental impact of reject brine

Reject brine has always been considered as waste by-product of the desalination processes that can not be recycled and must be disposed of. Its harmful effects on the surrounding environment have always been underestimated in spite of the high concentrations of chemicals and additives used in the pretreatment of the feed water. Numerous studies have evaluated the environmental impact of reject brine disposal on soil, groundwater and marine environment. The surface discharge of reject brine from inland desalination plants could have negative impacts on soil and groundwater (Rao et al, 1990; Mohamed et al, 2005; Al-Faifi et al, 2010). Other researchers have highlighted the impact of reject brine composition and conditions on marine life (Lattemann and Hopner, 2005; Sadhawani et al, 2008). Sánchez-Lizaso et al (2008) have reported that the high salinity associated with reject brine discharges has detrimental effects on sea grass structure and vitality.

Soil deterioration and groundwater contamination is a major concern when reject brine is discharged into concentration ponds, which is the most common means of brine disposal for inland desalination plants. Disposal of reject brine into unlined ponds could have significant environmental impacts and the improper disposal has the potential for polluting the groundwater resources and can have a profound effect on subsurface soil properties (Mohamed et al, 2005). However, the environmental implications related to brine discharge have not been adequately considered by the concerned authorities. Mohamed et al (2005) have conducted a comprehensive evaluation of the impact of land disposal of reject brine from desalination plants on soil and groundwater. The authors assessed the effect of reject brine disposed directly into surface impoundment (unlined pits) in a permeable soil with low clay content, cation exchange capacity and organic matter content. The study indicated that concentrate disposal in unlined pond or pits can pose a significant problem to soil and feed water and can increase the risk of saline brackish water intrusion into fresh water. The authors recommended considering proactive approaches such as using lining systems, long term monitoring programs, and field research to protect groundwater from further deterioration. They have also highlighted the importance of implementing and enforcing regulations and polices related to reject brine chemical composition and concentrate disposal.

Soil structure may deteriorate due to the high salinity of the reject brine, when calcium ions are replaced by sodium ions in the exchangeable ion complex (Al-faifi et al, 2010). This in turn results in reducing the infiltration rate of water and the soil aeration. Sodium does not reduce the intake of water by plants, but it changes soil structure and impairs the infiltration of water and hence affects plant growth (Hoffman et al, 1990; Maas, 1990). In addition, the elevated levels of sodium, chloride, and boron associated with reject brine can reduce plants productivity and increase the risk of soil salinization (Maas, 1990).

5. A new approach to reject brine management

The current options for reject brine management are rather limited and have not achieved a practical solution to this environmental challenge. There is an urgent need, therefore, for the development of a new process for the management of desalination reject brine that can be used by coastal as well as inland desalination plants. The chemical reaction of reject brine with carbon dioxide is a new approach that promises to be effective, economical and environmental friendly (El-Naas et al, 2010). The approach utilizes chemical reactions based on a modified Solvay process to convert the reject brine into useful and reusable solid product (sodium bicarbonate). At the same time, the treated brackish water can be used for irrigation. Another advantage is that the main gaseous reactant, carbon dioxide, can be pure or in the form of a mixture of exhaust or flue gases, which indicates that this approach can be utilized for the capture of CO_2 from flue gases or sweetening of natural gas. El-Naas et al (2010) reported that the reactions of CO_2 with ammoniated brine can be optimized at 20 °C and can achieve good conversion using different forms of carbon dioxide. Details of this promising approach are presented in the next sections.

5.1 Solvay process

The Solvay process was named after Ernst Solvay who was the first to develop and successfully use the process in 1881. It is initially developed for the manufacture of sodium carbonate (washing soda), where a saturated sodium chloride solution -in the form of concentrated brine- is contacted with ammonia and carbon dioxide to form soluble ammonium bicarbonate, which reacts with the sodium chloride to form soluble ammonium chloride and a precipitate of sodium bicarbonate according to the following reactions:

$$NaCl + NH_3 + CO_2 + H_2O \rightarrow NaHCO_3 + NH_4Cl \tag{5.1}$$

$$2NaHCO_3 \rightarrow Na_2CO_3 + CO_2 + H_2O \tag{5.2}$$

$$2NH_4Cl + Ca(OH)_2 \rightarrow CaCl_2 + 2NH_3 + 2H_2O \tag{5.3}$$

The overall reaction can be written as:

$$2NaCl + CaCO_3 \rightarrow Na_2CO_3 + CaCl_2 \tag{5.4}$$

The resulting ammonium chloride can be reacted with calcium hydroxide to recover and recycle the ammonia according to Reaction 5.3. Although the ammonia is not involved in the overall reaction of the Solvay process, it plays an essential role in the intermediate reactions, especially Reaction (5.1). The ammonia buffers the solution at a basic pH; without the presence of ammonia, the acidic nature of the water solution will hamper the precipitation of sodium bicarbonate.

The sodium bicarbonate ($NaHCO_3$), which precipitates from Reaction (5.1), is converted to the final product, sodium carbonate (Na_2CO_3) at about 200 °C, producing water and carbon dioxide as byproducts (Reaction 5.2). A well designed and operated Solvay plant can reclaim almost all its ammonia, and consumes only small amounts of additional ammonia to make up for losses. The only major feeds to the Solvay process are sodium chloride (NaCl) and limestone ($CaCO_3$), and its only major byproduct is calcium chloride ($CaCl_2$), which is usually sold as road salt or desiccant.

In industrial practice, Reaction (5.1) is carried out by passing concentrated brine through two towers, where the brine is ammoniated in the first tower by bubbling ammonia gas through the saturated brine. In the second column, carbon dioxide is bubbled up through the ammoniated brine to form sodium bicarbonate and ammonium chloride. The worldwide production of soda ash in 2005 has been estimated at about 42 billion kilograms (Kostick, 2005).

5.2 Thermodynamic analysis

The overall reaction in the Solvay process is not spontaneous as is, but it must go through the three steps given in Reactions 5.1, 5.2 and 5.3. The first step (Reaction 5.1) is the most important one, since it involves the initial contact of the three main reactants (CO_2, NaCl and NH_3). The prime target of the Solvay process is the formation of sodium carbonate, but for brine management the aim is to convert water-soluble sodium chloride into insoluble sodium bicarbonate that can be removed by filtration.

A chemical reaction and equilibrium software, HSC Chemistry (Roine, 2007) was used to carry out a thermodynamic analysis for Reaction (5.1) to determine the equilibrium composition at different temperatures and to estimate the heat of reaction as a function of temperature. For a fixed temperature and pressure the number of moles present at equilibrium for any species can be determined using the Gibbs free energy minimization method. The analysis indicates that Reaction (5.1) is spontaneous for the whole temperature range (0 to 90 °C) as indicated by the negative ΔG. At 20 °C, the values for ΔH and ΔG are -129.1 kJ/mol and -25.8 kJ/mol, respectively. The calculated thermodynamic properties for Reaction (5.1) are presented in Table 5.1. The reaction proceeds through the following two steps:

$$NH_4OH + CO_2 \rightarrow NH_4HCO_3 \tag{5.5}$$

$$NaCl + NH_4HCO_3 \rightarrow NaHCO_3 + NH_4Cl \tag{5.6}$$

Temperature (°C)	ΔH (kJ/mol)	ΔS (kJ/mol. °C)	ΔG (kJ/mol)
0.0	-123.7	-332.4	-32.9
10.0	-129.4	-353.4	-29.3
20.0	-129.1	-352.4	-25.8
30.0	-128.8	-351.5	-22.3
40.0	-128.6	-350.6	-18.8
50.0	-128.3	-349.7	-15.3
60.0	-128.0	-348.9	-11.8
70.0	-127.7	-348.0	-8.3
80.0	-127.4	-347.2	-4.8
90.0	-127.1	-346.4	-1.3

Table 5.1. Thermodynamic data for Reaction (5.1)

Given its highly negative ΔH and ΔG (Table 5.2), Reaction (5.5) is an exothermic reaction that takes place as soon as the CO_2 gets in contact with the ammoniated brine. Once ammonium bicarbonate is formed, it reacts with sodium chloride according to Reaction (5.6). As can be seen from Table 5.3, Reaction (5.6) is not as spontaneous as Reaction (5.5) and it is believed to be the rate limiting step.

Temperature (°C)	ΔH (kJ/mol)	ΔS (kJ/mol. °C)	ΔG (kJ/mol)
0.0	-127.6	-241.6	-61.7
10.0	-129.5	-248.4	-59.2
20.0	-131.5	-255.1	-56.7
30.0	-133.4	-261.5	-54.1
40.0	-135.3	-267.8	-51.5
50.0	-137.2	-273.8	-48.7
60.0	-139.2	-279.7	-46.0
70.0	-141.1	-285.5	-43.2
80.0	-143.1	-291.0	-40.3
90.0	-145.0	-296.5	-37.3

Table 5.2. Thermodynamic data for Reaction (5.5)

The thermodynamic analysis indicates that Reaction (5.6) is exothermic with a negative heat of reaction up to a temperature of 40 °C. Beyond this temperature, the reaction becomes endothermic as shown in Table 5.3. This phenomenon was observed experimentally in a semi-batch reactor study (El-Naas, 2010). The reactor temperature was monitored with time and found to increase up to 41 °C, then drop and stabilize at 30 °C. Although this sudden change in the heat of reaction may be attributed to the reactor dynamics, a similar finding was reported by Yeh and Bai (1999) who attributed it to variations in the concentration of NH_3 in the solution. This, however, is unlikely to be the case, since the heat of reaction obtained by the thermodynamic analysis (Table 5.3) is per mol of NH_3, and it is only a function of temperature. The phenomenon is believed to be due to the mechanisms of Reaction (5.6).

Temperature (°C)	ΔH (kJ/mol)	ΔS (kJ/mol. °C)	ΔG (kJ/mol)
0.0	-6.3	-11.8	-3.1
10.0	-4.6	-5.5	-3.0
20.0	-2.8	0.6	-3.0
30.0	-1.1	6.5	-3.0
40.0	0.7	12.2	-3.1
50.0	2.5	17.8	-3.3
60.0	4.2	23.2	-3.5
70.0	6.0	28.5	-3.8
80.0	7.9	33.8	-4.1
90.0	9.7	38.9	-4.4

Table 5.3. Thermodynamic data for Reaction (5.6)

5.3 Role of ammonia

Although ammonia is a major reactant in the first step of the Solvay process, it can be fully recovered in the process and, therefore, it is not seen in the overall reaction. Ammonia buffers the solution at a basic pH of greater than 9 and hence allows the precipitation of $NaHCO_3$, which is less water-soluble in basic solution than NaCl. Only a small amount of ammonia is needed to raise the pH to above 9; the increase of pH beyond this point is a little slower as shown in Figure 5.1. In the absence of ammonia, the acidic solution will deter the precipitation of sodium bicarbonate regardless of the concentrations of other salts. This reiterates the importance of ammonia as a catalyst in Reaction (5.1) and the importance of controlling sodium bicarbonate solubility in the overall process, which will be discussed in the next section.

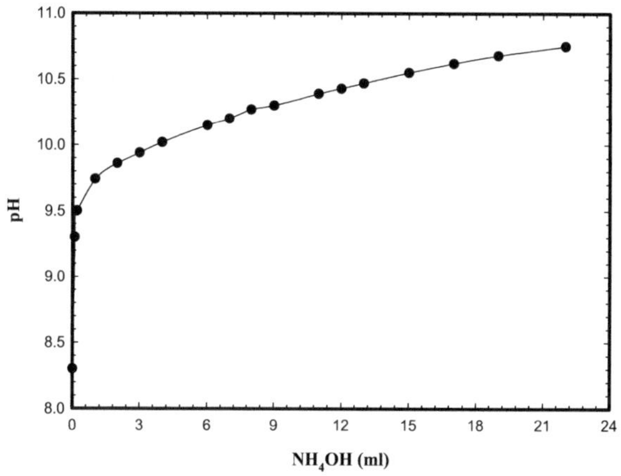

Fig. 5.1. Variation of solution pH with ammonia addition at 25 °C

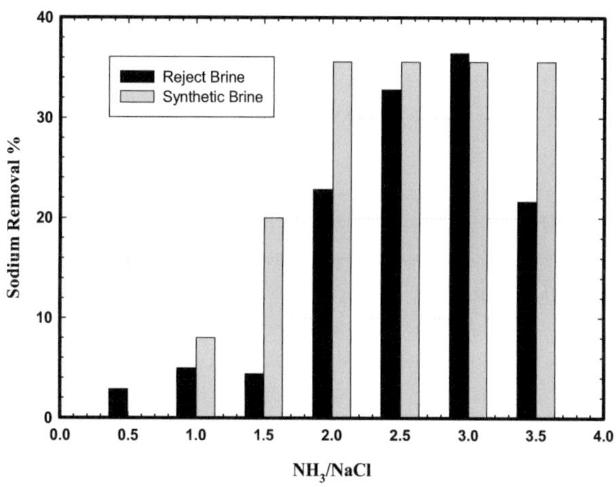

Fig. 5.2. Variation of sodium removal with $NH_3/NaCl$ molar ratio at 20 °C

It is important to note that the stoichiometric amount of ammonia required by Reaction (5.1) is one mole. However, in a real process excess ammonia may be needed for the reaction to reach completion. An experimental evaluation of the effect of excess ammonia on the removal of sodium at 20°C (El-Naas et al, 2010) indicated that the percent removal of sodium increased with increasing the $NH_3/NaCl$ ratio, reaching a maximum at 3 as shown in Figure 5.2. Similar experiments with synthetic brine solution, containing only NaCl in distilled water, in this study and in a previous study (Jibril and Ibrahim, 2001) revealed that the optimum sodium removal was achieved at a lower molar ratio ($NH_3/NaCl$) of 2. In both cases, the molar ratio is higher than that required stiochiometrically, which may be due to the fact that the reaction was carried out in a semi-batch reactor, where the CO_2 gas leaving the reactor stripped away some of the ammonia from the solution. This will not be the case for an industrial process, where the reactor will be run in a continuous mode and the ammonia is recycled within the system. As for the even higher molar ratio observed for the reject brine ($NH_3/NaCl=3$), it is believed to be due to the presence of other impurities in the brine.

Metal carbonates in the brine may compete for ammonia and reduce its availability for reaction with CO_2. Magnesium carbonate ($MgCO_3$), which is always present in the reject brine, consumes ammonia to form magnesium hydroxide and ammonium bicarbonate according to the following reaction:

$$NH_3 + MgCO_3 + 2H_2O \rightarrow NH_4HCO_3 + Mg(OH)_2 \quad (5.7)$$

Thermodynamic analysis of Reaction (5.7) indicates that this reaction is spontaneous for temperatures less than 22 °C. Thus one additional mole of ammonia is consumed by Reaction (5.7) to form magnesium hydroxide. This was confirmed experimentally, where milky colored turbidity was observed after mixing the reject brine with ammonium hydroxide.

It is worth noting here that after treatment of the reject brine through reactions with carbon dioxide, other ions such as Mg^{+2} and Ca^{+2} were significantly reduced at the end of the experimental runs. In fact, Mg^{+2}, Ca^{+2} and Sr^{+2} were reduced by more than 98%. Sodium (Na^+), which is the main focus of the treatment, was reduced by about 42% at the optimum conditions. This low reduction in sodium, however, is believed to only represent the conversion to insoluble sodium bicarbonate, which is removed by filtration. Since the amount of sodium in the filtrate comes from NaCl and soluble $NaHCO_3$, the true conversion can not be easily determined, and it is expected to be much higher than the 42%. Controlling the solubility of $NaHCO_3$, therefore, is a crucial step in optimizing the Solavy process for reject brine management.

5.4 Role of $NaHCO_3$ solubility

Sodium bicarbonate ($NaHCO_3$) is an important intermediate product in the Solvay process and its solubility plays an important role in the success of the process, since it determines the amount of the solid product that can be removed by filtration. For the process to achieve high conversion, the solubility of $NaHCO_3$ must be as low as possible. It is imperative, therefore, to evaluate factors that can limit or reduce its solubility. At room temperature, the solubility was determined experimentally to be about 9.75 g/100g and found to be negatively affected by the presence of other intermediates and reactants in Reaction (5.1) such as NaCl and NH_4HCO_3.

5.4.1 Effect of NaCl

The solubility of $NaHCO_3$ was found to decrease drastically with increasing the concentration of NaCl in the solution, from 9.75 g/100g at 0wt% NaCl to 3.6 g/100g at 10wt% NaCl as Shown in Figure 5.3. This is attributed to the presence of the sodium ion (Na^+) in the aqueous solutions of both salts. In aqueous solutions, both sodium chloride and sodium bicarbonate are present in their ionic format:

$$NaCl\ (a) \Leftrightarrow Na^+ + Cl^- \quad (5.8)$$

$$NaHCO_3\ (a) \Leftrightarrow Na^+ + HCO_3^- \qquad (5.9)$$

One would expect that increasing the concentration of the sodium ion (Na^+), by adding more NaCl into the solution, would force the equilibrium of Reaction (5.9) to the left and hence reduce the solubility of $NaHCO_3$. The solubility of NaCl in water at 25 °C is about 36 g/100g, which is almost four times that of $NaHCO_3$. The reduction in $NaHCO_3$ solubility with the presence of NaCl (Figure 5.3) seems to follow an exponential decay (**$y = 9.7e^{-0.095x}$**). According to this relation, the solubility of $NaHCO_3$ in a saturated NaCl solution will diminish to merely 0.3 g/100g. This highlights the necessity for using saturated brine in the Solvay process. It is to optimize the precipitation of $NaHCO_3$ by minimizing its solubility.

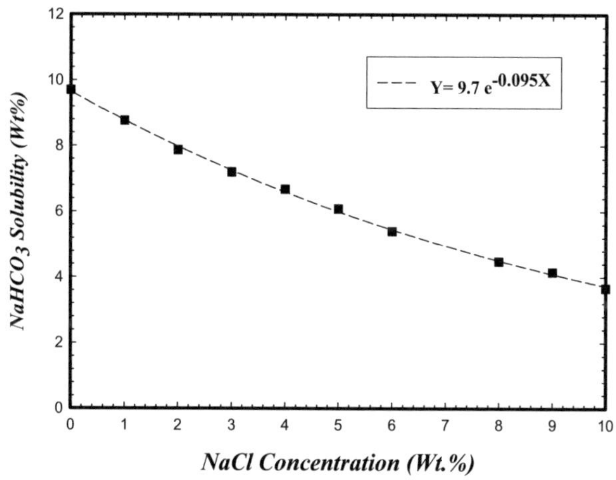

Fig. 5.3. Effect of NaCl on the solubility of $NaHCO_3$ at 25 °C

5.4.2 Effect of ammonium bicarbonate

Ammonium bicarbonate is another important intermediate in the formation of sodium bicarbonate according to Reactions 5.4 and 5.5. Its effect on the solubility of $NaHCO_3$ was evaluated for two aqueous solutions, containing 4% and 8% sodium chloride. The results are shown in Figure 5.4. Clearly, raising the concentration of ammonium bicarbonate seems to have a detrimental effect on the solubility of $NaHCO_3$. The rate of reduction in the solubility seems to be higher (about 33%) for the solution containing 8% NaCl. One may use similar argument to that used in the case of NaCl to explain this decline in the solubility. In this case, increasing the concentration of (HCO_3^-) by adding more ammonium bicarbonate would force the equilibrium in Reaction (5.11) below to the left and thus lower the solubility of $NaHCO_3$.

$$NH_4HCO_3\ (a) \Leftrightarrow NH4^+ + HCO_3^- \qquad (5.10)$$

$$NaHCO_3\ (a) \Leftrightarrow Na^+ + HCO_3^- \qquad (5.11)$$

The experimental results (Figure 5.4) indicate that for an aqueous solution containing 8% NaCl, the solubility of $NaHCO_3$ can be reduced to 0.0 g/100g with the addition of about 13wt% ammonium bicarbonate, which can definitely have significant effect on the possibility of using the Solvay process for reject brine management.

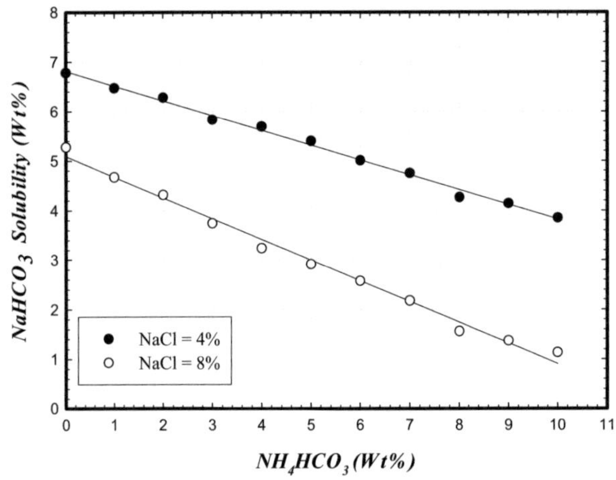

Fig. 5.4. Effect of NH_4HCO_3 on the solubility of $NaHCO_3$ at 25 °C

Ammonium chloride (NH_4Cl) is another byproduct formed in the Solvay process. Its effect on the solubility of $NaHCO_3$ was assessed in about the same way as that used with ammonium bicarbonate. The results, however, were not similar. The solubility of sodium bicarbonate does not seem to be affected by the presence of NH_4Cl regardless of the concentration of NaCl. This may be attributed to the fact that ammonium chloride is not involved in the formation of sodium bicarbonate and does not have any common ions with $NaHCO_3$; therefore, it does not affect its ionic equilibrium at these concentrations and temperature.

6. Industrial applications and CO_2 Capture

Application of the Solvay process for reject brine management has another important feature, which is the potential for carbon capture and storage (CCS). The process can be utilized for the removal of CO_2 from flue gases or for the sweetening of natural gas. Carbon dioxide is a major contributor to global warming and believed to have the greatest adverse impact on the observed greenhouse effect causing approximately 55% of global warming. The most common approach to CCS involves capturing CO_2 and then injecting it into rock layers in depleted or near-depleted oil and gas fields. The aim, off course, is to store the CO_2 and at the same time utilize it for Enhanced Oil Recovery (EOR). Although this option has gained the support of many industrialized and oil producing countries alike, it is not really problem-free and its long term effects are not yet known (El-Naas, 2008). Under typical storage conditions (1000 m below the surface), the density of CO_2 phase is approximately two-thirds that of the underground brine, which provides the driving force for escape (Bryant, 2007). Gradual seepage of CO_2 into the atmosphere may not pose much harm to human life, but it will certainly defeat the purpose of CCS.

Carbon dioxide reactions with ammoniated brine can offer a dual-purpose approach for the management of reject brine and capture of CO_2. The main unit of the process is the contact reactor, where the flue gases are contacted with the ammoniated reject brine. Other units include the ammoniating tank, where the high salinity water is mixed with ammonia gas; the ammonia recovery reactor, where the ammonia is recovered through reaction with calcium hydroxide; and a filter to separate the precipitated sodium bicarbonate from the rest of the solution. A schematic diagram of the process is shown in Figure 5.5. The carbon dioxide captured through this process is stored in the form of sodium bicarbonate.

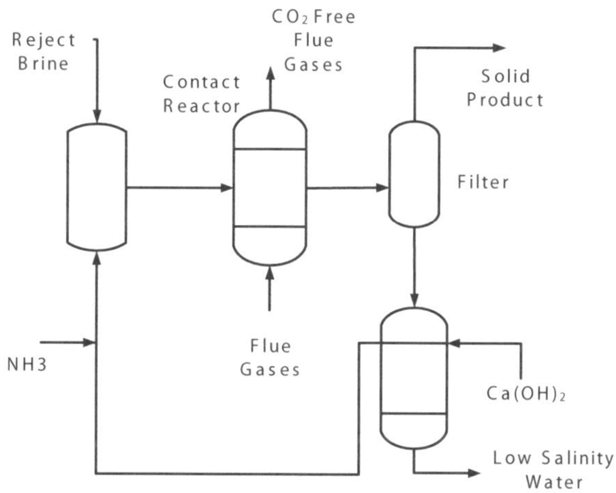

Fig. 5.5. A schematic diagram of a reject brine management process

The effectiveness of capturing CO_2 through the reaction with ammoniated brine was assessed experimentally. A gas mixture containing 10% CO_2 in methane was bubbled through one liter of ammoniated brine in three semi-batch bubble columns in series. The gas effluent of the first column was bubbled through the second and then the third. Half of the ammoniated brine was placed in the first column while the other half was divided equally between the other two columns. The total gas flow rate was controlled at 47 liter/hr using two mass flow controllers. The concentration of carbon dioxide and methane in the effluent gas stream were analyzed using a dual channel CO_2 and CH_4 infrared analyzer.

The experimental results for the CO_2 percent removal through the reaction with ammoniated reject brine solution are presented in Figure 5.6. It is evident that there is a considerable reduction in the CO_2 concentration in the effluent stream with 100% removal in the first two hours and more than 80% removal for the first five hours of run time. It is noticeable, nonetheless, that the percent removal is declining with time due to the consumption of the main reactants in the solution. Since the reactors were operated in the semi-batch mode, where only gases enter and leave the system, the other reactants in the ammoniated brine (NH_3 and NaCl) were consumed with time and hence less CO_2 was removed with time as shown in the figure. Although these results confirm the technical viability of the process for CO_2 capture and reduction of the reject brine salinity, more research is still needed to optimize the reactor design for continuous operation. An industrial process can be developed to offer an effective solution for the two major environmental challenges: reject brine management and CO_2 capture.

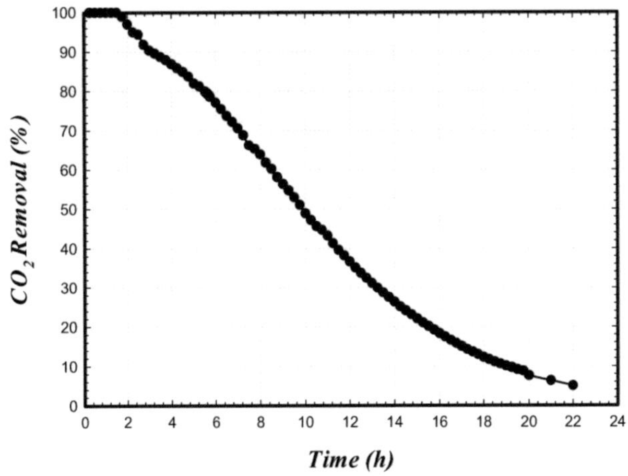

Fig. 5.6. CO_2 removal from a gas mixture containing 10% CO_2 in methane through reaction with ammoniated brine at 20 °C in a semi-batch three bubble columns in series.

7. Conclusions

Reject brine management represents a major environmental and economical challenge for most desalination plants. The current options for brine management are rather limited and have not achieved a practical solution to this environmental challenge. A new approach that involves reactions with CO_2 in the presence of ammonia has proven to be effective in reject brine management and capture of CO_2 .

8. References

Ahmed, M., W. H. Shayya, D. Hoey and J. Al-Handaly, "Brine disposal from reverse osmosis desalination plants in Oman and United Arab Emirates," Desalination 133, 135-147 (2001).

Ahmed, M., W. H. Shayya, D. Hoey, A. Maendran, R. Morris and J. Al-Handaly, "Use of evaporation ponds for brine disposal in desalination plants," Desalination, 130, 155-168 (2000).

Al-Faifi , H., A.M. Al-Omran, M. Nadeem, A. El-Eter , H.A. Khater , S.E. El-Maghraby, Soil deterioration as influenced by land disposal of reject brine from Salbukh water desalination plant at Riyadh, Saudi Arabia, Desalination 250 (2010) 479 - 484.

Yeh, A. C. and H. Bai, "Comparison of ammonia and monoethanolamine solvents to reduce CO greenhouse gas emissions", The Science of the Total Environment 228 (1999) 121-133.

El-Naas, M. H, A different approach for Carbon Capture and Storage (CCS), Research Journal of Chemistry and Environment, Volume 12, Issue 2, June 2008, Pages 3-4.

El-Naas, M. H., A. H. Al-Marzouqi, O. Chaalal, "A combined approach for the management of desalination reject brine and capture of CO_2", Desalination 251 (2010) 70–74.

Gilron, J., Y. Folkman, R. Savliev, M. Waisman, O. Kedem, "WAIV - wind aided intensified evaporation for reduction of desalination brine volume", Desalination, 158 (2003) 205.

Glater, J. and Y. Cohen, Brine disposal from land based membrane desalination plants a critical assessment, a report prepared for the Metropolitan Water District of Southern California, July 2003.

Hoffman, D., J.D. Rhodas, J. Letey and F. Sheng, Salinity management. In: G.J. Hoffman, T.A. Howell and K.H. Soloman, Editors, Management of Farm Irrigation Systems, American Society of Agricultural Engineers, New York (1990), pp. 667–715.

Jibril, B. and A. Ibrahim, "Chemical Conversions of salt concentrates from desalination plants" Desalination, 139 (2001) 287.

Khordagui, H.,Environmental aspects of brine reject from desalination industry in the SCWA region. ESCWA, Beirut, 1997.

Kostick, D., "Soda Ash", chapter in Minerals Yearbook, United States Geological Survey (2005).

Lattemann, S., T. Höpner, Environmental impact and impact assessment of seawater desalination, Desalination 220 (2008) 1–15.

Maas, E.V., Crop salt tolerance. In: K.K. Tanji, Editor, Agricultural Salinity Assessment and Management, Amercian Socity of Civil Enginers, New York (1990), pp. 262–303.

Mickley, M. C., Membrane Concentrate Disposal: Practice and Regulation,", Prepared for the U.S. Department of the Interior, Bureau of Reclamation, Technical Service Center, Water Treatment Engineering and Research Group, September 2001.

Mickley, M., Membrane Concentrate Disposal: Practices and Regulation, Second Edition. U.S. Department of the Interior, Bureau of Reclamation, Technical Service Center, Water Treatment Engineering and Research Group, April 2006.

Mickley, M., R. Hamilton, L. Gallegos and J. Truesdall, Membrane concentration disposal, American Water Works Association Research Foundation, Denver, Colorado, 1993.

Mohamed, A.M.O., M. Maraqa and J. Al Handhaly, Impact of land disposal of reject brine from desalination plants on soil and groundwater, Desalination 182 (2005), pp. 411–433.

Rao, N.S., R.T.N. Venkateswara, G.B. Rao and K.V.G. Rao, Impact of reject water from the desalination plants on ground water quality, Desalination 78 (1990), pp. 429–437.

Rhee, S-W, D. D. Reible and W. D. Constant, "Stochastic modeling of flow and transport in deep-well injection disposal systems," J. Hazardous Materials, 34, 313-333 (1993).

Roine, A., "Outokumpu HSC Chemistry for Windows", Ver. 6.12, User's guide, Outokumpu Research Oy, 2007.

Sadhwani, J. J., J. M. Veza, C. Santana, Case studies on environmental impact of seawater desalination, Desalination 185 (2005) 1–8.

Sánchez-Lizaso, J. L, J. Romero, J. Ruiz, E. Gacia, J. Buceta, O. Invers, Y. Torquemada, J. Mas, A.Ruiz-Mateoe, M. Manzanera, Salinity tolerance of the Mediterranean seagrass *Posidonia oceanica*: recommendations to minimize the impact of brine discharges from desalination plants, Desalination 221 (2008) 602–607.

Saripalli, K. P., M. M. Sharma and S. L. Bryant, "Modeling Injection Well Performance During Deep-Well Injection of Liquid Wastes," J. Hydrology, 227, 41-55 (2000).

Skehan, S and P. J. Kwiatkowski, "Concentrate Disposal via injection wells - permitting and design considerations," Florida Water Resources J., May 2000, 19-22.

12

DOE Method for Optimizing Desalination Systems

Amin Behzadmehr
Mechanical Engineering Department, University of Sistan and Baluchestan
I.R.Iran

1. Introduction

Fresh water production is one of the main concerns in the new century. Population grows fast and potable water resources are decreased. In the other hand energy crises would also be another issue that must be well addressed by the politicians and also scientists. Developing desalination plant with using renewable energy (particularly solar energy) is one of the important options to overcome these concerns. Thus many researchers have been working on different desalination plants to find the best conditions and to realize the most efficient performances for different cycles. Different approaches have been used to achieve the most efficient conditions or to find the optimum operation and design conditions. Some of the researchers used parametric study approach while many other adopted different conventional optimization algorithms for these tasks. The algorithms such as gradient based algorithm, genetic algorithm, search and pattern algorithm and neural network method have been used in the field of desalination. For instance; Ophir and Lokiec (2005) described the design principles of a MED plant and various energy considerations that result in an economical MED process and plant. Kamali and Mohebbinia (2007) showed that parametric study as one of the optimization methods on thermo-hydraulic data strongly helps to increase GOR value inside MED-TVC systems. Shamel and Chung (2006) used parametric study to find the optimum condition of a Reverse Osmosis (RO) system for sea water desalination. Metaiche et al (2008) developed optimization software, Desaltop, for RO system for water desalination. They used genetic algorithm to find suitable operating parameters and also to find appropriate type of membrane. Al-Shayji (1998) used neural network method for optimization of large-scale commercial desalination plants. Djebedjian et al. (2008) used genetic algorithm for optimization of a reverse osmosis desalination system. Mussati et al. (2003) used an evolutionary algorithm for the optimization of Multi Stage Flash (MSF) system. Finding the optimum conditions is a major challenge on the desalination plant studies. The plant performance depends on several different variables and constraints that need exhausting efforts to find the optimum conditions.
This chapter introduces Design of Experiment (DOE) method as a statistical tool for optimization of desalination systems. Thus two different desalination plants; Multi-Effect Desalination (MED) system and solar desalination using humidification–dehumidification cycle (SDHD) have been considered to show the ability of DOE method for optimizing such systems. These both desalination plants could use the low graded heating energy sources

such as solar energy. Thus it is very important to know the best thermodynamic conditions (variables) for which the desired objectives (objective functions) could be attained based on the technological and economical constraints. General perspective of effects of these thermodynamic conditions at different points of the plant on the rate of fresh water production and also the quantity and quality of heating energy sources would be very important and very useful for a plant designer. DOE method could show the optimal thermodynamic conditions of these systems.
Thus, first DOE method is briefly presented and then this method is adopted to investigate the MED plant and a solar desalination using humidification-dehumidification system.

2. Principles of Design of Experiments

Design of Experiment (DOE) is a statistical approach that could clearly show how much several parameters of a system as well as their interactions are important on the plant output and how these parameters could affect the objective function. DOE Method is capable to investigate simultaneously the effects of multiple parameters on an output variable (response). To illustrate statistically reasonable conclusions from the experiment, it is necessary to integrate an efficient statistical method into the methodology of experimental design. In the context of DOE in designing, one may encounter two types of plant variables or factors: qualitative and quantitative factors. For quantitative factors, the range of settings must be decided by designer. For instance, pressure, temperature or heat transfer surface are examples of quantitative factors. Qualitative factors are discrete in nature. For example, types of materials, nature of heating source, and types of equipments are examples of qualitative factors. A factor may take different levels, depending on the character of the factor- quantitative or qualitative. In general, compared to a quantitative factor, more levels are required by a qualitative factor. "Level" in this chapter refers to a specified value or setting of the factor that would be examined in the plant experiment. For instance, if the experiment is to be performed at three different pressures or using three different types of preheaters, then it could be said pressure or preheater has three levels. Three fundamental approaches on experimental design are replication, blocking, and randomization. The first two help to increase precision in the experiment; the last one is used to reduce bias. These three principles of experimental design can be used by industrial designer to improve the efficiency and performance of a product (see Behzadmehr et al. 2006a, 2006b). In addition these principles of experimental design are applied to decrease or even remove experimental bias. It should be mentioned that the large experimental bias could result in wrong optimal conditions or sometimes it could mask the effect of the really significant factors. This could cost lost of a primary factor for plant improvement.
Details and mathematical concepts of DOE are out of this chapter objective. The interested reader would find complete description of this method in the relevant text books such as Antony (2003) or book by Montogomery (2001).

3. Case I) Thermodynamic optimization of MED plant using DOE method

The multi-effect desalination (MED) plant is one of the most efficient thermal desalination processes currently in use. Development of MED in the last few years has brought this process to the point of competing economically and technically with the multi-stage flashing (MSF) process. MED process is based on the pressure reduction of water in each effect.

Many researchers have studied this process. Among them Sharmmiri and Safar (1999) discussed the general aspects of some commercial MED plants and also some of their specifications such as type of plant configuration, gain output ratio, number of effects, operating temperature and the construction material for the evaporator, condenser and preheaters. Ophir and Lokiec (2005) described the design principles of a MED plant and various energy considerations that result an economical MED process and plant. They also provided an overview of various cases of waste heat utilization, and cogeneration MED plants operating. Aybar (2004) considered a multi effect desalination system using waste heat of a power plant as energy source. A simple thermodynamic analysis of the system was performed with using energy and mass balance equations. Khademi et al. (2009) focused on the development of a steady-state model for the multi-effect evaporator desalination system. Hatzikioseyian et al. (2003) reviewed and developed a simulation program based on design parameters of the plant. They used mass and energy balance through each effect of the MED unit to predict the performance of the unit in terms of energy requirements. Kamali and Mohebbinia (2007) showed that parametric study as one of the optimization methods on thermo-hydraulic data strongly helps to increase GOR value inside MED-TVC systems. El-Nashar (2000) simulated part-load performance of small vertically stacked MED plants of the HTTF type using hot water as source of energy. Their model was validated with the experimental data obtained from an existing plant in operation. Narmine and El-Fiqi (2003) described a steady state mathematical model to analyze both multi-stage and multi-effect desalination systems. Relationships among the parameters which controlling the cost of fresh water production to the other operating and design parameters were presented. Parameters include plant performance ratio (PR), specific flow rate of brine, top brine temperature, and specific heat transfer area.

Here as an example the sensitivity of some important parameters on the minimum and maximum distilled water production is analyzed. Therefore the effects of feed water flow rate and its temperature, the number of effects, preheater temperature difference (performance of each preheater), and minimum pressure (at the end stage) are studied. Thus the mass, energy and salinity balance equations are solved for each effect to calculate temperature, pressure salinity of brine, enthalpies of outlet at each effect and mass flow rate of distilled water. For thermodynamic analysis and plant optimization, design of experiment method (DOE) is used to find the effective parameters on the minimum and maximum fresh water production.

3.1 Plant description

The multi effect desalination is an important process that has been used for desalination particularly in large scale plants. This method reduces considerably the production cost (Ophir and Lokiec 2005). The main parts of MED plant are: 1-Condenser, 2-Evaporator, 3-Collector (thermal source) and 4-Preheaters (heat exchanger). Each effect except the last one includes a preheater and an evaporator. The hot water circulates between the top effect evaporator and solar collector to supply thermal energy to the evaporator in the form of hot water flowing through the tube bundle of the first effect. The preheated sea water is sprayed on the tube bundle of first evaporator and vapour is generated. In the other effects, vapour is generated by both boiling and flashing processes. Pressure reduces on each effect to produce more vapours. Figure 1 shows a vertical configuration of MED plant.

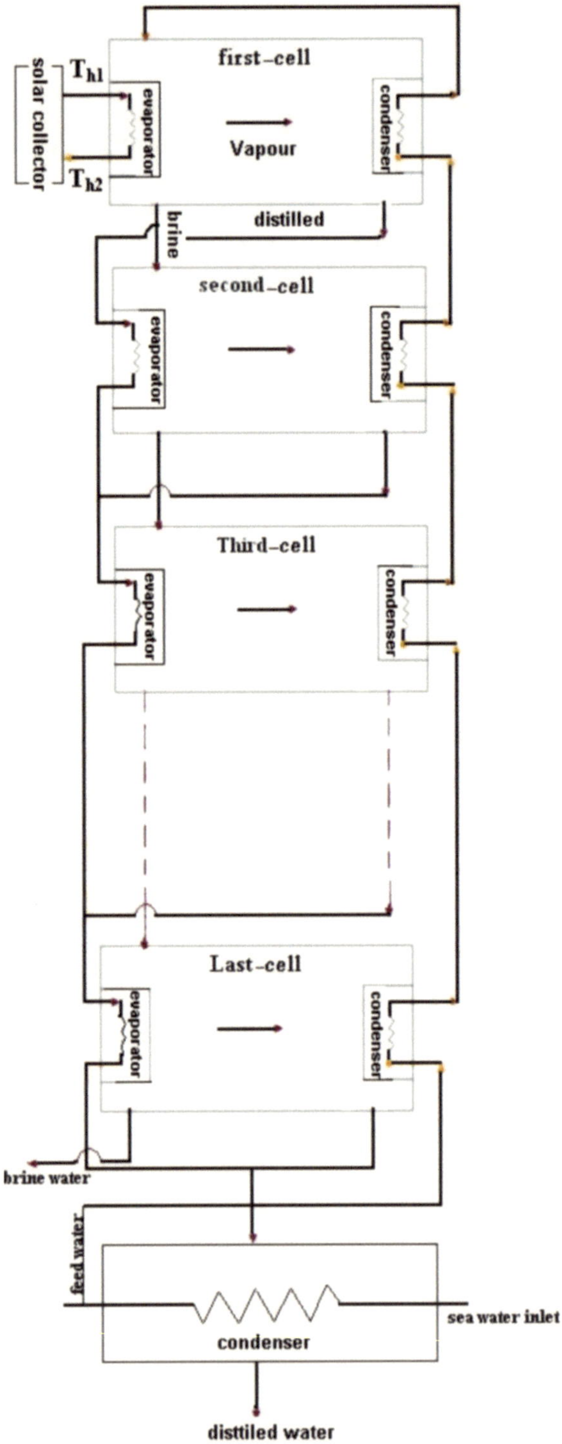

Fig. 1. Schematic diagram of MED plant

3.2 Mathematical modeling

A schematic of main multi effect desalination plant's parts is shown in Fig. 2. As seen in Fig. 2a the mass, energy and salinity balance equations for evaporator at the first effect (top effect) are as follow:

Mass balance equation:

$$\dot{m}_f = \dot{m}_b(1) + \dot{m}_v(1) \tag{1}$$

Balance of salinity:

$$\dot{m}_f X_{sw} = \dot{m}_b(1) X_b(1) \tag{2}$$

Balance of energy:

$$\dot{Q} + \dot{m}_f h_{fo}(1) = \dot{m}_b(1) h_b(1) + \dot{m}_v(1) h_v(1) \tag{3}$$

It is well known that the enthalpy of brine in each effect could be considered a function of salinity and temperature while the enthalpy of saturated vapour is only a function of temperature. Thus these equations are used to find temperature and salinity relationship. As shown schematically in Fig. 2b, preheaters consider being shell and tube heat exchanger. The mass and energy balance for all preheaters are as follow:

Mass balance equation

$$\dot{m}_d(i) = \dot{m}_v(i) \tag{4}$$

Energy balance equation:

$$\dot{m}_d(i) h_d(i) + \dot{m}_f h_{fo}(i) = \dot{m}_f h_{fi}(i) + \dot{m}_v(i) h_v(i) \tag{5}$$

Figure 2c shows the evaporator of the second effect. The mass, energy and salinity balance equations for this part are:

Mass balance equation:

$$\dot{m}_b(1) = \dot{m}_b(2) + \dot{m}_v(2)\,, \quad \dot{m}_{oe}(1) = \dot{m}_d(1) \tag{6}$$

Salinity balance equation:

$$\dot{m}_b(1) X_b(1) = \dot{m}_b(2) X_b(2) \tag{7}$$

Energy balance equation:

$$\dot{m}_b(1) h_b(1) + \dot{m}_d(1) h_d(1) = \dot{m}_b(2) h_b(2) + \dot{m}_v(2) h_v(2) + \dot{m}_{oe}(2) h_{oe}(2) \tag{8}$$

Energy balance equation:

$$\dot{m}_b(1) h_b(1) + \dot{m}_d(1) h_d(1) = \dot{m}_b(2) h_b(2) + \dot{m}_v(2) h_v(2) + \dot{m}_{oe}(2) h_{oe}(2) \quad (8)$$

Other evaporators are fed from both outlet of the previous evaporator and from the exit of distilled water side (mixture of vapour and condensed water) of the previous preheater. This is shown in Fig. 2d. The balance equations can be written as follows:

Mass balance equation:

$$\dot{m}_b(i-1)=\dot{m}_b(i)+\dot{m}_v(i)\,,\quad \dot{m}_{oe}(i)=\dot{m}_{ie}(i) \tag{9}$$

Balance of salinity:

$$\dot{m}_b(i-1)X_b(i-1)=\dot{m}_b(i)X_b(i) \tag{10}$$

Balance of energy equation:

$$\dot{m}_b(i-1)h_b(i-1)+\dot{m}_{ie}(i)h_{ie}(i)=\dot{m}_b(i)h_b(i)+\dot{m}_v(i)h_v(i)+\dot{m}_{oe}(i)h_{oe}(i) \tag{11}$$

Where:

$$\dot{m}_{ie}(i)=\dot{m}_{oe}(i-1)+\dot{m}_d(i-1),\qquad i=3,4,\ldots,n \tag{12}$$

$$h_{ie}(i)=\frac{\dot{m}_{oe}(i-1)h_{oe}(i-1)+\dot{m}_d(i-1)h_d(i-1)}{\dot{m}_{oe}(i-1)+\dot{m}_d(i-1)},\qquad i=3,4,\ldots,n \tag{13}$$

The last effect of MED plant just includes a condenser (see Fig. 2e).

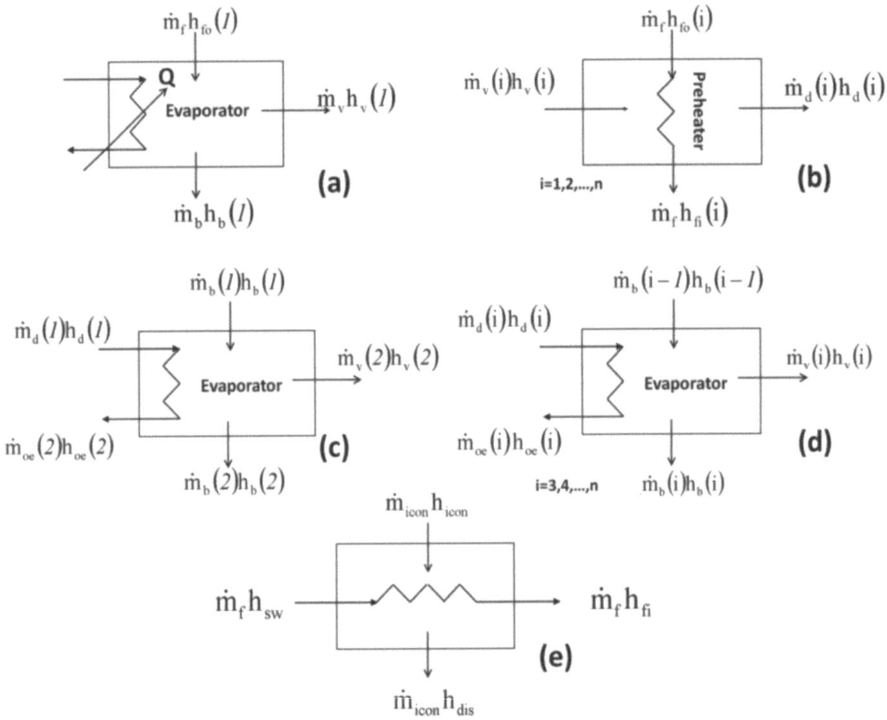

Fig. 2. Control volume of MED parts

The mass and energy balance in this stage are:
Mass balance equation:

$$\dot{m}_{icon} = \dot{m}_{dis} \tag{14}$$

Energy balance equation:

$$\dot{m}_{icon} h_{icon} + \dot{m}_{sw} h_{sw} = \dot{m}_f h_{fi}(n) + \dot{m}_{dis} h_{dis} \tag{15}$$

Where:

$$\dot{m}_{icon} = \dot{m}_{oe}(n) + \dot{m}_d(n) \tag{16}$$

$$h_{icon} = \frac{\dot{m}_{oe}(n)h_{oe}(n) + \dot{m}_d(n)h_d(n)}{\dot{m}_{oe}(n) + \dot{m}_d(n)} \tag{17}$$

The equations (1)-(17) are simultaneously solved to predict the thermodynamic properties of water and steam (temperature, pressure, enthalpy and salinity) at each part. The thermodynamic properties of distilled water, water vapour and brine are calculated for known parameters and P_{out} in the simulation code. Since these parameters have been specified in the simulation code, the input heating energy ($\dot{Q}$) must be limited to a particular range based on these parameters in order to achieve the balanced thermodynamic conditions at all parts of MED plant. Therefore for given input parameters minimum and maximum values for the heating energy is calculated for which equations (1)-(17) would be thermodynamically balanced with real physical conditions. More details of numerical procedure and validation could be found in the work by Kazemian et al. (2010).

3.3 Results and discussions

The effects of different parameters such as feed water flow rate, temperature of feed water, number of effects, temperature difference of preheaters and output pressure on the rate of fresh water production have been studied. Therefore to be more efficient the test conditions are design based on the method of design of experiment (DOE). DOE is performed on k parameters at two or more than two levels to understand their direct effects and also their interactions on the desired responses (Montogomery 2001). First a 2^k factorial design is chosen to construct the tests table. Five parameters are selected to study their effects on the minimum and maximum distilled water. These parameters are: feed water flow rate (A), temperature of feed water (B), number of effects (C), temperature difference of preheater (D) and output pressure (E).

Factors	Parameters	Level 1	Level 2	Level 3
A	Feed water (kg/s)	30	60	90
B	Temperature of seawater (°C)	25	30	32.5
C	Number of effects	12	15	18
D	Temperature difference of each pre-heater (°C)	1	3	5
E	Out-put pressure (MPa)	0.003	.004	0.005

Table 1. Parameters and their three levels value for 3^k factorial model of minimum distillate

Source	Sum of Squares	df	Mean Square	F Value	p-value	
Model	2076.82	20	103.841	846.4712	< 0.0001	Significant
A- m_f	597.2135	1	597.2135	4868.251	< 0.0001	Significant
B- T_{sw}	3.868855	1	3.868855	31.5374	0.0002	Significant
C-n	549.1412	1	549.1412	4476.385	< 0.0001	Significant
D- ΔT_{pr}	539.6624	1	539.6624	4399.118	< 0.0001	Significant
E- P_{out}	42.44459	1	42.44459	345.9918	< 0.0001	Significant
AB	0.617234	1	0.617234	5.031455	0.0464	Significant
AC	87.78436	1	87.78436	715.5839	< 0.0001	Significant
AD	86.38329	1	86.38329	704.1629	< 0.0001	Significant
AE	6.796065	1	6.796065	55.39888	< 0.0001	Significant
BC	1.162114	1	1.162114	9.473101	0.0105	Significant
BD	0.561878	1	0.561878	4.580208	0.0556	Non significant
BE	4.281921	1	4.281921	34.90455	0.0001	Significant
CD	126.191	1	126.191	1028.66	< 0.0001	Significant
CE	5.539345	1	5.539345	45.15458	< 0.0001	Significant
DE	1.563831	1	1.563831	12.74774	0.0044	Significant
ABE	0.684065	1	0.684065	5.576231	0.0377	Significant
ACD	20.19142	1	20.19142	164.5926	< 0.0001	Significant
ACE	0.884603	1	0.884603	7.210942	0.0212	Significant
BCD	0.431494	1	0.431494	3.517369	0.0875	Non significant
BDE	1.416334	1	1.416334	11.5454	0.0060	Significant

Table 2. analysis variance of 3^k factorial model for minimum distillate

A 2^k factorial with two levels for the minimum distilled water has been performed to see if there are any non significant parameters. It should be mentioned that the signification of the parameters are quantified by the p-value, a p-value less than 0.05 indicates significance (Montogomery 2001) and are specified as significant parameter. The results show that the effects of these parameters on the minimum distilled water are significant (see Fig. 3).

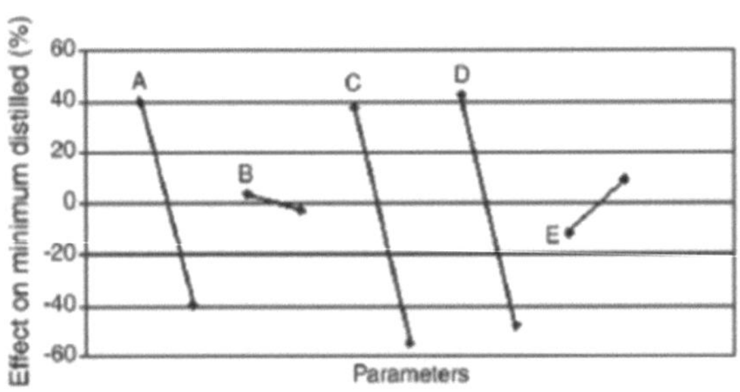

Fig. 3. Response of first DOE for minimum distilled water

Thus in order to have more accuracy a new DOE with three levels is performed to study the effects of these parameters on the minimum distilled water a 3^k factorial test table is designed which is shown in table 1.
Therefore 243 (3^5) tests have been executed to find the response of the objective function (minimum distilled water) on the variations of these parameters. Analysis variance of the 3^k factorial tests is shown in table 2. Then a regression has been performed on the results of factorial to show and also to predict the effects of these parameters on the minimum water distilled. Equation (19) is the regression function estimated from DOE analysis of minimum amount of distilled water.

$$\begin{aligned}\sqrt{(\dot{m}_d)_{min}} = & -11.01593 - 0.023307\, m_f + 0.35466\, T_{sw} + 0.014194\, n + 2.64044\, \Delta T_{pr} \\ & + 3042.42454\, P_{out} + 1.73109\times 10^{-3}\, m_f \times T_{sw} - 1.177766\times 10^{-3}\, m_f \times n \\ & + 1.39631\times 10^{-4}\, m_f \times \Delta T_{pr} + 6.08346\, m_f \times P_{out} + 3.14011\times 10^{-3}\, T_{sw} \times n \\ & - 0.090929\, T_{sw} \times \Delta T_{pr} - 95.27274\, T_{sw} \times P_{out} + 0.049661\, n \times \Delta T_{pr} - 26.34805\, n \times P_{out} \\ & - 749.91251\, \Delta T_{pr} \times P_{out} + 1.71863\times 10^{-5}\, m_f \times T_{sw} \times n - 1.19483\times 10^{-4}\, m_f \times T_{sw} \times \Delta T_{pr} \\ & - 0.34827\, m_f \times T_{sw} \times P_{out} + 5.14699\times 10^{-4}\, m_f \times n \times \Delta T_{pr} + 0.28864\, m_f \times \Delta T_{pr} \times P_{out} \\ & + 0.19816\, m_f \times n \times P_{out} - 9.56007\times 10^{-4}\, T_{sw} \times n \times \Delta T_{pr} + 0.18593\, T_{sw} \times n \times P_{out} \\ & + 24.71972\, T_{sw} \times \Delta T_{pr} \times P_{out} + 1.57852\, n \times \Delta T_{pr} \times P_{out}\end{aligned} \quad (19)$$

For given values of the parameters the prediction contours of minimum water distilled can be plotted by this equation. In order to see the precision of the predicted results by these contours, comparisons are done with the results obtained directly from the simulation code. As seen in table 3, within the range of performed tests, these results are very close while out of the range of executed tests the concordance between the results is acceptable (6.59%).

	Prediction	Actual	Error%
In the range	8.958	8.896	0.69
Out of the range	34.398	36.8249	6.59

Table 3. Predicted error for the minimum distilled water

The same approach is also adopted for the maximum distilled water. The results of 2^k factorial tests for the maximum amount of distilled water shows that the effect of parameter B (T_{sw}) on the maximum amount of distilled water is negligible (see in Fig. 4). Thus another tests routine with factorial is performed. The parameters and their levels (three for each) are shown in table 4. The tests table for analysis of variance of maximum amount of distilled water consists of 81 tests (3^4) which is presented in table 5.

Factors	Parameters	Level 1	Level 2	Level 3
A	Feed water (kg/s)	30	50	70
B	Number of effects	12	15	18
C	Temperature difference of each pre-heater (°C)	2	3	4
D	Output pressure (MPa)	0.003	.0045	0.006

Table 4. Parameters and their three levels value for 3^k factorial model of maximum distillate

Source	Sum of Squares	df	Mean Square	F Value	p-value Prob > F	
Model	4974.632	11	452.2393	3444.212	< 0.0001	significant
A-mf	2447.064	1	2447.064	18636.61	< 0.0001	Significant
B-n	661.6533	1	661.6533	5039.089	< 0.0001	Significant
C-deltpr	1583.357	1	1583.357	12058.69	< 0.0001	Significant
D-Pout	0.278922	1	0.278922	2.124243	0.1495	Non- significant
AB	70.48853	1	70.48853	536.834	< 0.0001	Significant
AC	168.8939	1	168.8939	1286.28	< 0.0001	Significant
AD	0.03011	1	0.03011	0.229317	0.6335	Non- significant
BC	29.61552	1	29.61552	225.549	< 0.0001	Significant
CD	9.119239	1	9.119239	69.45126	< 0.0001	Significant
ABC	3.161682	1	3.161682	24.07907	< 0.0001	Significant
ACD	0.970699	1	0.970699	7.392755	0.0083	Significant

Table 5. analysis variance of 3^k factorial model for maximum distilled water

The estimated function resulted from DOE analysis for maximum distilled water is as follow:

$$
\begin{aligned}
(\dot{m}_d)_{max} = & -9.98114\times10^{-3} + 0.02892\,m_f + 1.11264\times10^{-3}\,n + 3.24331\times10^{-4}\,\Delta T_{pr} \\
& - 0.7575\,P_{out} + 5.17374\times10^{-3}\,m_f\times n - 0.012606\,m_f\times\Delta T_{pr} \\
& - 21.07516\,m_f\times P_{out} - 1.28832\times10^{-4}\,T_{sw}\times n + 0.34856\,\Delta T_{pr}\times P_{out} \\
& + 6.04926\times10^{-3}\,m_f\times n\times\Delta T_{pr} + 6.70371\,m_f\times\Delta T_{pr}\times P_{out}
\end{aligned}
\tag{20}
$$

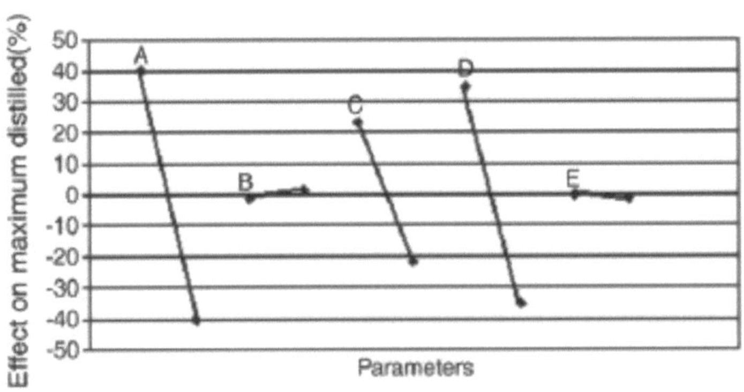

Fig. 4. Response of first DOE for maximum distilled water

To show the precision of this equation comparisons are done with the direct results of the simulation code. As seen in table 6 within the range of performed tests, the results are very close while at the out of range the concordance between the results is acceptable (4.33%).

Error%	actual	Prediction	
1.27	13.5522	13.724	In the range
4.33	44.9896	43.0413	Out of the range

Table 6. Predicted error for maximum distilled water

Thus at this step the contours for prediction of minimum and maximum distilled water could be presented and discussed.

As mentioned the regression functions are obtained by using the responses of the parameters on the objective function. These functions are composed of the effective parameters and their interactions. The contours of responses on each parameter could be plotted using these equations. These contours are an excellent tools to show the effect of each parameter rather than calculating by the simulation code.

Several contours are shown in Figs 5-10 (these results were presented by Kazemian et al. 2010). It is shown that the amount of feed water mass flow rate has a significant effect on increasing the amount of minimum distilled water. As seen in Fig. 5, for a given feed water mass flow rate, increasing the temperature of the inlet feed water slightly augments the amount of minimum distilled water.

There are two parallel phenomena to increase the amount of minimum distilled water by increasing the numbers of effects (see Fig. 6). The top brine temperature is increased by increasing the effects, so the salt concentration in the first effect is increased and more vapours is produced. On the other hand, the pressure and temperature differences of effects are decreased by increasing the effects. Therefore the salt concentration differences at each effect would be decreased and the amount of distilled water would be increased. The influence of increasing temperature difference of preheaters on the minimum distilled water production is fairly the same as the one that was seen by increasing the numbers of effects (Fig. 7).

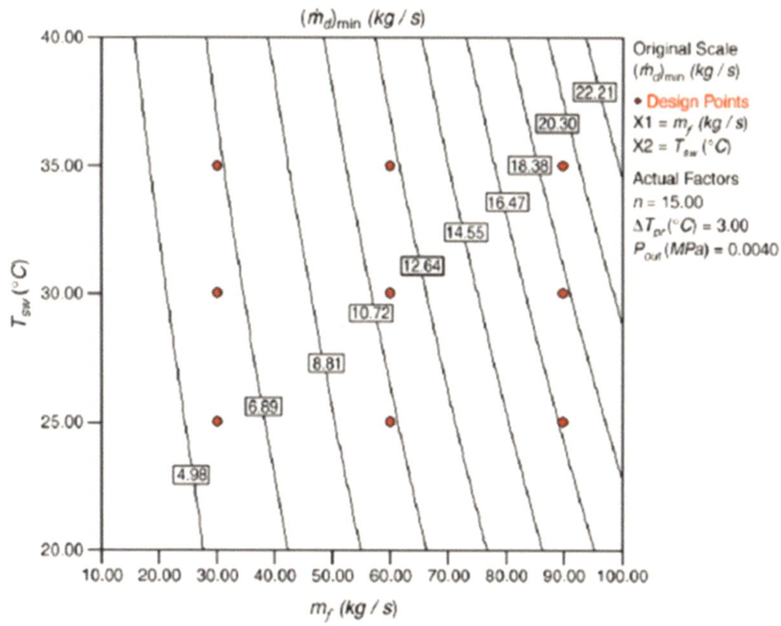

Fig. 5. Contour of minimum distilled water for N=15, ΔT_{pr}=3°C, P_{out}=0.004MPa

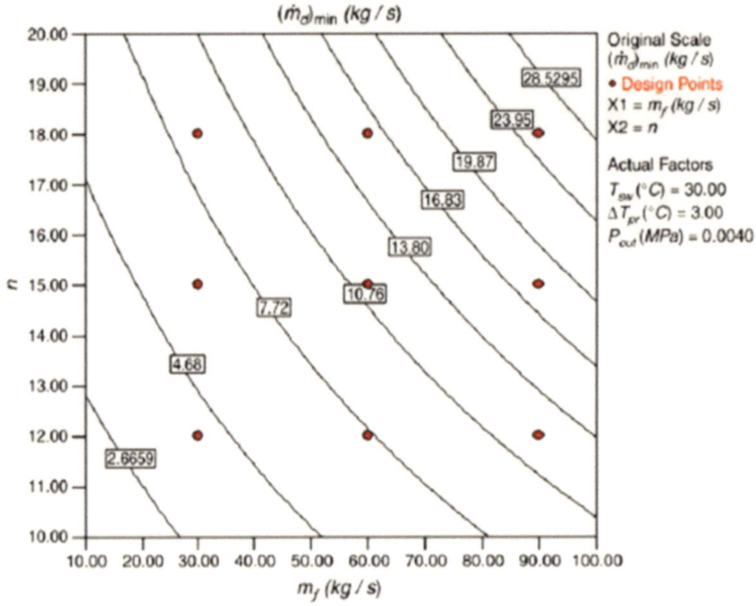

Fig. 6. Contour of minimum distilled water for T_{sw}=30°C, ΔT_{pr}=3°C, P_{out}=0.004MPa

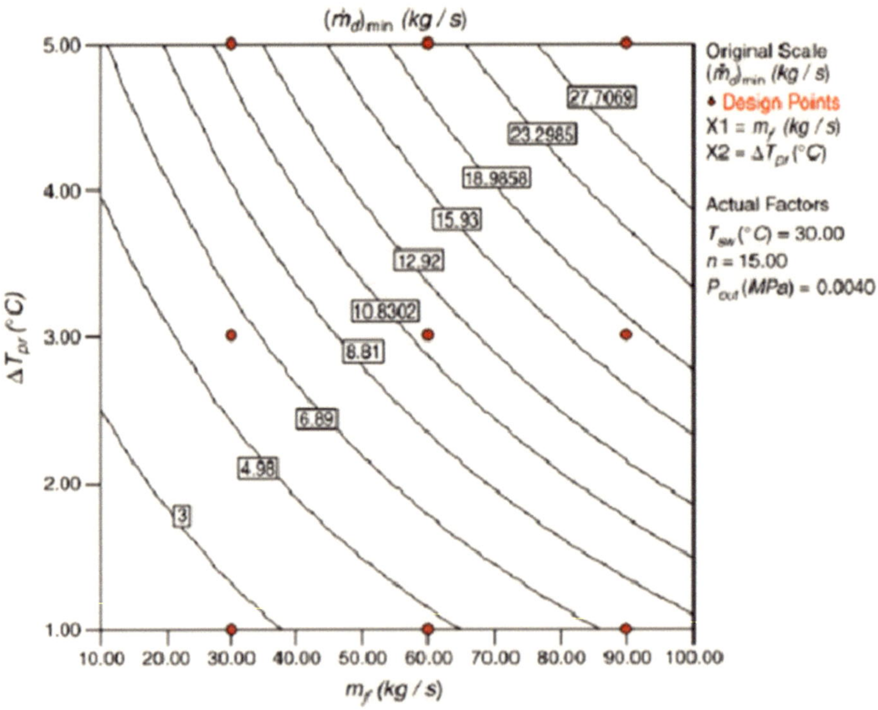

Fig. 7. Contour of minimum distilled water for N=15, T_{sw}=30°C, P_{out}=0.004MPa

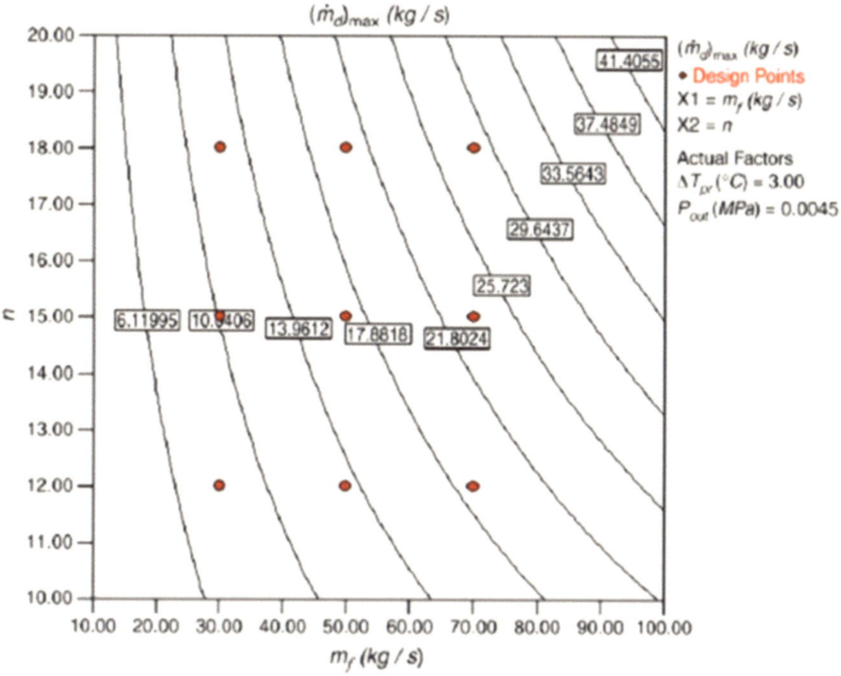

Fig. 8. Contour of maximum distilled water for T_{sw}=30°C, ΔT_{pr}=3°C, P_{out}=0.004MPa

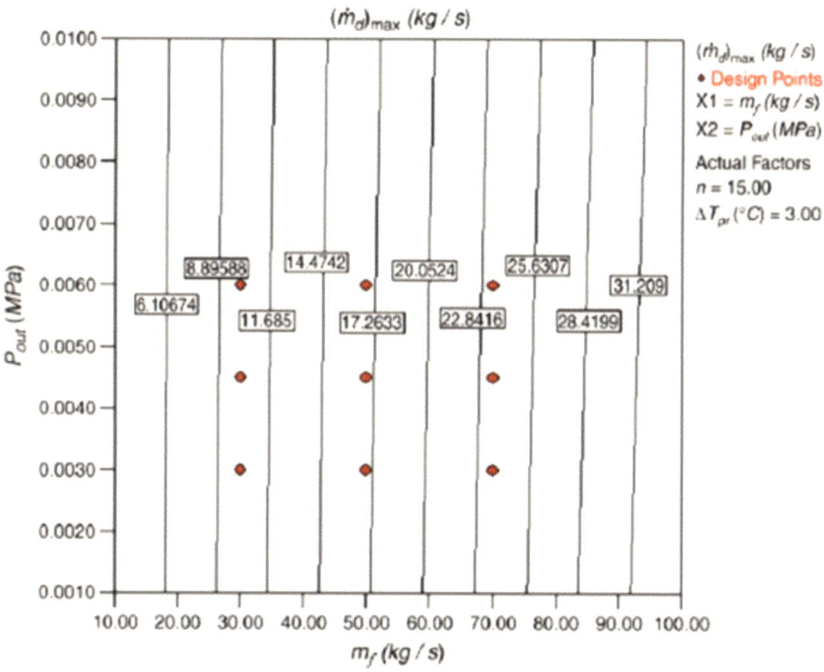

Fig. 9. Contour of maximum distilled water for N=15, ΔT_{pr}=3°C

In the case of maximum distilled water, the top brine temperature and also salt concentration of the first effect are increased by the numbers of effects. Consequently the amount of vapour on the first effect is augmented. Therefore it causes to enhance the amount of maximum distilled water which can be seen in Fig. 8. The temperature difference of condenser is reduced by augmenting pressure output, thus the top brine temperature is decreased and the heating energy is increased. On the other hand increasing of heating energy is restricted with the maximum distilled water. So as seen in Fig. 9 the sensitivity of pressure output on the maximum distilled water is negligible.
The influence of increasing temperature difference of preheaters on the maximum distilled water is fairly similar to the one that was seen in the minimum distilled water approach (see Fig. 10).

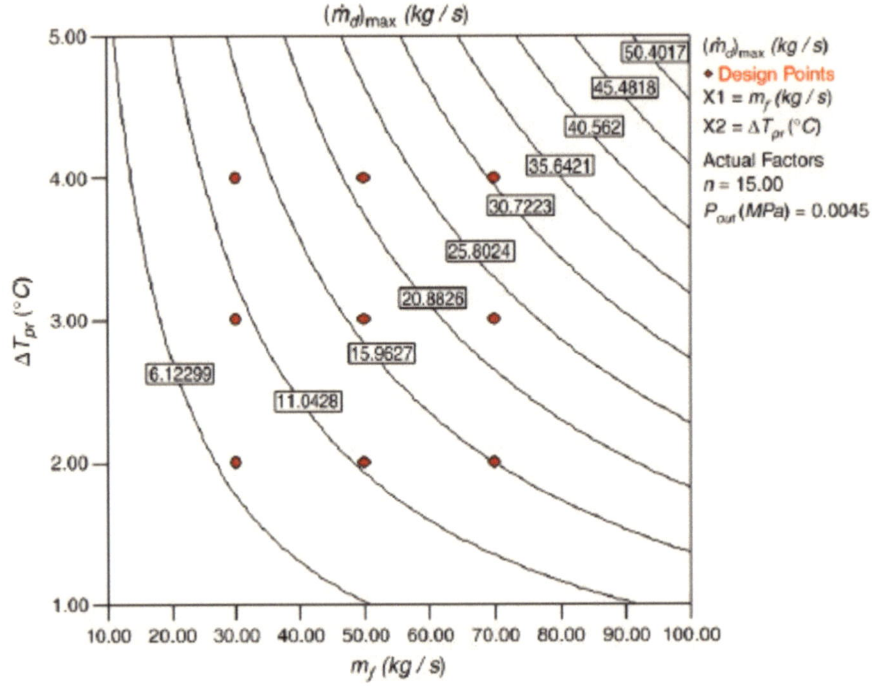

Fig. 10. Contour of maximum distilled water for N=15, P_{out}=0.0045MPa

4. Case II) Optimization of a SDHD plant using DOE method

One of the plants that potentially could supply fresh water at the humid regions (such as the Persian Gulf region) is solar desalination with using humidification-dehumidification (SDHD) process. Because of importance of this system, it is chosen to investigate using DOE method as the second case study. This process is, mainly, based on the ability of air to be saturated with water. Thus it would be very efficient at the regions for which the relative humidity of air is significant.
Many studies have been carried out on the various types of humidification-dehumidification (HD) cycle desalinations. These studies investigated different methods of increasing the production of desalinated water and on augmenting performance of the

plants. Goosen, et al. (2000) with the aid of HD process, examined some economic and thermodynamic aspects of solar desalination. Their report was based on this fact that commercial production of solar desalination is economically and efficiently advantageous. Parekh et al. (2004) carried out a comprehensive study on the background of solar desalination using humidification-dehumidification (SDHD) systems. They studied development of solar stills historically, and concluded that frequent use of the latent heat of condensation is the major factor of the development of these systems. They concluded that most of the researchers have indicated the effect of inlet air flow rate in the cycle is insignificant. However, the effect of feed water flow rate on the efficiency of a SDHD unit has been described as significant. Al Hallaj et al. (1998) undertook an experimental study on a SDHD unit. In their unit the air circulates by natural or forced convection and is humidified by the constant water obtained either from a collector (indoor type) or from an electrical heater (outdoor type).Their results in indoor and outdoor conditions showed factors of performance and daily production of desalinated water. In outdoor conditions, the results showed higher production of desalinated water compared to that of solar stills, whereas the effect of air velocity was formerly regarded only in lower performance temperatures. Nafey, et al.(2004) carried out an experimental work on SDHD process. Their plant consists of humidification and dehumidification towers, which are located next to flat-plate solar collectors (for air heating) and water concentrator (for heating water). They found that the effect of air velocity is insignificant, while great influence of inlet water and air temperatures on the production of desalinated water is observed. They predicted the fresh water production numerically. Their experimental and theoretical findings are in good agreement (Nafey, et al 2004). Multi-effect humidification-dehumidification (MEHD) is another interesting plant that was studied by Chafik.(2002), Ben Amara et al.(2004). This technique includes humidification and air heating in several stages which leads to an increase in the moisture density in air flow. Hou, et al.(2005) mentioned that in most of the previous studies concerning SDHD technology, obtaining optimal conditions of design, has been a difficult and complicated procedure. Using Pinch method, they proposed a design for optimizing the performance of SDHD process. Results show that as the temperatures of the sprayed (humidifier tower) and cooling water (in condenser) are known, there is an optimal rate of flow for the ratio of water to dry air. Recently, Farsad et al (2010) numerically investigated SDHD process. They showed that based on the conditions and desired fresh water production, condenser characteristic and humidifier characteristic has an optimum configuration (performance-economy).

This section intends to investigate the optimum condition of several parameters on the fresh water production of SDHD plant using DOE method.

4.1 Plant description

The analyzed system consists of main sections including air and water solar collectors, condenser and humidifier tower. Figure 11 illustrates a schema of this system. Feed water (brackish water, turbid water, flowing water with high heaviness and seawater) flows into the condenser through point number one. Normally, the model of condenser for air-water flows is a type of extended surface heat exchanger. In this system, extended surface heat exchanger is used because heat transfer coefficient at the air side is far smaller than that of liquid side, and more heat transfer surface is needed at the air side.

Therefore, the humid air starts flowing at the air side of the vanes which are fixed on the tubes. Feed water runs through the tubes and is preheated by the recovery of condensation latent heat. Then it is heated in the solar collector and after that flows into the humidifying

tower. Humidifying tower consists of packing. Packing material is selected from wooden chips which are laid on each other separately by some distribution porous plates. Hot water stream flows into the humidifier through the top and is sprayed on the packing.
Thus, the contact level of air and water is increased and heat and mass transfer between the two streams augments. Humid air at the exit of humidifier then passes through the condenser to be cold and fresh water to be distilled.

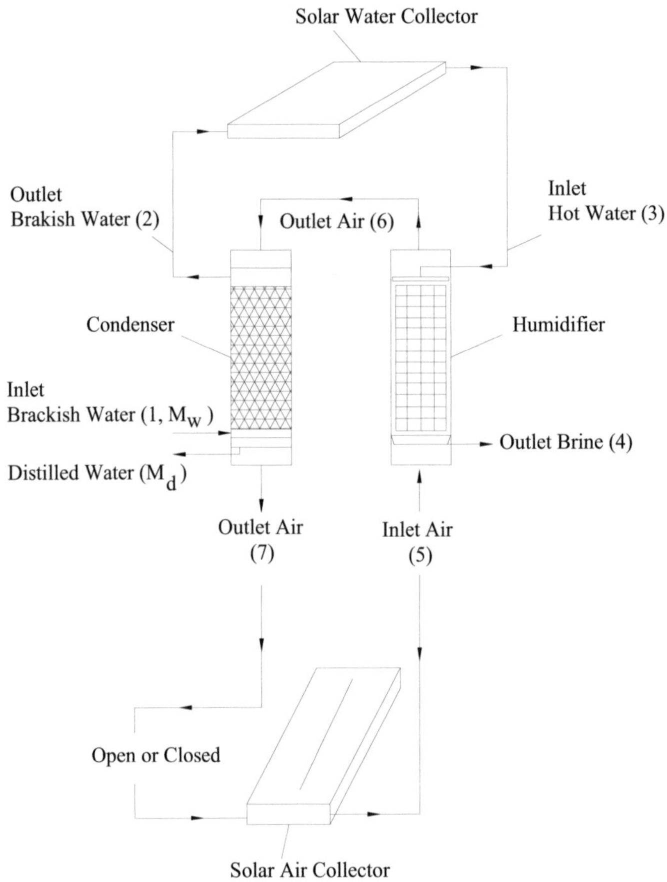

Fig. 11. Schematic of SDHD plant

4.2 Mathematical formulation

(Energy and mass balance equations have been considered for all parts of the cycle. Few assumptions that believed do not have significant effect on the analysis, is considered for simplicity of calculation. These are as follows:

1. The process is assumed to be in steady state condition.
2. Heat loss is neglected.
3. Since the operating pressure is close to the atmospheric pressure air and water vapor are assumed behave as ideal gas.
4. Saturated air at the exit of humidifier and also at the exit of condenser.
5. Kinetic and potential energy changes are neglected.

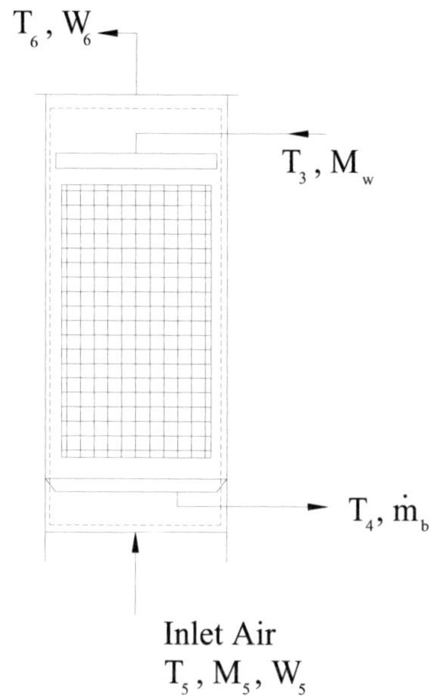

Fig. 12. Humidifier

Accordingly, mass and energy balance equations in the humidifier (Fig.12) are defined as:

$$\dot{m}_a h_{a5} + \dot{m}_{v5} h_{v5} + \dot{m}_{w3} h_{f3} = \dot{m}_a h_{a6} + \dot{m}_{v6} h_{v6} + \dot{m}_{b4} h_{f4} \tag{21}$$

$$\dot{m}_{v6} + \dot{m}_{b4} = \dot{m}_{v5} + \dot{m}_{w3} \tag{22}$$

$$M_5(h_6 - h_5) = \text{Ka V}\left[\frac{(h_3 - h_6) - (h_4 - h_5)}{\ln \frac{h_3 - h_6}{h_4 - h_5}}\right] \tag{23}$$

In the above equation KaV, the humidifier characteristic, could be determined by the following imperial equation (Nafey et al. 2004):

$$\frac{KaV}{M_w} = 0.07 + A.N\left(\frac{M_w}{M_5}\right)^{-n} \tag{24}$$

where A and n are constant value for a kind of packing material (see Table 7).
Humidity ratio is characterized as a function of atmospheric pressure, steam partial pressure and dry bulb temperature.

$$w_n = \frac{m_{vn}}{m_a} = 0.622 \frac{P_{vn}}{P - P_{vn}} \tag{25}$$

Relative humidity is also defined as follow:

$$\Phi_n = \frac{P_{vn}}{P_{gn}} \tag{26}$$

n	A	Type of Packing
0.62	0.060	A
0.62	0.070	B
0.60	0.092	C
0.58	0.119	D
0.46	0.110	E
0.51	0.100	F
0.57	0.104	G
0.47	0.127	H
0.57	0.135	I

Table 7. Constant value of n and A used in Eq.24 (Frass 1989)

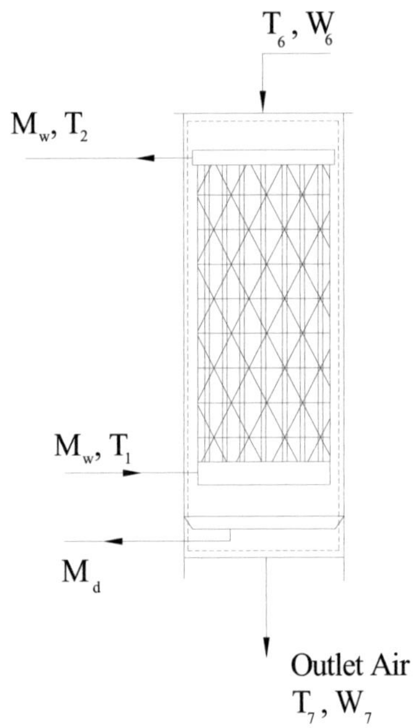

Fig. 13. Condenser (dehumidifier)

The energy and mass balance equations for the condenser which is shown in Fig. 13 are defined as:

$$\dot{m}_a h_{a6} + \dot{m}_{v6} h_{v6} + \dot{m}_{w1} h_{f1} = \dot{m}_a h_{a7} + \dot{m}_{v7} h_{v7} + \dot{m}_d h_{f7} + \dot{m}_{w2} h_{f2} \quad (27)$$

$$\dot{m}_d = \dot{m}_{v6} - \dot{m}_{v5} \qquad \& \qquad \dot{m}_{w1} = \dot{m}_{w2} = \dot{m}_{w3} = M_w \quad (28)$$

$$Q_c = M_w\, C_{pw} (T_2 - T_1) = U_{cond}\, A_{cond} LMTD \quad (29)$$

LMTD is condenser's logarithmic average temperature difference which is described by:

$$\mathrm{LMTD} = \frac{(T_6 - T_2) - (T_7 - T_1)}{\ln \frac{(T_6 - T_2)}{(T_7 - T_1)}} \tag{30}$$

Enthalpy and humidity ratio for saturation can be obtained from the following relationship.

$$h = 0.00585\, T^3 - 0.497\, T^2 + 19.87\, T - 207.61 \tag{31}$$

$$W = 2.19\, T^3 (10^{-6}) - 1.85\, T^2 (10^{-4}) + 7.06\, T (10^{-3}) - 0.077 \tag{32}$$

Heating input energy at the flat-plat solar collector is calculated by:

$$Q_u = F_R A_c \left[I\, \tau \alpha - U_L (T_i - T_a) \right] \tag{33}$$

These equations have been solved simultaneously to find the plant performance. Details of numerical procedure and validation could be found in the work by Farsad et al. (2010).

5. Results and discussions

The adopted mathematical formulation and numerical procedure could determine the thermodynamic properties of air and water streams throughout the cycle and fresh water production for inlet air and water conditions. Therefore air and water flow rates, temperature and, inlet relative humidity and input heating energy (solar collectors) are considered as variable to see their effects on the fresh water production.

Design of experiment (DOE) is performed on k parameters at two or more than two levels to understand their direct effects and also their interactions on the desired responses.

Therefore, at first a 2^k factorial approach with two levels is chosen to see if there are any non significant parameters on the fresh water production. Therefore 64 (2^6) tests have been executed to find the response of objective function (fresh water) on the variations of these parameters. Providing the P-value model shows that all the parameters are effective in water production and are evaluated as significant in the table. Therefore, to have more accuracy a new DOE with three levels (capturing nonlinear effects) is performed to study the effects of these parameters on the distilled water production. Therefore the parameters are written in three levels (see table 8) and 3^k factorial model is designed for the tests. Thus 729 (3^6) tests have been performed to see the effects of these parameters on the fresh water productions. The results from the Analysis of Variance using backward elimination regression method are displayed in table 9. Then a regression has been performed on the

Factors	Parameters	Level 1	Level 2	Level 3
A	Inlet Water Temperature (°C)	15	20	25
B	Inlet Air Temperature (°C)	5	20	35
C	Input Heat Flux (kW)	50	75	100
D	Acond Ucond (kW/°C)	8	13	18
E	Mass Flow Rate Of Water (kg/s)	0.4	0.9	1.4
F	Mass Flow Rate Of Air (kg/s)	0.4	0.8	1.2

Table 8. Parameters and their three levels value for 3^k factorial model of fresh water production.

Source	Sum of Squares	*df*	Mean Square	*F* Value	*p*-value	
Model	324.6028	27	12.02233	210.7277	0.0001	significant
A-T_1	13.83002	1	13.83002	242.4131	0.0001	significant
B-T_5	19.65184	1	19.65184	344.4581	0.0001	significant
C-Q	75.669	1	75.669	1326.329	0.0001	significant
D-$A_{cond}U_{cond}$	16.12721	1	16.12721	282.6782	0.0001	significant
E-M_w	15.04927	1	15.04927	263.7842	0.0001	significant
F-M_5	30.03497	1	30.03497	526.454	0.0001	significant
AB	0.911795	1	0.911795	15.98197	0.0004	significant
AC	2.526584	1	2.526584	44.28605	0.0001	significant
AD	0.343341	1	0.343341	6.018092	0.0206	significant
AE	1.104146	1	1.104146	19.35351	0.0001	significant
AF	5.187897	1	5.187897	90.93363	0.0001	significant
BC	0.953295	1	0.953295	16.70938	0.0003	significant
BF	11.33596	1	11.33596	198.697	0.0001	significant
CD	2.717269	1	2.717269	47.62838	0.0001	significant
CE	19.13845	1	19.13845	335.4595	0.0001	significant
DE	12.8603	1	12.8603	225.4157	0.0001	significant
EF	26.65787	1	26.65787	467.26	0.0001	significant

Table 9. Analysis of variance of 3^k factorial model for fresh water production

results of factorial to show and also to predict the effects of these parameters on the fresh water production. Equation (34) is the regression function estimated from DOE analysis of 3^k factorial model to predict distilled water (M_d).

$$\begin{aligned}
\ln(M_d) = {} & 4.04483 - 0.098587(T_1) - (7.19727\times10^{-3})(T_5) + 0.019074(Q) \\
& + 0.043618(A_{cond}U_{cond}) + 1.14683(M_w) - 0.80018(M_5) + (1.11087\times10^{-3})(T_1\times T_5) \\
& + (9.40156\times10^{-4})(T_1\times Q) + 0.031299(T_1\times M_w) - 0.083390(T_1\times M_5) \\
& - (2.97633\times10^{-4})(T_5\times Q) - (4.84324\times10^{-3})(T_5\times M_w) + 0.031011(T_5\times M_5) \\
& + (8.14539\times10^{-3})(Q\times M_w) + (6.66603\times10^{-3})(Q\times M_5) \\
& + 0.045739(A_{cond}U_{cond}\times Mw) + 1.70131(M_w\times M_5) - (1.79173\times10^{-4})Q^2 \\
& - (2.09743\times10^{-3})(A_{cond}U_{cond})^2 - 2.06950(M_w)^2 - 0.68742(M_5)^2
\end{aligned} \tag{34}$$

For given values of the parameters the prediction contours of water production can be plotted by using this equation. In order to see the precision of the predicted results by these contours, comparisons have been done with the results obtained directly from the simulation code. As seen in table 10, within the range of performed tests, these results are very close while out of the range of executed tests the concordance between the results is acceptable (8.78%).

	Response	Prediction	Actual	Error %
Within the range	M_d(kg/s)	98.9881	101.9117	2.87
Out of the range	M_d(kg/s)	91.9274	100.77	8.78

Table 10. Error of predicted fresh water production by the regression equation.

As mentioned the regression functions are obtained by using the responses of the parameters on the objective function (fresh water production). These functions are composed of the effective parameters and their interactions. These contours are an excellent tool to show the effect of each parameter simultaneously rather than calculating one by one by the simulation code.

To show this ability, for instance, Figs. 14-17 present the effects of some of the parameters on the fresh water production. Fig. 14 presents the effect of inlet air and water temperature on the fresh water production for give conditions (Q, M_w, M_5, $A_{cond}U_{cond}$). It shows that with decreasing the inlet water temperature and increasing the air inlet temperature distilled water production enhances. The effects of inlet water temperature and total heat flux on the fresh water production is shown in Fig.15. As shown decreasing the inlet water temperature reduces the necessary input energy. Interesting information is found in Fig.16; the effects of water inlet temperature and water mass flow rate on the distilled water production. As seen, for given conditions there are two different inlet water temperatures that could produce similar fresh water production (because of its different effects on the humidifier and

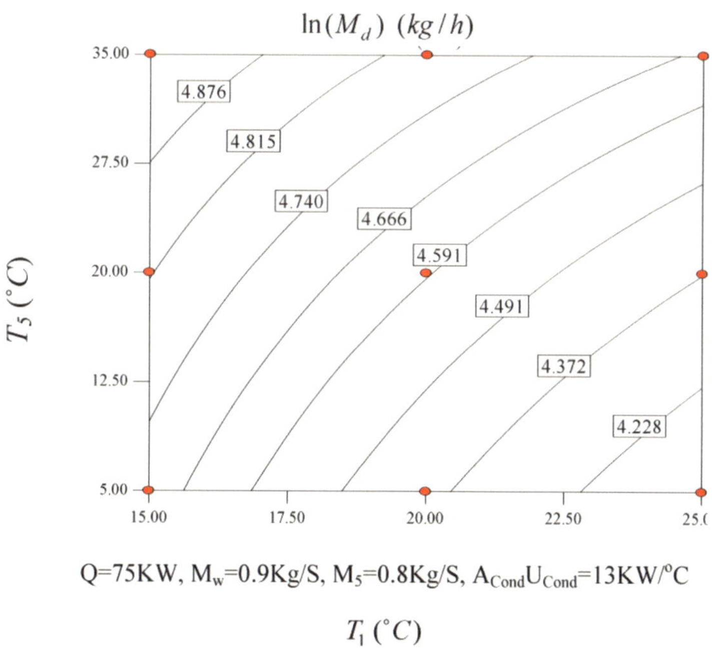

Fig. 14. Contour of variation of inlet air and water temperatures on the fresh water production.

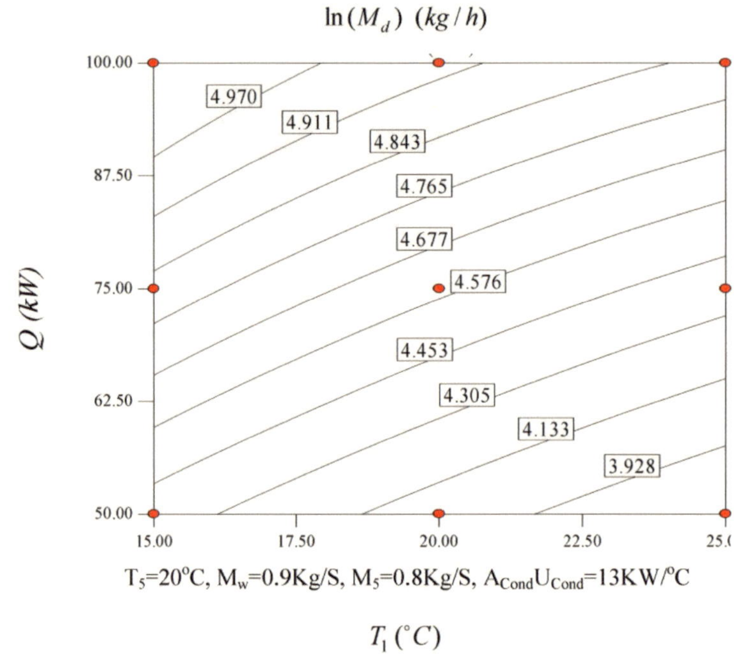

Fig. 15. Contour of feed water temperature and the given total heat flux of the cycle on the fresh water production.

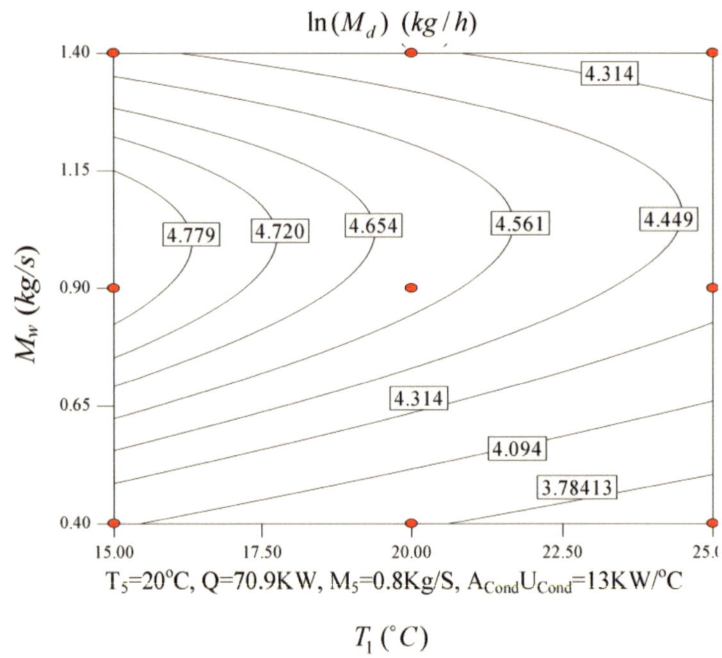

Fig. 16. Contour of the inlet water temperature and its mass flow rate on the fresh water production.

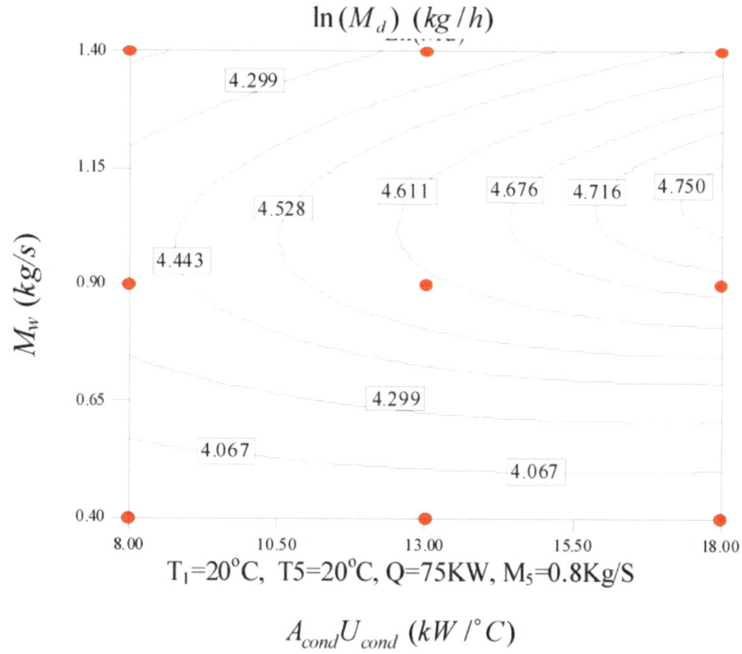

Fig. 17. Contour of condenser characteristic parameter and the feed water flow rate on the fresh water production.

condenser). Another contour that could show the effect of condenser's design parameter on the fresh water production is presented in Fig.17. As shown there are different condenser characteristic that could produce particular distilled water.

6. Conclusion

This chapter introduces Design of Experiment (DOE) method as a statistical tool for optimization of desalination systems. Two different desalination plants; Multi-Effect Desalination system and solar desalination using humidification-dehumidification cycle have been numerically investigated to show the ability of DOE method for optimizing such systems. Thus several different contours that could help a designer to achieve the best thermodynamic conditions in these systems are presented and discussed. It is shown that DOE method is capable to well determine the optimum conditions for such systems.

Nomenclature:

A_c	Solar collector area	n	Number of effect	b	Brine
A_{cond}	Condenser heat transfer area	P	Pressure	con	Condenser
a	Area per volume of humidifier	Q	Input heating energy	d,dis	Distilled water
C_p	Specific heat	T	Temperature	e	evaporator
F_R	Solar collector heat	U_l	Overall loss coefficient	f	Feed water

	removal factor		of the collector		
h	Enthalpy	U_{cond}	Overall heat transfer coefficient of the condenser	O,out	Outlet
I	Solar irradiance	V	Volume of humidifier	pr	Preheater
K	Mass transfer coefficient	X	Salt concentration	sw	seawater
M	Mass flow rate	**Subscript**		v	vapor
$\dot{m}$	Mass flow rate	a	air	w	water

7. References

Al Hallaj, S.; Farid, M.M. & Tamimi, A.R. (1998). Solar desalination with a humidification-dehumidification cycle: performance of the unit, *Desalination*, Vol.120, pp.273-280, ISSN: 0011-9164

Al-Shammiri, M. & Safar, M. (1999). Multi-effect distillation plants: state of the art, *Desalination*. Vol.126, pp. 45-59, ISSN: 0011-9164

Al-Shayji, K.A.M. (1998). Modeling simulation, and optimization of large-scale commercial desalination plants. *Dissertation submitted to the Faculty of the Virginia Polytechnic Institute and State University in partial fulfilment of the requirements for the degree of Doctor of Philosophy in Chemical Engineering*.

Antony, J. (2003). *Design of Experiments for Engineers and Scientists*, Elsevier Science & Technology Books, ISBN: 978-0-7506-4709-0

Aybar, H. (2004). Desalination system using waste heat of power plant, *Desalination*. Vol.166, pp.167-170, ISSN: 0011-9164

Behzadmehr, A.; Piaud, J. B.; Oddo, R. & Mercadier, Y. (2006). Aero-acoustical effects of some parameters of a backward- curved centrifugal fan using DOE, *ASRAEH, HVA&R Research*, Vol.12, No.2, pp.353-365, ISSN: 1078-9669

Behzadmehr, A.; Mercardier, Y. & Galanis, N. (2006). Sensitivity Analysis of Entrance Design Parameters of a Backward-Inclined Centrifugal Fan using DOE Method and CFD Calculations, *ASME Transaction, Journal of Fluid Engineering*, Vol. 128, pp.446-453, ISSN: 0098-2202.

Ben Amara, M.; Houcine, I.; Guizani, A. & Maalej, M. (2004). Experimental study of a multiple-effect humidification solar desalination technique, *Desalination*, Vol.170, pp.209-221, ISSN: 0011-9164

Chafik, E. (2002). A new seawater desalination process using solar energy, Desalination, Vol. 153, pp. 25-37, ISSN: 0011-9164

Djebedjian, B.; Gad, H.; Khaled , I. & Rayan, M.A. (2008). Optimization of Reverse osmosis desalination system using genetic algorithm technique. *Twelfth International Water Technology Conference, IWTC12, Alexandria, Egypt.*

El-Nashar, A. (2000). Predicting part load performance of small MED evaporators - a simple simulation program and its experimental verification, *Desalination*. Vol.130, pp. 217-234, ISSN: 0011-9164

Farsad, S.; Behzadmehr, A. & Sarvari, S.H. (2005). Numerical analysis of solar desalination using humidification-dehumidification cycle, *Desalination and Water Treatment*, Vol.19, pp.294-300, ISSN: 1944-3994

Frass, P. (1989). *Heat exchanger design*, Wiley, John & Sons, ISBN: 0471628689

Goosen, M.F.A.; Sablani, S.S.; Shayya, W.H.; Paton, C. & Al-Hinai, A. (2000). Thermodynamic and economic considerations in solar desalination, *Desalination*, Vol.129, pp. 63-89, ISSN: 0011-9164

Hatzikioseyian, A.; Vidali, R. & Kousi P. (2003). Modelling and thermodynamic analysis of a multi effect desalination (MED) plant for seawater desalination, *Improving of Human Potential 2003 (IHP) Research Results at Plataforma Solar de Almeria within the Year 2003 Access Campain Plataforma Solar de Almeria (PSA)-CIEMAT*, Almeria, Spain, July 2003, pp.17-25

Hou, S.; Ye, S. & Zhang, H. (2005). Performance optimization of solar humidification–dehumidification desalination process using Pinch technology, *Desalination*, Vol. 183, pp.143-149, ISSN: 0011-9164

Kamali, R. & Mohebbinia, S. (2007). Optimization of the tube size and the arrangement of evaporator tube bundle to improve the performance of MED-TVC systems, *11th International Water Technology Conference. Egypt, 440.*

Kazemian, E.; Behzadmehr, A. & Sarvari, S.M.H. (2010), Thermodynamic optimization of multi effect desalination plant using DoE method, *Desalination*, Vol. 257 pp 195-205, ISSN: 0011-9164

Khademi, M.H. ; Rahimpour, M.R. & Jahanmiri, A. (2009).Simulation and optimization of a six-effect evaporator in a desalination process, *Chem. Eng. Process*. Vol. 48, pp. 339-34, ISSN: 0255-2701

Metaiche, M.; Palmeri, J. & David, P. (2008). Development of Optimization Software of Ro Systems For Water Desalination: 'Desaltop'. The 3rd International Conference on Water Resources and Arid Environments and the 1st Arab Water Forum.

Montogomery, C. Douglas. (2001). *Design and analysis of experiments*, 5th Ed., John Wily &Sons, New York, ISBN 0471316490.

Mussati, S.F.; Aguirre, P.A. & Scenna, N.J. (2003). A hybrid methodology for optimization of multi-stage flash-mixer desalination systems, *Latin American Applied Research*, Vol.33, pp. 141-147, ISSN 0327-0793

Nafey, A.S.; Fath, H.E.S.; El Heaby, S.O. & Soliman, M. (2004).Solar desalination using humidification–dehumidification processes. Part II. An experimental investigation, *Energy Conversion Man.*, Vol.45, pp.1263-1277, ISSN: 0196-8904

Nafey, A.S.; Fath, H.E.S.; El Heaby, S.O. & Soliman, M. (2004). Solar desalination using humidification dehumidification processes. Part I. A numerical investigation, *Energy Conversion Man.*, Vol. 45, pp.1243-1261, ISSN: 0196-8904

Narmine, H.A. & El-Fiqi, A.K. (2003). Thermal performance of seawater desalination systems, *Desalination*. Vol.158, pp. 127-142, ISSN: 0011-9164

Ophir, A. & Lokiec, F. (2005). Advanced MED process for most economical sea water desalination, *Desalination*. Vol., 182, pp. 187–198, ISSN: 0011-9164

Parekh, S.; Farid, M.M.; Selman, J.R. & Al Hallaj, S.A. (2004).Solar desalination with a humidification-dehumidification technique—a comprehensive technical review, *Desalination*, Vol.160, pp. 168-186, ISSN: 0011-9164

Shamel, M. & Chung, O.T. (2006). Drinking water from desalination of seawater: optimization of reverse osmosis system operating parameters, *Journal of Engineering Science and Technology*. Vol. 1, pp.203- 211, ISSN: 1823-4690

13

Impacts of Brine Discharge on the Marine Environment. Modelling as a Predictive Tool

Pilar Palomar and Iñigo. J. Losada
Environmental Hydraulics Institute "IH Cantabria", (Universidad de Cantabria)
Spain

1. Introduction

Desalination is a rainfall independent source of water for security long term water supplies. Is is expected that in the medium term desalination would be an optimum to apply to different uses of human consumption, such as irrigation.

Desalination is any of the several processes involved in removing dissolved minerals (especially salt) from seawater, brackish water, or treated wastewater. A number of technologies have been developed for desalination, including thermal processes and membrane technologies. In the present chapter we will focus on seawater desalination, with the aim of obtaining fresh water for human supply, irrigation or industrial facilities.

Seawater desalination has gained importance in coastal countries where conventional water sources are insufficient or overexploited. It can be considered an inexhaustible natural source that generates a high quality product and guarantees demand supply. On the other hand, desalinated water is expensive (due to high energy consumption) and the brine discharged into the sea has negative effects on some important marine ecosystems.

1.1 Environmental Impact by type of desalination project.

The main environmental impacts of desalination projects are associated with construction, marine structures, waste water disposal and energy consumption. The importance of these impacts depends on the type of technology used in salt separation.

MSF thermal plants work with small conversion rates (10% - 20%), so they need greater amounts of feedwater to produce the same volume of fresh desalinated water. The consequences are: a higher water intake, pipes and outfall structures, increased energy losses in pipes and more concentration of chemical additives required. Energy consumption with this technology is very high, which means a higher fuel consumption (Afgan et al, 1998), and thus, emissions of greenhouse gases. The waste water effluent has a slight hypersalinity with respect to the seawater receiving body. However, it has a significant thermal and chemical pollution capability, thus affecting water quality. In general, MSF brine is less dense than seawater, so it floats and rises to the surface, reducing impact risk on benthic ecosystems, but increasing the risk of contamination of recreational or commercial fishing areas. The combustion processes that take place in the plant generate emissions of air pollutants. Finally, visual impact is also significant because of the large amount of piping, tanks and chimneys associated with such plants.

RO plants work with conversion rates of 40 - 50%, so that the need of feedwater is smaller, as are the environmental impacts associated to it. Energy consumption is high but much lower than in MSF plants. The waste effluent or brine has no chemical or thermal pollution, but the salt concentration is very high, making it denser than seawater and thus increasing the risk of negative effects on stenohaline benthic ecosystems. RO plants do not include combustion processes resulting in no air pollution. Its visual impact is less because the plants are usually compact. However, an additional solid waste is generated by RO plants compared to those of MSF, since membranes need to be changed at a certain frequency and at the moment they are not reusable (Hoepner, 1999).

1.2 Desalination impacts on the marine environment.

Among the most important and significant **impacts of seawater desalination projects** are those associated with marine structures construction, as the water intake and outlet:

- Impacts on the water quality and on the benthic organisms present in the receiving water body, due to dredging of trenches and placement of new infrastructures.
- Impacts on navigation and fishing because of the presence of new infrastructures.
- Impacts on the coastal dynamics of beaches by the presence of structures in the active beach profile zone, which may affect longshore and cross-shore sediment transport.

The second and third impacts can be avoided by locating the marine structures in zones with no interference with other applications or processes, and informing the competent authorities of these activities. The following pages are dedicated to impacts and prevention and mitigation measures related to marine dredging and location of pipes.

To place underwater pipelines (associated with water intake and outfall), seabed dredging and trenching are conducted. The **impacts associated with dredging** are:

- Occupation and physical destruction of benthic ecosystems located in the dredging area.
- Effects on water quality_due to increase in suspended particles and turbidity (suspended solid concentration in the water column).
- Reduction in the percentage of light passing through the water column and reaching the seabed Gacía et al, (1999). This reduction can affect benthic primary producers. Some scientific studies carried out with *Posidonia oceanica* seagrasses show that suspended solid concentrations higher than 20mg/l adversely affect their growth.
- Burial of benthic organisms by suspended solids sedimentation. These particles may be transported by ambient currents and therefore affect benthic organisms even far away from the dredging area.

Regarding **marine water intakes**, the main impacts include:

- The risk of saltwater intrusion into nearby fresh groundwater aquifers, in case of subsurface water intakes.
- Regarding, open seawater intakes, the main impacts include:
  - Needs for more concentration of chemical additives in the pre-treatment phase, due to lower quality of feedwater.
  - Negative impacts on habitats which are in the vicinity of the intake due to the extraction of large quantities of water.
  - Impingement: Pinning of larger organisms on screen mesh by the withdrawn water flow, causing physical damage (peeling) and disorientation, due to the extraction of huge seawater flows through the screens.

- Entrainment: Passage of smaller organisms (often passive life stages, but also small fishes) living in the vicinity of the intake, through the screen mesh (Hogan, 2008).

The impacts associated with brine discharges into seawaters are related to:

- Effects on Water Quality due to potentital chemical pollution, anoxia at the sea bottoms and turbidity because of the presence os hipersaline effluent.
- Impacts on plankton by causing a drop in osmotic pressure (breaking the osmotic equilibrium between plankton organisms and seawater) and hence causing negative effects in primary production.
- Impacts on fish fauna. These communities, thanks to their mobility can swim far away from the turbidity and emissions associated with the brine and cleaning water discharges. However, extinction of the larvae and younger individuals (Einav & Lokiec, 2003) has been detected near MSF brine discharges. In the case of discharges by high velocity jets, a significant alteration of local hydrodynamics in the environment can affect sensitive fish species, especially the smaller individuals, creating confusion and increasing their vulnerability to predators. To reduce this impact, a jet discharge velocity of 3 -3.5 m/s should not be exceeded.
- Effects on coral reefs, which are very sensitive to changes in environmental conditions (chemical pollution, hydrodynamic alterations, temperature, salinity, etc.), and thus, brine disposal may have significant negative effects.
- Impacs on seagrasses and algae due to turbidity of the brine presence, which affects seagrasses by reducing the percentage of light filtered through the water column that reaches the seabed, thus affecting seagrass photosynthesis (Gacía et al, 2007).
- Impacts on seagrasses due to the presence of the hypersaline brine effluent, depending on the sensitivity of the species. Studies on marine angiosperms have detected a low tolerance to salinity and temperature changes in the conditions of the receiving environment. As an example, in the Mediterranean Sea there are ecologically important angiosperms (Gacía et al, 2007), as is the case of *Posidonia oceanica, Cymodocea nodosa, Zostera noltii,* with high ecological value, which are stenohaline species, and hence sensitive to salinity variations.

At the moment, there are no regulations limiting the physical parameters and chemical concentrations of brine effluents resulting from desalination processes (Palomar & Losada, 2009). The lack of legislation and the vulnerability and ecological importance of marine ecosystems justify the diverse studies carried out over the last years regarding the impact of hypersaline discharges in the marine environment.

Table 1 shows salinity thresholds, established by different authors, for some of the main Mediterranean Sea ecosystems and species.

In order to minimize the impacts of brine discharges on water quality and marine ecosystems, the following prevention and **mitigation measures** are proposed:

- Brine disposal should be placed in non-protected areas or in areas under anthropic influence.
- The brine discharge system should be placed in areas of high turbulence (Hoepnet & Windelberg, 1996), where ambient currents and waves facilitate brine dilution into the receiving water body. Ambient conditions, including slope, water column stratification and bottom currents are essential in far field dilution. If the discharge zone is deeper than the area to be protected, the latter should not be affected, since brine flows down slope to the bottom.

ECOSYSTEMS / SPECIES	CRITICAL SALINITY LIMITS	SOURCE
Posidonia oceanica seagrasses	Should not exceed 38.5psu in more than 25% of measurements: $S_{25,lim}$=38.5 Should not exceed 40psu in more than 5% of measurements: $S_{25,lim}$=40	(Sánchez Lizaso et al, 2008)
Cymodocea nodosa seagrasses	Should not exceed 39.5psu in more than 25% of measurements: $S_{25,lim}$=39.5 Should not exceed 41psu in more than 5% of measurements: $S_{25,lim}$=41	Spanish Ministry of the Environment
Caulerpa prolifera algae	Threshold established around 50-60psu	(Terrados & Ros, 1992)
Zostera noltii seagrasses	Threshold established around 41psu	(Fernández &Sánchez, 2006)
Mussels	Threshold established around 50-70psu	(Iso et al, 1994)

Table 1. Suggested limits in saline concentration for different ecosystems and species present in the Mediterranean Sea. Salinity in "psu", practical salinity units.

- The brine discharge configuration should consider the particular characteristics of the discharge area and the degree of dilution necessary to guarantee compliance with environmental quality standards and the protection of marine ecosystems located in the area affected by the discharge.
- If there are any protected ecosystems along the seabed in the area surrounding the discharge zone, it is recommended to avoid direct surface brine discharge systems because the degree of dilution and mixing is very weak.
- To maximize brine dilution, multiport jet diffuser discharge systems are recommended.

The following sections are focused on brine discharge, as one of the most important environmental impacts of desalination plant projects. Descriptions of the behaviour of brine in the near and far field regions, disposal systems and experimental and numerical modelling are included.

2. Brine discharge into seawaters

2.1 Behaviour of the brine: near and far field regiones.

Two regions with a different effluent behaviour should be considered when studying the discharge of brine into receiving water body: the near and the far field regions.

The **Near field region** is located in the vicinity of the discharge point and is characterised by initial mixing, which mainly depends on the brine discharge configuration design and the effluent and ambient properties. Higher dilution rates are reached at the near field, due to the turbulence effects created by the shear layer because of the differences of velocity between the jet and the ambient body. Flow and mixing characteristics are dominated by small scales (~metres and ~minutes). Normally, the brine discharge system is designed to maximize dilution in the near field region.

The **Far field region** is located further away from the discharge point, where the brine turns into a gravity current that flows down the seabed. Mixing depends on the ambient conditions (bathymetry, currents, waves, etc.) and the differences in density between the hypersaline plume and receiving waters. The water column appears stratified and the pycnocline difficults mixing between the hypersaline plume and seawater. The brine dilution ratio is very small in this region and tends to take an almost constant value. Flow and mixing characteristics are dominated by large scales (~kilometers and ~hours).

Figure 1 shows a diagram of the different behaviour areas of a brine jet discharge: ① jet ascending trajectory: the inclined jet is discharged with a certain velocity, so momentum (impulse) significantly influences its, ascending trajectory opposite to gravity force. At some distance from the discharge point, the buoyant force (weight) equals the momentum and the jet reaches its maximum height. From this point buoyancy is the dominant force and the jet descends ② to impact the bottom, where it undergoes an additional dilution due to turbulence phenomena and flow expansion. The region between the bottom impact zone and the far field region ③ is a transition zone, where flow behaves as a "spreading layer". In the far field region, brine behaves as a gravity current ④

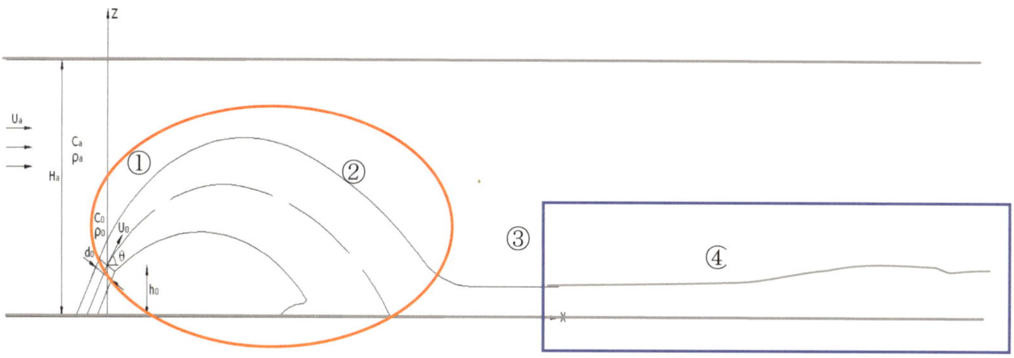

Fig. 1. Near and far field regions in a jet discharge, comparing brine and waste water effluents.

Figure 2 shows photographs of a brine single jet discharged from the SWRO Maspalomas desalination plant, located in Gran Canaria Island (Spain). Brine is coloured with rhodamine in order to study *ad hoc* the behaviour of the effluent discharged, in the near and far field regions. Pictures belong to the Instituto Canario del Agua, S.A. and area related to a Venturi research project (Portillo, 2009).

2.2 Brine discharge systems.

There are different management possibilities for the brine waste effluent generated in the desalination process:

- Discharge directly into the sea through some discharge configuration
- Discharge combined with other effluents (e.g., power plant cooling water or sewage treatment effluent).
- Dry out.

In most cases, especially in large desalination plants, the brine is discharged into seawater, because other alternatives are technically, socially, economically or environmentally not feasible.

There are different discharge configurations for brine discharges, the optimal one depending on the brine physical and chemical properties, the discharge location, the

Fig. 2. Pictures from an ad hoc brine discharge dyed by rhodamine in Maspalomas beach. Near (upper panel) and Far field (lower panel) regions can be observed.

ambient conditions and the presence of stenohaline protected species that can be particularly vulnerable to brine. Among others, the most common discharge systems are: direct surface disposal through gravel beaches, through watercourses, etc., overflow spill in a cliff, submerged single or multiple jets by outfalls, and discharge on a breakwater.

Figure 3 shows pictures of some types of brine discharge configurations:

The design of the discharge system determines the degree of brine dilution in the near field region, where density differences (between brine and seawater) and momentum (depending on the discharge system) control the geometry and mixing processes of the brine effluent. This dilution influences the salinity of the gravity current in the far field region and, consequently increasing risk of impact on benthic communities located far away from the discharging point.

Faced with the expected increase in flow rate of brine discharged into the Mediterranean Sea and the negative impact on the marine environment, the Spanish Center of Studies and Experimentation of Publish Works (CEDEX) carried out an experimental investigation on scaled physical models to determine the most effective dilution brine discharge systems in the near field region. Several systems were tested (Ruiz Mateo, 2007). According to previous studies, CEDEX concluded that the system generating the greatest dilution is the submerged

Fig. 3. Photographs of brine discharge configurations located in Spain. .A) Dicharge trough a submerged outfall. B) Surfce discharge .C) Discharge trough multiple jets (CEDEX).

multiport diffuser outfall with an angle of discharge of approximately 65°. In contrast, physical model tests simulating a surface discharge directly on a watercourse flowing into the sea revealed that, except in the collapse zone, mixing and dilution are very weak. According to this, the brine effluent rapidly turns into a negatively buoyant plume with a very high salt concentration that flows down the seabed, as a gravity current, in the far field region. Surface discharge tests indicate a dilution degree of about 4 at the end of the near field under stagnant ambient conditions.

3. Brine discharge modelling.

3.1 Introduction

Water quality modelling can simulate the behaviour of brine discharges, thus it is an essential prediction tool in the environmental assessment of desalination projects.

Regarding the goal of polluted effluent discharge models: considering the properties of the brine effluent and the discharge configuration, the model predicts brine disposal evolution under ambient conditions in the receiving water body. Simulation leads to prediction of the performance of quality standards (EQS) in the receiving waters and to guarantee that critical salinity limits will not be exceeded.

There are two types of modelling techniques:

- Experimental modelling: scaled physical models.
- Numerical modelling.

The following sections describe the main characteristics of these techniques. We will focus especially on numerical modelling since it is less expensive and, if correctly calibrated, can be generalized and used on any type of discharge configuration or ambient scenarios.

3.2 Experimental physical modelling.

Experimental modelling consists in performing laboratory experiments using scale physical models, which are a copy of the real case being tested, i.e.: the prototype, but normally at a smaller scale. Experimental tests can be carried out on any effluent, discharge configuration and ambient conditions.

The model and the prototype maintain the relative proportions (the scale factor) and they are scaled in terms of both geometry and forces. In order to guarantee the correspondence between the model and the prototype behaviour, the following conditions must be achieved:

1. **Geometric similarity** exists between model and prototype if the ratio of all corresponding dimensions in the model and prototype are equal. Dimension scales are defined by the formulas: $n_L = \frac{L_{model}}{L_{prototype}} = \frac{L_m}{L_p}$ where n_L is the scale factor for length and $n_L^2 = \frac{A_{model}}{A_{prototype}} = \frac{L_m^2}{L_p^2}$. All corresponding angles are the same.
2. **Kinematic similarity** is the similarity of time and geometry. It exists between model and prototype if the paths of moving particles are geometrically similar and if the ratio of the particles velocities are similar. Scales include ratios of discharge, acceleration: $n_a = \frac{a_{model}}{a_{prototype}} = \frac{L_m / T_m^2}{L_p / T_p^2}$, velocity: $n_v = \frac{v_m}{v_p} = \frac{L_m / T_m}{L_p / T_p}$ and time: $n_t = \frac{t_m}{t_p}$, with the relations: $n_v = (n_L)^{0.5}$ y $n_v = (n_t)^{0.5}$
3. **Dynamic similarity** includes geometrically and kinematically similar systems, if the ratios of all forces in the model and prototype are the same. The force ratio: $\frac{F_m}{F_p} = \frac{M_m a_m}{M_p a_p} = \frac{\rho_m L_m^3}{\rho_p L_p^3} * \frac{n_L}{n_T^2}$.

The forces acting on the fluid are: inertial gravity, viscosity, surface tension, elasticity and pressure, with different scales. In order to achieve dynamic similarity, the most influential forces are identified and secondary forces are neglected. Deviations between the model and the prototype are called "scale effects." In the case of moving fluids, the inertial ones are the predominant forces. The relationship of inertial forces and others leads to dimensionless numbers. The ratio between inertia and viscous forces is defined by the dimensionless Reynolds number. If its value is sufficiently high, the viscous forces can be neglected, thus the brine effluent behaviour depends mainly on the Densimetric Froude number, which is defined as the ratio between the inertial and the gravity forces: $F_{rd} = \frac{u}{\sqrt{g \frac{\Delta\rho}{\rho} D}} = \frac{u}{\sqrt{g_o' D}}$,

being u:: velocity; D: diameter of the orifice and $g_o' = g \frac{\rho_o - \rho_A}{\rho_{ref}}$: reduced gravitational buoyancy acceleration. ρ_o, ρ_a: effluent and ambient fluid density.

Traditionally, measures taken in laboratory experiments have been done with conventional techniques such as video, photography, conductivity meters, peristaltic pumps, etc., in order to typify the main characteristics of the effluents being discharged. In experimental tests, brine is usually dyed with rhodamine to distinguish its presence easily and describe its behaviour qualitatively. Geometric and kinematic similarities are guaranteed by scaling magnitudes in the model and the prototype, and dynamic similarity is considered to be achieved when the Densimetric Froude number remains the same. A high Reynolds number: Re> 1500 (Jirka, 2004) is required for the assumption of fully turbulent flow and neglected viscous forces.

Figure 4 shows photographs of physical model tests of a brine single jet discharge (Portillo, CEDEX). Rhodamine colouring makes it possible to observe the brine, which is denser than the receiving water and thus sinks to the bottom.

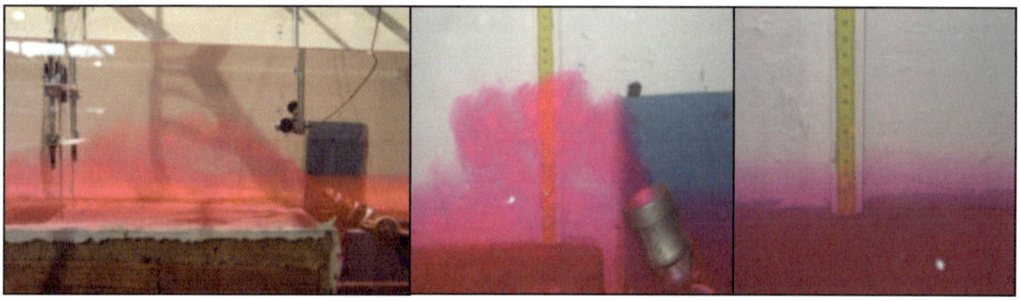

Fig. 4. Physical model test of a brine single jet discharge. Figure 4A shows the jet flow path in the near field region. Figure 4B) shows a detail of the jet orifice and flux exit. Figure 4.C) shows the brine hypersaline plume which is typical of the far field region.

Figure 5 shows a tank and precision conductivity meters to measure salinity at the bottom layer of the water column in the receiving body. Tank walls are white in order to correctly observe the rhodamine coloured brine.

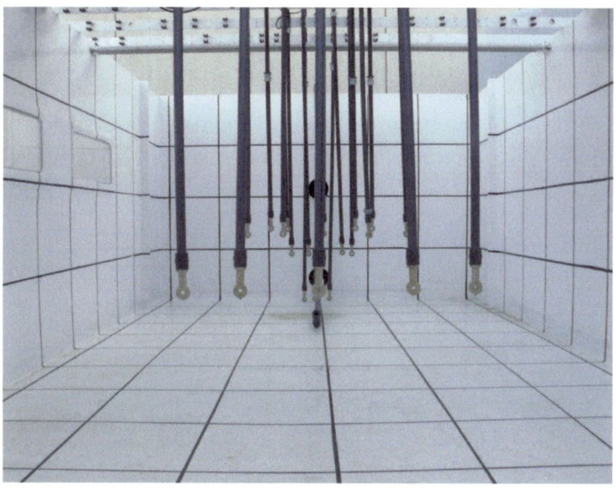

Fig. 5. Tank and gauges for brine discharge physical experiments.

Scaled physical models with conventional measurement techniques are generally used for a qualitative description of the effluent being discharged, and for determining the approximate geometry, dimensions and dilution degree of the effluent. Typically, quantitative measures are taken at control points (e.g. maximum rise height, impingement point distance and centerline dilution, in the case of jet discharges). Table 2 (Ruiz Mateo, 2007) shows, as an example, the approximate dilution rates obtained at the end of the near field region. Tests were carried out at the CEDEX Laboratory, in Spain.

APPROXIMATE VALUES OF DILUTION IN THE NEAR FIELD AREA Hypersaline effluent discharge		
BRINE DISCHARGE SYSTEM		DILUTION IN THE NEAR FIELD REGION
Discharge on gravel beaches		2,5
Discharge on mouth of channels flowing to seawaters		4
Discharge on a breakwater of a sheltered dock		6
Discharge by and horizontal submerged jet		10
Overflow spillway in a cliff discharge (influenced by the discharge height and depth available)		18
Discharge by single jet outfalls (Minimum dilution at the impact point)	Submerged 65° inclined jet, on the bottom.	30
	Submerged vertical jet, at surface level	8,7
	Submerged horizontal jet, at surface level.	10
	Above surface vertical jet	9
	Above surface horizontal jet.	23
Discharge by multiple jets diffuser outfalls	One orifices per diffuser	24
	Two orifices in opposite directions	30

Table 2. Estimated dilutions of the brine effluent in the near field region under different discharge configurations. Results obtained by scaled physical laboratory tests (Ruiz Mateo, 2007).

The experimental results obtained from conventional techniques are generally used to calibrate simple formulas based on dimensional analysis which describe the flux approximately. Some of the main dimensional analysis formulas for a single jet discharge characterization are (Pincince & List, 1973):

$$\frac{y_t}{DF} = C_1 ; \quad \frac{X_i}{DF} = C_2 ; \quad \frac{S_i}{F} = C_3$$

Being:

y_t : maximum rise height (maximum height of the top boundary or upper edge of the jet).

X_i : horizontal distance of the centerline peak at the impact (impingement) point

S_i : minimum centerline dilution at the impact point.

D: diameter of the orifice.

F: Densimetric Froude number.

$C_1, C_2.C_3$: experimental constants or coefficients obtained from laboratory physical scale models.

New and more sophisticated measuring techniques for laboratory experiments have been developed in the last years using advanced optical technology as Laser Induced Fluorescence (LIF) and Particle Image Velocimeter (PIV). With these techniques the concentration and velocity fields can be completely characterized. Results can also be used to calibrate and validate complex CFD (Computational Fluid Dynamics) numerical models. Table 3 shows the experimental coefficient values obtained by experimental research, focused on negatively buoyant jet discharges into stagnant environment:

NEGATIVELY BOUYANT SINGLE JET IN STAGNANT ENVIRONMENT.					
RESEARCH	α	Nº Froude	$\frac{y_t}{D}$	$\frac{x_i}{D}$	S_i
Zeitoun et al (1970) *Conventional techniques*	30º	25-60	1.04F	3.48	-
	45º	25-60	1.56F	3.33	-
	60º	25-60	2.13F	3.19	1.12F
Roberts et al, (1997) *Optical techniques*	60º	18-36	2.2F	2.4F	1.6F+/-12%
Cipollina et al (2009) *Convencional techniques*	30º	18-32	1.08	3.03	-
	45º	18-32	1.61	2.82	-
	60º	18-32	2.32	2.25	-
Kikkert et al (2007) (LA) *Optical techniques*	30º	27-50	1.07	3.18	1.51
	45º	27-50	1.71	3.332	1.71
	60º	27-50	2.2	2.79	1.81
Shao et al (2010) *Optical Techniques*	30º	18-36	1.05	3	1.45
	45º	18-36	1.47	2.83	1.26

Table 3. Experimental coefficients for dimensional analysis formulas for single port hyperdense jets (α : discharge angle).

3.3 Numerical modelling.

Water quality modelling is a mathematical representation of the physical and chemical mechanisms determining the development of pollutant concentrations discharged into the seawater receiving body. It involves the prediction of water pollution using mathematical simulation techniques and determines the position and momentum of pollutants in a water body taking into account ambient conditions.

Water quality modelling applied to brine discharges solves the hydrodynamics and transport equations adapted to a negatively buoyant effluent. The equations can be set up by a Lagrangian or Eulerian system. In the first case, the effluent brine is represented by a collection of particles moving in time and changing their properties. In the second case, the space is represented by a mesh of fixed points defined by their spatial coordinates, on which differential equations are solved.

Figure 6 shows the modelling scheme for designing brine discharges (Palomar et al, 2010).

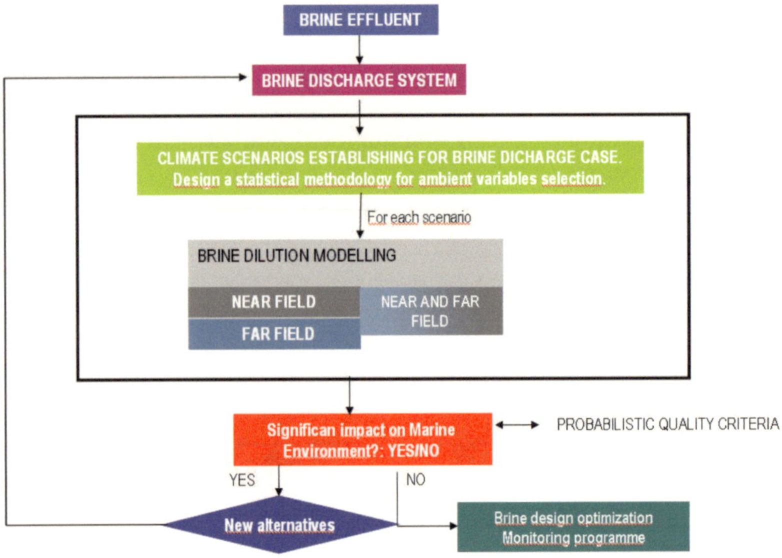

Fig. 6. Scheme of brine discharge modelling.

3.3.1 Symplifying assumptions within modelling.

Simplifying assumptions which are generally taken in the modelling of brine discharges are (Doneker & Jirka, 2001):

1. Incompressible fluid (pressure does not affect density of the fluid).
2. Reynolds decomposition: $f(t) = \overline{f}(t) + f'(t)$ the instantaneous value of a magnitude is the sum of a time-averaged component and a random (instant, turbulent) component.
3. Boussinesq approximation: density differences between effluent discharges and the water receiving environment are small and are important only in terms of the buoyancy force.
4. Turbulence closure model based on Boussinesq turbulent viscosity theory, $\rho\overline{u'_i u'_j} = \rho\mu_{ei}\dfrac{d\overline{U}_i}{dx_j}$. Turbulent terms are proportional to the average value of the magnitude, with an experimental proportionality coefficient (eddy viscosity). In recent years, more rigorous and sophisticated closure models, such as the k-ε model, are being applied.
5. Molecular diffusion is negligible compared to turbulent diffusion in the effluent.
6. There are no fluid sources or drain.

3.3.2 Governing equations.

Once the simplifying assumptions have been applied, the partial differential equations to be solved in brine discharge modelling are:

Equation of Continuity (Mass Conservation)

It is a statement of mass conservation. For a control volume that has a single inlet and a single outlet, the principle of mass conservation states that, for steady-state flow, the mass flow rate into the volume must equal the mass flow rate out of it. It relates velocity and density of the fluid.

$$\frac{\partial \bar{u}_i}{\partial x_i} = 0 \text{ Cartesian coordintes: } \left(\frac{\partial \bar{u}}{\partial x} + \frac{\partial \bar{v}}{\partial y} + \frac{\partial \bar{w}}{\partial z} \right) = 0$$

Equation of momentum conservation

The momentum equation is a statement of Newton's Second Law and relates the sum of the forces acting on a fluid element (incompressible) to its acceleration or momentum change rate: $\Sigma \bar{F} = \frac{d\bar{p}}{dt}$. Total force is the sum of surface forces (viscous stresses) acting by direct contact, and volume forces (inertial) acting without contact

$$\frac{D\vec{u_i}}{Dt} = -\frac{1}{\rho_o}\vec{\nabla} p - g\delta_{i3} + \mu_{ei}\nabla^2 \vec{u_i} \text{ Cartesian coordinates:}$$

$$\text{X Axis:} \rightarrow \rho_o \left(\frac{\partial \bar{u}}{\partial t} + \bar{u}\frac{\partial \bar{u}}{\partial x} + \bar{v}\frac{\partial \bar{u}}{\partial y} + \bar{w}\frac{\partial \bar{u}}{\partial z} \right) = -\frac{\partial p}{\partial x} + \mu_{ex}\left(\frac{\partial^2 \bar{u}}{\partial x^2} + \frac{\partial^2 \bar{u}}{\partial y^2} + \frac{\partial^2 \bar{u}}{\partial z^2} \right)$$

$$\text{Y Axis} \rightarrow \rho_o \left(\frac{\partial \bar{v}}{\partial t} + \bar{u}\frac{\partial \bar{v}}{\partial x} + \bar{v}\frac{\partial \bar{v}}{\partial y} + \bar{w}\frac{\partial \bar{v}}{\partial z} \right) = -\frac{\partial p}{\partial y} + \mu_{ey}\left(\frac{\partial^2 \bar{v}}{\partial x^2} + \frac{\partial^2 \bar{v}}{\partial y^2} + \frac{\partial^2 \bar{v}}{\partial z^2} \right)$$

$$\text{Z Axis} \rightarrow \rho_o \left(\frac{\partial \bar{w}}{\partial t} + \bar{u}\frac{\partial \bar{w}}{\partial x} + \bar{v}\frac{\partial \bar{w}}{\partial y} + \bar{w}\frac{\partial \bar{w}}{\partial z} \right) = -\frac{\partial p}{\partial z} + \mu_{ez}\left(\frac{\partial^2 \bar{w}}{\partial x^2} + \frac{\partial^2 \bar{w}}{\partial y^2} + \frac{\partial^2 \bar{w}}{\partial z^2} \right) - g\rho$$

Transport equation (Conservation of Solute mass)

For a control volume, changes in concentration (salinity) are due to: advective transport of fluid containing the substance, solute mass flow by diffusion, and destruction or incorporation of the substance in the fluid.

$$\text{Cartesian coordinates: } \frac{\partial \bar{c}}{\partial t} + \bar{u}\frac{\partial \bar{c}}{\partial x} + \bar{v}\frac{\partial \bar{c}}{\partial y} + \bar{w}\frac{\partial \bar{c}}{\partial z} = \frac{\partial}{\partial x}\left(\varepsilon_x \frac{\partial \bar{c}}{\partial x} \right) + \frac{\partial}{\partial y}\left(\varepsilon_y \frac{\partial \bar{c}}{\partial y} \right) + \frac{\partial}{\partial z}\left(\varepsilon_z \frac{\partial \bar{c}}{\partial z} \right)$$

Equation of State.

For an incompressible fluid, relates temperature, salinity and density. Normally the empirical equation of the UNESCO is used. Salinity is expressed in "psu (practical salinity units) and is calculated through fluid conductivity:

$$\begin{aligned} \rho(T,S) &= 999.842594 + 6.793952 \cdot 10^{-2}T - 9.09529 \cdot 10^{-3}T^2 + 1.001685 \cdot 10^{-4}T^3 - 1.120083 \cdot 10^{-6}T^4 + \\ &+6.536332 \cdot 10^{-9}T^5 + (0.824493 - 4.0899 \cdot 10^{-3}T + 7.6438 \cdot 10^{-5}T^2 - 8.2467 \cdot 10^{-7}T^3 + 5.3875 \cdot 10^{-9}T^4)S + \\ &+(-5.72466 \cdot 10^{-3} + 1.0227 \cdot 10^{-4}T - 1.6546 \cdot 10^{-6}T^2)S^{1.5} + 4.8314 \cdot 10^{-4}S^2 \end{aligned}$$

Variables in the equations are:
p : Fluid pressure at position (x, y, z).
(u,v,w) : Time averaged velocity components.
ρ : Effluent density at position (x,y,z).
μ_{ei} : Fluid dynamic viscosity of the fluid.
ν : Eddy viscosity
ε_i : Turbulent diffusion coefficient.
c : Pollutant concentration, in this case: salinity, at position (x,y,z).
U_o ; V_o ; Q_o ; ρ_o : velocity, volume, flow and density of the effluent at discharge.
U_A ; V_A ; Q_A ; ρ_A : velocity, volume, flow and density of the receiving seawater body.
D : diameter of the orifice.
$g_o' = g\frac{\rho_o - \rho_A}{\rho_{ref}}$: reduced gravitational buoyancy acceleration.
The variables "x" time averaged are expressed through an upper dash.

3.3 Model types according to mathematical approach.

There are three basic approaches for solving the equations according to the hypothesis and simplifications assumed, resulting in three types of physical and mathematical models to describe the behaviour of a discharge (Doneker &Jirka, 2001):

- Models based on a dimensional analysis of the phenomenon.
- Models based on integration of differential equations along the cross section of flow.
- Hydrodynamics models.

A) Models based on a dimensional analysis of the phenomenon.

The length scale models, derived from a dimensional analysis of the phenomenon, are the simplest models because they accept important simplifying assumptions.
Dimensional analysis is used to form reasonable hypotheses about complex physical situations that can be tested experimentally and to categorize types of physical quantities and units based on their relations to or dependence on other units, or their dimensions if any.
In dimensional analysis, variables with a higher influence in the phenomenon are considered, setting up the value of the ones with less influence, to reduce the independent variables under consideration. Selected independent variables are related through "flux" magnitudes, which represent the major forces determining effluent behaviour. For the discharging phenomenon, the main fluxes are:

- **Kinematic flux of mass**: $Q_0 = \frac{\pi}{4} D^2 U$. Dimension $\left[L^3 / T\right]$. Represents effluent flow discharged into the receiving environment.
- **Kinematic flux of momentum**: $M = UQ$. Dimension: $\left[L^4 / T^2\right]$. It represents the energy transmitted during the discharge of the effluent.
- **Kinematic flux of buoyancy**: $J = g'Q$ in dimension $\left[L^4 / T^3\right]$. Represents the effect of gravity on the effluent discharge.

Fluxes are combined with each other and with other parameters that influence discharge behaviour (ambient currents, density stratification, jet vertical angle, etc.) to generate length scale magnitudes that characterise effluent behaviour. The value of the length scales depends, anyhow, on the role of the forces acting on the effluent and varies along the trajectory of the effluent. The main length scales for a round buoyant jet are (Roberts et al, 1997):

Flux-momentum length scale. $l_Q = \frac{Q}{M^{1/2}}$: a measure of the distance over which the volume flux of the entrained ambient fluid becomes approximately equal to the initial volume flux.

Momentum-Buoyancy length scale. $l_M = \frac{M^{3/4}}{J^{1/2}}$: a measure of the distance over which the buoyancy generated *momentum* is approximately equal to the initial volume flux.

Assuming full turbulent flow (thus neglecting viscous forces), any dependent variable will be a function of the fluxes: Q, M, J. The dependent variables of interest may be expressed in terms of length scales, with a proportionality coefficient, which is obtained from laboratory experiments.

$$y_t, X_{i,} S_i = f_1(Q,M,J) = f_2(l_Q, l_M)$$

Considering $l_Q << l_M$, assuming Boussinesq hypothesis for gravity terms and using the equivalent expression obtained by substituying the values of M and J in the l_M expression: $l_M = \left(\frac{\pi}{4}\right)^{1/4} \bullet DF$, the variables of interest will depend on the diameter orifice and the Densimetric Froude number:

$$\frac{y_t}{DF} = C_1 ; \quad \frac{X_i}{DF} = C_2 ; \quad \frac{S_i}{F} = C_3$$

Being:

y_t : maximum rise height (maximum height of the top boundary or upper edge of the jet).
X_i : horizontal distance of centerline peak at the impact (impingement) point
S_i : minimum centerline dilution at the impact point.
U: discharge velocity.
D: diameter of the orifice.
F: Densimetric Froude number.
$C_1, C_2.C_3$: experimental constants or coefficients obtained from laboratory physical scale models (for a stagnant environment, different discharge angles, etc.).

As already explained, the dimensional analysis derives from highly simplified formulas for the characterization of the flow because governing equations are reduced to semi-empirical expressions of length scales. Since this method does not solve rigorous equations of the phenomenon, its reliability would depend on the range and quality of the experimental tests performed.

Some examples of the length scale models for brine discharge modelling are those showed in section 3.2, with the experimental coefficients obtained by several authors and showed in Table 3. Dimensional analysis formulas are also those used for CORMIX1 (Doneker & Jirka,

2000), and CORMIX2 (Akar & Jirka, 1991) subsystems of the CORMIX software (Doneker & Jirka, 2001).

B) Models based on the integration of differential equations.

Governing equations of flow are in this case integrated over the cross section, transforming them into simple ordinary differential equations which are easily solved with numerical methods, as Runge Kutta formula. These integration models are mainly used for jets and gravity current modelling.

Integration of the equation requires assumption of an unlimited receiving water body and consequently boundary effects cannot be modelled. Because of this, even if these models give detailed descriptions of the jet effluent, results are valid only in the effluent trajectory prior to the impact of the jet on the bottom, and whenever the effluent does not previously reach the surface or impact with obstacles or lateral boundaries. Since the results of the integrated equation refer to magnitudes in the brine effluent axis, calculations of these values in cross-sections require assuming a distribution function, generally Gaussian, and experimentally determining the basic parameters. Effluent diffusion is controlled in these models through simple "entrainment" formulas with coefficients obtained experimentally.

Commercial models of this type are: CORJET (Jirka, 2004, 2006) of CORMIX software; JetLag of VISJET software (Lee & Cheung, 1990) and UM3 of VISUAL PLUMES (Frick, 2004), all of them available for negatively buoyant discharges.

Some of the advantages of integration models are (Palomar & Losada, 2008): equation solving and calibration are quite easy and need few input data for modelling. Among the disadvantages is the unlimited receiving water, which limits brine discharges modelling to the near field region.

C) Hydrodynamic models

Hydrodynamics three-dimensional models are the most general and rigorous models for effluent discharge simulation. They solve differential hydrodynamics and transport equations with complete partial derivates. These models require a great number of initial data but can consider more processes and variables such as: boundary effects, bathymetry, salinity/ temperature (density) water columns stratification, ambient currents at different depths, waves, tides, etc.

Among their advantages are: more rigorous and complex phenomena modelling, possibility of continuous simulation of the near and far field region, simulation of any discharge configuration and ambient conditions.

At present, these models are not completely developed and have some limitations such as: coupling between the near and far field regions, because of the different spatial and time scales; need of a large amount of initial data; difficulty in calibration of the model and long computational time.

Hydrodynamics three dimensional models are: COHERENS software (Luyten et al, 1999), DELFT3D], etc.

3.4 Commercial tools for brine discharge modelling.

Nowadays there are many commercial tools for discharge modelling and some of them are adapted to simulate negatively buoyant effluents, as that of brine. These tools solve the numerical equations with approaches such as those explained in the previous section, considering the most relevant processes and determining the geometry and saline concentration evolution of the effluent.

CORMIX, VISUAL PLUMES and VISJET are some of the most notable commercial software for brine discharge modelling. The models predict brine behaviour, including trajectory, dimensions and dilution degrees, considering the effluent properties (e.g., flow rate, temperature, salinity, etc.), the disposal configuration and the ambient conditions (e.g., local water depth, stratification, currents, etc.). Commercial models are often used by promoters to design the discharge and by environmental authorities to predict potential marine impacts.
Figure 7 shows images and schemes of numerical results obtained by commercial software:

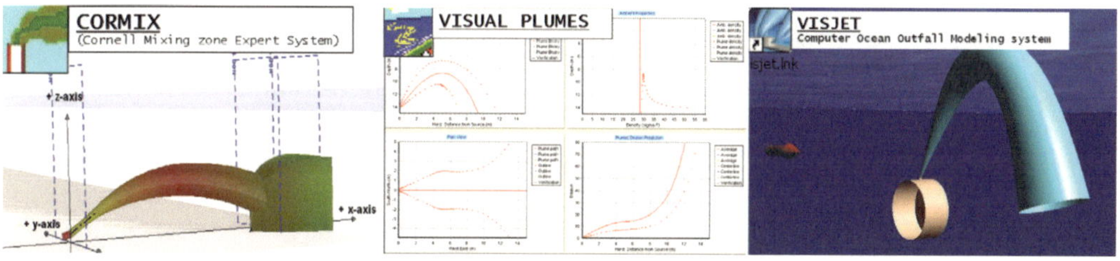

CORMIX, VISUAL PLUMES and VISJET include several models to simulate brine discharges through different types of discharge configuration. Table 4 shows the software models adapted to negatively buoyant effluents modelling:

CORMIX software	VISUAL PLUMES software	VISJET software
CORMIX 1: submerged and emerged single port jet. CORMIX 2: submerged multiport jets D-CORMIX: Direct surface discharge CORJET: submerged single and multi-port jets	UM3: submerged jets single and multi-port	JetLag; submerged jets single and multi-port
OTHER MODELS OF THE COMMERCIAL SOFTWARE		
CORMIX3: for positively buoyant effluents	DKHW, RSB: only positively buoyant effluents	

Table 4. Software models for brine discharge modelling.

3.4.1 CORMIX software.

CORMIX software (Cornell Mixing Zone Expert System) (Doneker & Jirka, 2001) was developed in the 1980s at Cornell University as a project subsidized by the Environmental Protection Agency (EPA). Since it was supported by EPA, it has become one of the most popular programs for discharge modelling.
CORMIX is defined as a Hydrodynamic Mixing Zone Model and Decision Support System for the analysis, prediction, and design of aqueous toxic or conventional pollutant discharges into diverse water bodies. It is an expert system, which also includes various subsystems for simulating the discharge phenomenon.
The subsystems: CORMIX 1, 2 and 3 are based on dimensional analyses of the phenomenon while the model CORJET is based on the integration of differential equations. CORMIX can simulate disposals of effluents with positive, negative and neutral buoyancy, under different types of discharge (single port and multiple port diffusers, emerged and submerged jets,

surface discharges, etc.) and ambient conditions (temperature/salinity, currents direction and intensity, etc.).
CORMIX is a steady state model, therefore time series data and statistical analyses cannot be considered.

CORMIX1: SUBMERGED SINGLE PORT DISCHARGES.

CORMIX1 (Doneker & Jirka, 1990) is the CORMIX subsystem applicable to single port discharges. Regarding negatively buoyant effluents, CORMIX1 can simulate submerged and emerged jets.
The model is based on a dimensional analysis of the phenomenon. The subsystem calculates flows, length scales and dimensionless relationships, and identifies and classifies the flow of study in one of the 35 flux classes included in its database. Once the flow has been classified, simplified semi-empirical formulas are applied in order to calculate the main features of the brine effluent behaviour.
CORMIX1 can make a roughly approximation of the brine effluent's behaviour in the near and the far field regions. CORMIX1 simulates the interaction of the flow with the contours and if no interaction is detected, it applies the model CORJET. CORMIX1 includes some terms to consider the COANDA attachment effect.
The main assumptions of CORMIX1 are:

- Since calculation formulas are mainly empirical, reliability depends on the quality and approach of the case study to the experiments used to calibrate the formulas.
- Unrealistically sharp transitions in the development of flow behaviour, for example: from the near to the far field region.
- "Black box" formula based on volume control for the characterization of some flux regions.
- Water body geometry restrictions: rectangular, horizontal and flat channel receiving water bodies. Limitations related to the port elevation with respect to the position of the pycnocline in a stratified water column.
- Unidirectional and steady ambient currents
- If flow impacts the surface, depending on water depth, CORMIX1 makes the simplification of flow homogenized in the water column, etc.

The initial data for CORMIX1 are: temperature, salinity or density of the effluent, pollutant concentration, jet discharge velocity or brine flow, diameter of the orifice, discharge angle, local water depth, port elevation, ambient salinity and temperature or ambient density, ambient current velocity and direction, among others.
One of the main limitations of CORMIX1 is the lack of validation studies for negatively buoyant effluents. Studies presented in the CORMIX1 manual only include the case of a vertical submerged jet discharged in a dynamic receiving water body, and the validation is restricted to trajectories, but not dilution rates. Other shortcoming is that in many cases the flux classification assumed by CORMIX1 does not match with the type of flow observed in the laboratory experiments. It is also important to be careful when using CORMIX1 since it is very sensitive to changes of input data and occasionally small changes in the data values lead to a misclassification of the flow in another flux class, resulting a completely different behaviour.
Some recommendations for using CORMIX1 in brine discharge modelling are: if a single jet with no interaction with the contours is to be designed, it is recommended to utilize the CORJET module instead of CORMIX1, or utilize both and compare the results to ensure that

the classification of the flow is correct and the results are consistent. Given the strong simplifying assumptions imposed and the lack of validation data, CORMIX1 should be avoided for simulations of single port brine discharges impacting the surface.

CORMIX 2: SUBMERGED MULTI-PORT DISCHARGES

CORMIX2 (Akar & JIrka, 1991) is the CORMIX subsystem applicable to submerged multiport discharges.

The model is based on a dimensional analysis of the phenomenon. The subsystem calculates flows, length scales and dimensionless relationships, and identifies and classifies the flow of study in one of the 31 flux classes included in its database. Once, the flow has been classified, simplified semi-empirical formulas are applied to characterize brine behaviour.

CORMIX2 can make a rough approximation of the brine effluent behaviour in the near and far field regions. CORMIX2 simulates the interaction of the flow with the contours and if no interaction is detected, it applies the model CORJET. CORMIX1 includes some terms to consider the COANDA attachment effect. One of the most important advantages of CORMIX2 is the possibility of modelling merging phenomena when contiguous jets interact.

The main assumptions of CORMIX2 are:

- If CORMIX2 detects merging between contiguous jets, it assumes the hypothesis of a equivalent slot diffuser, in which the discharge from the diffuser of equally spaced ports is assumed to be the same as a line slot discharge with the same length, brine flow rate and momentum as the set of ports. This assumption makes the model to consider a two-dimensional flow, with a uniform distribution across the section.
- As CORMIX1: since the calculation formulas are mainly empirical, reliability depends on the quality and the approach of the case studies of the experiments used to calibrate the formulas. Unrealistically sharp transitions in the evolution of flow behaviour and simplified receiving water body and "Black box" formulas are applied.
- Although CORMIX2 supposedly simulates a large variety of diffuser multi-port configurations (unidirectional, staged, alternating diffusers; same direction and fanned out jets), important assumptions are made, all cases leading to two types: a unidirectional diffuser with perpendicular jets and a diffuser with vertical jets. This fact causes important errors in the case of negatively buoyant effluents.

CORMIX2 initial data are: temperature, salinity or density of effluent, pollutant concentration, jet discharge velocity or brine flow, discharge angle, diameter of the orifices, port elevation, diffuser length, port spacing, number of ports, local water depth, ambient salinity and temperature and current velocity and direction, among others. An important shortcoming of CORMIX2 is the assumption applied to bilateral or rosette discharges, in which CORMIX2 considers the jets merging in a unique vertical single jet. This assumption is roughly correct for positively buoyant effluents whereas it is not valid for negatively buoyant effluents, leading to completely wrong results. The equivalent slot diffuser hypothesis leads in some cases to unrealistic results.

The limitations are similar to those of CORMIX1 in relation to receiving water body geometry simplifications, lack of validation studies for hyperdense effluents, or sensitivity to initial data variations.

Some recommendations for using CORMIX2 in brine discharge modelling are: given the strong simplifying assumptions imposed and the lack of validation data, CORMIX2 subsystem should be avoided in the case of flux interacting with contours. Due to the invalid hypotheses assumed, CORMIX2 cannot be used with bidirectional and alternating diffusers, rosettes and unidirectional diffuser with jets forming less than 60º. The typical diffuser configuration with bidirectional jets forming 180º should be modelled by CORMIX2 considering separately each diffuser side.

CORJET: CORNELL BUOYANT JET INTEGRAL MODEL

CORJET is a model of CORMIX applicable to submerged single port (Jirka, 2004) and multi port discharges (Jirka, 2006).
It is a three dimensional eulerian model based on the integration of the differential equations of motion and transport through the cross section, obtaining the evolution of the jet axis variables. The integration of the differential equations transforms them into an ordinary equation system, which is solved with a four order Runge Kutta numerical method. Integration requires assuming an unlimited receiving water body and sections self similarity. Regarding the variables distribution in the jet cross section, CORJET assumes Gaussian profiles since it has been experimentally observed in round jets.
Since the model assumes unlimited environment, it cannot simulate the interaction of the jet with the contours, thus the scope is limited to the near field zone, before the impingement of the jet with the bottom. The COANDA effect and intrusion are not modelled by CORJET. As CORMIX1 and CORMIX2, CORJET validation studies are very scarce and limited to the jet path with few dilution data (Jirka, 2008). Regarding the diffuser configuration, CORJET can only model unidirectional jets perpendicular to the diffuser direction, with the same diameter orifices, equal spaces, and with the same port elevation and discharge angle.
CORJET initial data are similar to those indicated for CORMIX1 and CORMIX2, with the advantage of a more detailed description of the flux, with the evolution of the variables of interest (axis trajectory (x,y,z), velocity, concentration, etc.)
For calculating the jet upper edge position it is recommended to add to the maximum height axis (zmax), the radius, calculated with the formulas $r=\sqrt{2}b$ or $r=2b$, "b" being the radial distance in which the concentration is 50% and velocity amounts to 37% of axis concentration and velocity respectively. The $r=\sqrt{2}b$ value stands for the radial distance in which the concentration is 25% and velocity is 14% of that in the jet axis. The value $r=2b$ stands for the radial distance in which the concentration is 6% and velocity is 2% of that in the jet axis. The user must verify that the jet does not impact the surface by calculating this addition.
Since CORJET cannot simulate COANDA effects it is recommended not to simulate jets with a discharge angle smaller than 30º and zero port height. Since it does not either model reintrusion phenomena, discharge angles larger than 70º should not be simulated with CORJET.

3.4.2 VISUAL PLUMES software.

VISUAL PLUMES (Frick, 2004) is a software developed by the Environment Protection Agency (EPA), which includes several models to simulate positively, negatively and neutrally buoyant effluents discharged into water receiving bodies.
VISUAL PLUMES considers the effluent properties, the discharge configuration and the ambient conditions (temperature, salinity and currents whose intensity and direction can be variable through the water column). It is limited to the near field region modelling and does not simulate the interaction of the flow with the contours. VISUAL PLUMES can consider time series data, simulating discharges under scenarios which change over time.

"UM3" MODEL (UPDATED MERGE 3D): SINGLE AND MULTI-PORT DIFFUSER.

UM3 is the only model of VISUAL PLUMES applicable to negatively buoyant effluents. It is a three dimensional lagrangian model which simulates the behaviour of submerged single or multi port jet discharges into stagnant or dynamic environments. It is based on the integration of motion and transport differential equations, and shows the evolution of the variables along the jet axis. As CORJET, UM3 also assumes an unlimited receiving water body and sections self similarity, but it considers a uniform ("top hat") distribution of the variables across the section.

UM3 includes the possibility of simulating a tide effect on the behaviour of the discharge. The water column can be separated into layers with different temperature and salinity values, and velocity or intensity of currents.

As a model based on the integration of differential equations, it cannot simulate COANDA effects, reintrusion phenomena or interaction of the flow with the contours, so its scope is limited to the point before jets impinge with the bottom. Regarding the diffuser configuration, UM3 can only model unidirectional jets perpendicular to the diffuser's direction, with the same diameter orifices, equal spaces, and with the same port elevation and discharge angle.

No validation data have been found in the literature for negatively buoyant effluents modelled with UM3.

Some recommendations are: the user must enter at least two levels (surface and depth) to run the model; UM3 does not break when the jet impacts the bottom so the user must be careful to reject results beyond this point. UM3 considers a uniform distribution of magnitudes in the cross section, thus if UM3 dilutions are compared with CORJET axis dilutions, the following formula must be applied: $D_{axis} = D_{Top-Hat} / 1.7$.

3.4.3 VISJET Software.

VISJET software (Innovative Modeling and Visualization Technology for Environmental Impact Assessment) has been developed by the University of Hong Kong.

JETLAG MODEL (LAGRANGIAN JET MODEL): SINGLE AND MULTI-PORT DIFFUSERS.

JetLag is a three dimensional lagrangian model which simulates single and multi-port submerged jet discharges. It can simulate positively, negatively and neutrally buoyant effluents, considering stagnant or dynamic water environments.

JetLag does not strictly resolve the mathematical governing equations, but makes an approximation of the physical processes, considering entrainment phenomena, in each slice in which the jet has been previously discretized. It assumes section self similarity and considers a uniform ("Top Hat") distribution of the variables in the cross section.

Among its possibilities, it can consider tidal effects on the effluent behaviour. Water column can be discretized into layers, with different temperature or salinity values, and ambient currents. JetLag allows different designs for each jet, i.e.: a different diameter in each orifice, different port elevations, angles of discharge, velocity, etc., in each jet. This fact is due to the fact that JetLag calculates each jet independently.

JetLag cannot simulate the COANDA effect, the intrusion phenomenon or the interaction of the flow with the contours. Because of this, JetLag is limited to the point before the jet impacts the bottom. An important shortcoming of Jetlag, which the users should take into

account, is that the model does not consider the merging between jets although it seems to do that. Thus, the choice of diffuser type is not relevant since JetLag always calculates each jet individually as a single port. JetLag cannot consider time series.
Some recommendations for using JETLAG in brine discharge modelling are: the user must enter at least two vertical levels in the discretization of the vertical column. Because Jetlag only simulates single individual jets and cannot calculate merging between jets, it should not be used for multi-port diffuser modelling. The user must calculate the upper edge of the jet and calculate if it impacts the surface (invalidating the model) since JetLag only fails when the axis impacts the surface. JetLag results can be directly compared with UM3 since both assume a uniform distribution.

3.5 Research related to brine discharge behaviour and modelling: State of art.

The first research related to brine discharge behaviour started in the 1940s in the United States, and increased radically during the 1960 and 1970 decades.
Regarding the description of the near field region, Turner, 1996, carried out a dimensional analysis of the phenomenon and established length scales for jet characterization, considering those variables with strongest influence. Some years later, Turner conducted physical (scale) laboratory tests to determine experimental coefficient values for the maximum rise height of a negatively buoyant vertical jet in stagnant waters. Other authors, such as Holly et al, 1972, followed this line, but extended the studies to other geometrical jet characteristics. Zeitoun et al, 1970, studied the influence of the discharge angle on jet behaviour for 30º, 45º, 60º and 90º angles, obtaining the highest dilution with 60º angles. Since then 60º has been established as the optimum angle for hyperdense jet discharges. Gaussian profiles along jet cross sections were also observed by Zeitoun. Pincince & List, 1973, based on Zeitoun´s results, studied the effect of dynamic environments in a 60º jet, concluding that they increase dilution. Chu, 1975, proposed a theoretical model. Fisher et al, 1979, described the three fluxes which are the base of dimensional analysis in relation to round buoyant jets. Roberts & Toms, 1987, studied the behaviour of vertical and 60º jets into stagnant and dynamic receiving environments. A significant quantity of laboratory tests were carried out obtaining experimental coefficients for dimensional analysis formulas. Roberts et al, 1997, developed new experiments using optical Laser Fluorescence induced (LIF) techniques for a more rigorous study of a 60º hyperdense jet, discharged on a stagnant environment.
Cipollina et al, 2005, developed a numerical model for hyperdense jets discharged into a stagnant environment, based on the integration of differential equations. Jirka, 2004, proposed a more complex eulerian three dimensional integration model for stagnant and dynamic environments. This same author (Jirka, 2006) extended his model to multiport discharges, considering the interaction or merging of jets. Jirka, 2008, introduced the effect of the bottom slope on jet behaviour. Cipollina et al, 2009, presented new experimental coefficients for dimensional analysis formulas.
During the last decade, several authors have performed experimental research using advanced optical techniques, as LIF and PIV, in order to acquire a better knowledge of jet velocity and concentration fields. Ferrari, 2008, studied 60º and 90º jets in stagnant and wavy environments. Chen et al, 2008, also considered the effect of waves on jets.
Kikkert & Davidson, 2007, proposed an analytical model for single jet modelling and calibrated it with experimental coefficients obtained from physical scale tests, using LIF and

LA techniques. Kikkert compared his results with those of other authors. Papanicolaou et al, 2008, reviewed the entrainment state of the art and proposed new values for negatively buoyant effluents. Gungor & Roberts, 2009, studied the behaviour of a vertical jet in a dynamic environment. Recently, Shao, 2010, carried out physical scale experiments with 30° and 60° jets, taking measurements with PIV and LIF optical techniques, and obtained experimental coefficients for dimensional analysis formulas. Plum, 2008, applied the commercial CFD software FLUENT for brine modelling, analysing different turbulence models.

Regarding the far field region, where brine forms a gravity current, the first important research was carried out by Ellison & Turner, 1959, who developed a two dimensional integration model with a simple entrainment formula. The authors experimentally proved that, at some distance from the discharge point, the plume takes a Richardson number constant value. Fietz & Wood, 1967, considered a three dimensional plume and analyzed the influence of the discharge. Alavian, 1986, proposed a three-dimensional integration model and distinguished between supercritical and subcritical behaviours. Garcia, 1996, presented an interesting two dimensional integration model based on the eddy viscosity formula for entrainment. Raithby et al, 1988, applied a more complex turbulence model in a three-dimensional hydrodynamic model, calibrating it with experimental results.

Regarding entrainment phenomena research, Turner, 1986, studied the mixing associated to turbulence movement and the effect of viscosity in effluent mixing and behaviour. Kaminski et al, 2005, experimentally and theoretically studied turbulent entrainment in jets with arbitrary buoyancy. Papanicolau et al, 2008, studied the entrainment phenomenon in negatively buoyant jets.

Alavian at al, 1992, expanded their study to a three-dimensional flow moving in a stratified environment. Tsihrintzis & Alavian, 1986, experimentally obtained an equation for calculating the plume width in a laminar regime. Christodoulou & Tzachou, 1979, simulated the behaviour of three-dimensional gravity currents in scaled tanks and obtained formulas for calculating the velocity, the width and the thickness of the gravity current. Cheong & Han, 1997, studied the influence of the bottom slope in plume behaviour. Bournet et al, 1999, applied different turbulence closure models, performing laboratory experiments and obtaining coefficients for dimensional analysis formulas.

Ross et al, 2001, presented a model based on integration equations to simulate a gravity current on a sloping bottom, and supported it with laboratory data, including geometry and dilution. Özgökmen & Chassignet, 2002, studied the behaviour of a plume, varying the parameters of interest and considering small-scale turbulence.

Bombardelli et al., 2004, studied three-dimensional gravity currents using CFDs (Computational Fluid Mechanics) models, capturing small-scale turbulent phenomena, and comparing the results obtained using different commercial software. Oliver et al, 2008, discussed the mixing of a hypersaline plume with ambient fluid using a closure model for turbulent terms. Joongcheol Paik et al, 2009, used a three dimensional RANS equations model to simulate a two-dimensional plume, comparing experimental data with numerical results using different turbulence closure models.

Dallimore et al, 2003. used an underflow model coupled to a three dimensional hydrodynamic model, comparing numerical results with field data. Martin & García, 2008, conducted an experimental research combining optical PIV/LIF measurements to study gravity currents. Recently, Hodges et al, 2010, modelled a real case of a brine discharge gravity current from a desalination plant in Texas (U.S).

3.6 Shortcomings and research line proposal.

The following paragraphs illustrate the main shortcomings detected in the different fields related with brine discharge modelling and the knowledge of impact on the marine environment, proposing some research lines.

As regards the **effects on the marine environment**, it is necessary to establish critical salinity limits, in statistical terms, for ecologically important species which are sensitive to hyper-salinity and are located in areas of frequent brine discharges. It is also important to carry out additional studies regarding the synergistic effects of different effluent discharges, as is the case of brine mixed with cooling water or seawater waste effluents.

Regarding **regulations,** a new legislation regulating brine discharges, which includes emission limit values and quality standards in the environment is still necessary. Regulation defining dimensions of the mixing zone would be also interesting.

Regarding **brine discharge systems,** some discharge configurations such as direct surface disposal, discharge on gravel beaches, on the mouth of channels flowing to seawaters, discharge on a breakwater sheltered dock or overflow spillway in a cliff discharge, among others, are in need of further investigations. Research must be focused on quantitative descriptions, including dilution rates and modelling.

As regards **methodologies,** new ones are needed for brine discharge systems design and marine impact assessment, that describe all the aspects that need to be taken into account.

Regarding **brine discharge modelling,** the following research is proposed to improve current knowledge on the matter (Palomar & Losada, 2010):

- Methodology to describe the marine climate and selection of the ambient scenarios in statistical terms, including the most frequent and unfavourable conditions.
- Further investigation of the entrainment phenomenon for negatively buoyant effluents.
- Recalibration of numerical models with experimental coefficients obtained from experimental measurements carried out with the most rigorous and precise optical techniques developed in the last years.
- To improve the knowledge of the gravity currents behaviour and to develop tools for three dimensional numerical modelling, considering the effect of bathymetry, waves, bottom currents, environment stratification, etc.
- To study the possibility of coupling near and far field processes modelling.

To improve the knowledge in some of these areas, several investigation projects are being developed. One of the most important in the Mediterranean area is the project "Horizon 2020 initiative" which aims to eliminate pollution in the Mediterranean by the year 2020 by tackling the sources of pollution, including brine from desalination plants. In Spain, the following projects, financed by the Ministry of Environment, are being developed to improve brine discharge knowledge and methodologies:

- **ASDECO project** (Automated control system for Desalination dilution), the objectives of which are: to design, develop and validate a prototype of the Automatic Control of Toxic Desalination; analyzing real-time ocean-meteorological data of the receiving environment and effluent data (all recorded by the system itself ASDECO), focusing on its application in brine discharge environmental monitoring plans.
- **VENTURI project** (Portillo, 2010), , which aims to test the efficiency in the dilution degree of Venturi systems as compared to conventional broadcasters, for single port submerged jet discharges, while "ad hoc" studying the near and far field regions of a brine discharge in the Canary Islands (Atlantic Ocean).

- **MEDVSA project** (Palomar et al, 2010) which aims to develop a methodology in order to improve brine discharge system design to reduce the impacts of brine discharges on the marine environment. The objective is to make compatible the use of desalination as an important water resource in some Spanish coastal areas, with the protection of marine areas, while following Sustainable Development principles. Two important Spanish Research Centres: IH Cantabria and CEDEX are collaborating in the R&D project "MEDVSA" development. It includes the following tasks: experimental research (Scale physical models), numerical research for near and far field simulations (including commercial tool analysis, online MEDVSA tools, using CFDs for near field modelling and ROMS application for far field simulation, etc.); climate scenario research; numerical tool validation (field works and experimental tests); a methodological guide and dissemination and training. Regarding commercial tools analysis, CORMIX, VISUAL PLUMES and VISJET, focused on negatively buoyant effluents, have been analyzed in detail. As a result, Technical Specification Cards have been developed, including: theoretical basis, simplifying assumptions, modelling options, possibilities and limitations and recommendations for implementation and management. After having analyzed in detail the brine discharge commercial simulation tools and having reviewed the existing literature on the matter, different codes are being programmed in order to have freely accessible tools, with codes similar to those of the commercial software. These tools will be calibrated and validated with the results of new laboratory tests. Technical Specification Cards and MEDVSA online tools are available and can be downloaded from the Web Page of the MEDVSA project.

4. Recommendations on the design and modelling of brine discharges into the sea.

In order to improve design of brine discharge systems, the following paragraphs propose some recommendation for reducing marine environment impacts faced to these disposals (Palomar & Losada, 2010):

- Brine disposal should be placed in non-protected areas or in areas under anthropic influence.
- The brine discharge system should be placed in areas of high turbulence, where ambient currents and waves facilitate brine dilution into the receiving water body. Ambient conditions, including slope, water column stratification and bottom currents are essential in far field dilution. If the discharge zone is deeper than the area to be protected, the latter should not be affected, since brine flows down slope to the bottom.
- The brine discharge configuration should consider the particular characteristics of the discharge area and the degree of dilution necessary to guarantee compliance with environmental quality standards and the protection of marine ecosystems located in the area affected by the discharge.
- If there are any protected ecosystems along the seabed in the area surrounding the discharge zone, it is recommended to avoid direct surface brine discharge systems because the degree of dilution and mixing is very weak.
- To maximize brine dilution, jet discharge configurations, through outfall structures, are recommended to be installed. It can be a solution when there are ecologically important stenohaline species near the discharge area. The following requirements are recommended to optimize jet discharges:

- The densimetric Froude number at the discharge must always be higher than 1, even so the installation of valves is recommended.
- Jet discharge velocity should be maximized to increase mixing and dilution with seawater in the near field region. The optimum ratio between the diameter of the port and brine flow rate per port is set so that the effluent velocity at discharge is about 4 – 5 m/s.
- Nozzle diameters are recommended to be bigger than 20cm, to prevent their clogging due to biofouling.
- To maximize mixing and dilution with submerged outfall discharges, a jet discharge angle between 45° and 60° with respect to the seabed is advisable, under stagnant or co-flowing ambient conditions. In case of cross-flow, vertical jets (90°) reach higher dilution rates (Roberts et el, 1987)- Avoid angles exceeding 75° and below 30 °.
- Diffusers (ports) should be located at a certain height (elevation) above the seabed, avoiding the brine jet interaction with the hypersaline spreading layer formed after the jet impacts the bottom. This port height can be set up between 0.5 and 1.5 m.
- The discharge zone is recommended to be deep enough to avoid the jet from impacting the surface under any ambient conditions.
- Avoid designs with several jets in a rosette.
- Riser spacing is recommended to be large enough to avoid merging between contiguous jets along the trajectory, because this interaction will reduce the dilution obtained in the near field region and also because the modelling tools to simulate this merging are less feasible.

- If it is necessary to build a submarine outfall, and it passes through interesting benthic ecosystems, a microtunnel to locate the pipeline should be constructed.
- As a prevention measure, modelling tools should be used for modelling discharge and brine behaviour into seawaters, under different ambient scenarios.
- An interesting alternative is to discharge brine into closed areas with a low water renovation rate, or areas receiving wastewater disposals. This mixture is favourable since it reduces chemicals concentration and anoxia in receiving waters.
- An environmental monitoring plan must be established, including the following controls: feedwater and brine flow variables, surroundings of the discharge zone, receiving seawater bodies and marine ecosystems under protection located in the area affected by the brine discharge.

Regarding **brine discharge modelling** (Palomar & Losada, 2010):

- Modelling data must be reliable and representative of the real brine and ambient conditions. Their collection should be carried out by direct measurements in the field. The most important data in the near field region are: 1) brine effluent properties: flow rate, temperature and salinity, or density, and 2) discharge system parameters. In the far field region, mixing is dominated by ambient conditions: bathymetry, density stratification in the water column, ambient currents on the bottom, etc.
- In the case of using CORMIX1 or CORMIX2 for brine discharge modelling, it must be taken into account that both are based on dimensional analysis and thus reliability depends on the quality of the laboratory experiments on which they are based, and on the degree of assimilation to the real case to be modelled. The scarcity of validation studies for negatively buoyant effluents in CORMIX1 and CORMIX2, is one of the main shortcomings of these commercial tools.

- For each simulation case, it is recommended to use different models and to compare the results to ensure that jet dimensions and dilution are being correctly modelled. It is also recommended to run the case under different scenarios, always within the range of realistic values of the ambient parameters.
- With respect to brine surface discharges, most of the commercial codes: RSB and PSD of VISUAL PLUMES or CORMIX 3 of CORMIX focus on positively buoyant discharges. D-CORMIX is designed for hyperdense effluent surface discharges but has not yet been sufficiently validated and therefore cannot be considered feasible at the moment.
- For far field region behaviour modelling, hydrodynamics three-dimensional or quasi-three dimensional models are recommended. At present, these models have errors linked to numerical solutions of differential equations, especially in the boundaries of large gradient areas, such as the pycnocline between brine and seawater in the far field region. These errors can be partially solved if enough small cells are used in the areas where large gradients may arise, but it significantly increases the modelling computation time.
- It is necessary to generate hindcast databases of ambient conditions in the coastal waters which are the receiving big volumes of brine discharges, considering those variables with a higher influence in brine behaviour. Analysis of this database by means of statistical and classification tools will allow establishing scenarios to be used in the assessment of brine discharge impact.

5. Conclusion

Desalination projects cause negative effects on the environment. Some of the most significant impacts are those associated with the construction of marine structures, energy consumption, seawater intake and brine disposal.

This chapter focuses on brine disposal impacts, describing the most important aspects related to brine behaviour and environmental assessment, especially from seawater desalination plants (SWRO). Brine is, in these cases, a hypersaline effluent which is denser than the seawater receiving body, and thus behaves as a negatively buoyant effluent, sinking to the bottom and affecting water quality and stenohaline benthic marine ecosystems.

The present chapter describes the main aspects related to brine disposal behaviour into the seawater, discharge configuration devices and experimental and numerical modelling. Since numerical modelling is currently and is expected to be in the future, a very important predictive tool for brine behaviour and marine impact studies, it is described in detail, including: simplifying assumptions, governing equations and model types according to mathematical approaches. The most used commercial software for brine discharge modelling: CORMIX, VISUAL PLUMES y VISJET are also analyzed including all modules applicable to hyperdense effluent disposal. New modelling tools, as MEDVSA online models, are also introduced.

The chapter reviews the state of the art related to negatively buoyant effluents, outlining the main research being carried out for both the near and far field regions. To overcome the shortcomings detected in the analysis, some research lines are proposed, related to important aspects such as: marine environment effects, regulation, disposal systems, numerical modelling, etc. Finally, some recommendations are proposed in order to improve the design of brine discharge systems in order to reduce impacts on the marine environment. These recommendations may be useful to promoters and environmental authorities.

6. References

Afgan, N.H; Al Gobaisi, D; Carvalho, M.G. & Cumo, M. (1998). Sustainable energy management. *Renewable and Sustainable Energy Review, vol 2*, pp. 235–286.

Akar, P.J. & Jirka, G.H. (1991). CORMIX2: An Expert System for Hydrodynamic Mixing Zone Analysis of Conventional and Toxic Submerged Multiport Diffuser Discharges, *U.S. Environmental Protection Agency (EPA), Office of Research and Development..*

Alavian, V. (1986). Behaviour of density current on an incline. *Journal of Hydraulic Engineering, vol 112*, No 1.

Alavian, V; Jirka, G.H; Denton, R.A; Jhonson, M.C; Stefan, G.C (1992). Density currents Entering lakes and reservoirs. *Journal of Hydraulic Engineering, vol 118*, No 11.

ASDECO project: Automated System for desalination dilution control: http://www.proyectoasdeco.com/

Bombardelli, F.A; Cantero, M.I; Buscaglia, G.C & García, M.H. (2004). Comparative study of convergence of CFD commercial codes when simulating dense underflows. Mecánica *computacional. Vol 23*, pp.1187 -1199.

Bournet, P.E; Dartus, D; Tassin.B & VinÇon-Leite.B. (1999). Numerical investigation of plunging density current. *Journal of Hydraulic Engineering. Vol. 125*, No. 6, pp. 584-594.

CEDEX: Spanish Center of Studies and Experimentation of Publish Works www.cedex.es

Chen, Y.P; Lia, C.W & Zhang, C.K. (2008). Numerical modelling o a Round Jet discharged into random waves. *Ocean Engineering, vol. 35*, pp.77-89.

Christodoulou, G.C & Tzachou, F.E, (1997). Experiments on 3-D Turbulent density currents. *4th Annual Int. Symp. on Stratified Flows, Grenoble, France.*

Chu, V.H. (1975). Turbulent dense plumes in a laminar crossflow. *Journal of Hydraulic research*, pp. 253-279.

Cipollina, A; Brucato, A; Grisafi, F; Nicosia, S. (2005). "Bench-Scale Investigation of Inclined Dense Jets". *Journal of Hydraulic Engineering, vol 131, n 11*, pp. 1017-1022.

Cipollina, A; Brucato,A & Micale, G. (2009). A mathematical tool for describing the behaviour of a dense effluent discharge. *Desalination and Water Treatment, vol 2*, pp. 295-309.

Dallimore, C.J, Hodges, B.R & Imberger, J. (2003). Coupled an underflow model to a three dimensional Hydrodinamic models. *Journal of Hydraulic Engineering, vol 129, nº10*, pp. 748-757

Delft Hydraulics Software. Delft Hydraulics part of Deltares. Available from: http://delftsoftware.wldelft.nl/.

Doneker, R.L. & Jirka, G.H. (1990). Expert System for Hydrodynamic Mixing Zone Analysis of Conventional and Toxic Submerged Single Port Discharges (CORMIX1). *Technical Report EPA 600-3-90-012. U.S. Environmental Protection Agency (EPA).*

Doneker, R.L & Jirka, G.H. (2001). CORMIX-GI systems for mixing zone analysis of brine wastewater disposal. Desalination *(ELSEVIER), vol 139*, pp. 263–274. www.cormix.info/.

Einav, R & Lokiec, F. (2003). Environmental aspects of a desalination plant in Ashkelon. Desalination (ELSEVIER), vol 156, pp. 79-85.

Ellison, T.H& Turner, J.S. (1959). Turbulent entraintment in stratified flows. *Journal of Fluids Mechanics, vol 6, part 3,* pp.423-448.

Environmental Hydraulics Institute "IH Cantabria" & Centre of Studies and Experimentation of Public Works (CEDEX). Project MEDVSA (A methodology for the design of brine discharges into the seawater), *funded by the National Programme for Experimental Development of the Spanish Ministry of the Environment and Rural and Marine Affairs (2008-2010). Available from:* www.mevsa.es. *Technical specification cards*

Environmental Hydraulics Institute: IH Cantabria. University of Cantabria, Spain. www.ihcantabria.com

Fernández Torquemada, Y & Sánchez Lisazo, J.L. (2006). Effects of salinity on growth and survival of Cymodocea *nodosa* (Ucria) *Aschernon* and *Zostera noltii Hornemann". Biology Marine Mediterranean, vol 13 (4),* pp 46-47.

Ferrari, S & Querzoli, G. (2004). Sea discharge of brine from desalination plants: a laboratory model of negatively buoyant jets. *MWWD 2004, 3th International Conference on Marine Waste Water Disposal and Marine Environment.* Ferrari, S;

Fietz, T.R & Wood, I. (1967). Three-dimensional density currents. *Journal of Hydraulics Division.* Proceedings of the American Society of Civil Engineers.

Frick, W.E. (2004). Visual Plumes mixing zone modelling software. *Environmental & Modelling Software (ELSEVIER),* vol 19, pp 645-654. http://www.epa.gov/ceampubl/swater/vplume/

Gacia, E.; Granata, T.C. & Duarte, C.M. (1999). An approach to measurements of particle flux and sediment retention within seagrass *Posidonia oceanica* meadows. *Aquatic Botany, vol 65,* pp. 255–268.

Gacia, E.; Invers, O.; Ballesteros, E.; Manzanera M. & Romero, J. (2007). The impact of the brine from a desalination plant on a shallow seagrass *(Posidonia oceanica)* meadow. *Estuarine, Coastal and Shelf Science, vol 72, Issue 4,* pp.579-590.

García, Marcelo. (1996). Environmental Hydrodynamics. Sante Fe, Argentina: Publications Center, Universidad Nacional del Litoral.

Gungor, E & Roberts, P.J. (2009). Experimental Studies on Vertical Dense Jets in a Flowing Current. *Journal of* Hydraulic *Enginnering, vol 135, No11,* pp.935-948.

Hauenstein, W & Dracos, TH. (1984). Investigation of plunging currents lacustres generated by inflows. *Journal of* Hydraulic *Research, vol 22,* No 3.

Hodges, B.R; Furnans, J.E & Kulis, P.S. (2010). Case study: A thin-layer gravity current with implications for desalination brine disposal. *Journal of Hydraulic Engineering (in press).*

Hogan, T. (2008). Impingement and Entrainment: Biological Efficacy of Intake Alternatives. *Desalination Intake Solutions Workshop Alden Research Laboratory.*

Holly, Forrest M., Jr., & Grace, John L., Jr. (1972). Model Study of Dense Jet in Flowing Fluid. *Journal of the Hydraulics* Division, *ASCE, vol. 98,* pp. 1921-1933.

Höpner, T & Windelberg, J. (1996). Elements of environmental impact studies on the coastal desalination plants. *Desalination (ELSEVIER), vol 108,* pp. 11-18.

Hópner, T. (1999). A procedure for environmental impact assessment (EIA) for seawater desalination plants. *Desalination (*ELSEVIER*), vol 124,* pp. 1-12.

Hyeong-Bin Cheong & Young-Ho Han (1997). Numerical Study of Two-Dimensional Gravity Currents on a Slope. *Journal of Oceanography, vol. 53*, pp. 179 - 192.

Iso, S; Suizu, S & Maejima, A. (1994). The Lethal Effect of Hypertonic Solutions and Avoidance of Marine Organisms in relation to discharged brine from a Desalination Plant. *Desalination (ELSEVIER), vol 97*, pp. 389-399.

Jirka, G-H. (2004). Integral model for turbulent buoyant jets in unbounded stratified flows. Part I: The single round jet. *Environmental Fluid Mechanics (ASCE), vol 4*, pp.1–56.

Jirka, G. H. (2006). Integral model for turbulent buoyant jets in unbounded stratified flows. Part II: Plane jet dynamics resulting from multiport diffuser jets. *Environmental Fluid Mechanics, vol. 6*, pp.43–100.

Jirka, G.H. (2008). Improved Discharge Configurations for Brine Effluents from Desalination Plants. *Journal of Hydraulic Enginnering*, vol 134, nº1, pp 116-120.

Joongcheol Paik, Eghbalzadeh, A; Sotiropoulos, F. (2009). Three-Dimensional Unsteady RANS Modeling of Discontinous Gravity Currents in rectangular Domains. *Journal of Hydraulic Engineering, vol 135*, n 6, pp. 505-521.

Kaminski, E; Tait, S & Carzzo, G. (2005). Turbulent entrainment in jets with arbitrary buoyancy. *Journal of Fluid Mechanics, vol. 526*, pp. 361-376.

Kikkert, G.A; Davidson, M.J; Nokes, R.I (2007). "Inclined Negatively Buoyant Discharges". *Journal of Hydraulic* engineering, *vol 133, pp.545 – 554.*

Lee, J.H.W. & Cheung, V. (1990). Generalized Lagrangian model for buoyant jets in current. *Journal of* Environmental *Engineering (ASCE), vol 116 (6)*, pp. 1085-1105.

Luyten P.J., Jones J.E., Proctor R., Tabor A., Tett P. & Wild-Allen K. (1999). COHERENS: A Coupled Hydrodynamical-Ecological Model for Regional and Shelf Seas: User Documentation. *MUMM Report, Management Unit of the Mathematical Models of the North Sea, 914.* www.mumm.ac.be/coherens/.

Martin, J.E; García, M.H. (2008). Combined PIV/LIF measurements of a steady density current front. *Exp Fluids, 46*, pp.265-276

Oliver, C.J; Davidson, M.J & Nokes, R.I. (2008). K-ε Predictions of the initial mixing of desalination discharges. Environmental *Fluid Mechanics, vol 8*: pp.617-625

Özgökmen, T.M. & E.P. Chassignet (2002). Dynamics of two-dimensional turbulent bottom gravity currents. Journal *of Physical Oceanography, vol 32/5*, pp.1460-1478.

Palomar, P & Losada, I.J. (2008). Desalinización de agua marina en España: aspectos a considerar en el diseño del sistema de vertido para protección del medio marino. *Public civil works Magazine (Revista de Obras Públicas). Nº 3486, pp. 37-52.*

Palomar, P & Losada, I.J. (2009). Desalination in Spain: Recent developments and Recommendations. *Desalination (ELSEVIER), vol 255*, pp. 97-106.

Palomar, P; Ruiz-Mateo, A; Losada, IJ; Lara, J L; Lloret, A; Castanedo, S; Álvarez, A; Méndez, F; Rodrigo, M; Camus, P; Vila, F; Lomónaco, P & Antequera, M. (2010). *"MEDVSA: a methodology for design of brine discharges into seawater" Desalination and Water Reuse, vol. 20/1*, pp. 21-25.

Palomar, P & Losada, I.J. (2010). "Desalination Impacts on the marine environment". *Book. The Marine Environment: ecology, management and conservation". Edit. NOVA Publishers.* Submitted.

Papanicolau, P, Papakonstantis, I.G; & Christodoulou, G.C.(2008). On the entrainment coefficients in negatively buoyant jets. *Journal of Fluid Mechanics, vol. 614,* pp. 447-470.

Pincice, A.B., & List, E.J. (1973). Disposal of brine into an estuary. *Journal Water Pollution, vol 45 (11),* pp. 2335-2344.

Plum, B.R. (2008). Modelling desalination plant outfalls. *Final Thesis Report. University of New South Wales at the Australia Defence Force Academy.*

Portillo, 2009. Instituto Tecnológico de Canarias. S.A. "Venturi Projects" , funded by the National Programme for Experimental Development of the Spanish Ministry of the Environment and Rural and Marine Affairs (2008-2010).

Querzoli, G. (2006) An experimental investigation of interaction between dense sea discharges and wave motion. *MWWD 2006, 4th International Conference on Marine Waste Water Disposal and Marine Environment.*

Raithby, G.D; Elliott, R.V; Hutchinson, B.R (1988). Prediction of three-dimensional thermal Discharge flows. *Journal of Hydraulic Division, ASCE 114(7),* 720–737

Roberts, P.J.W & Toms, G. (1987). Inclined dense jets in a flowing current. *Journal of Hydraulic Engineering, vol 113,* nº 3.

Roberts, P.J.; Ferrier, A & Daviero, G. (1997). Mixing in inclined dense jets. *Journal of Hydraulic Engineering (ASCE), vol. 123, No 8,* pp. 693-699.

Ross, A; Linden, F y Dalziel S.B. (2001). A study os three-dimensional gravity currents on a uniform slope. *Journal of Fluid Mechanics, vol 453,* pp.239-261.

Ruiz Mateo, A. (2007). Los vertidos al mar de las plantas desaladoras" (Brine discharges into seawaters). *AMBIENTA magazine, vol .51,* pp. 51-57.

Sánchez-Lizaso, J.L.; Romero, J.; Ruiz, J.; Gacia, E.; Buceta, J.L.; Invers, O.; Fernández Torquemada, Y.; Mas, J.; Ruiz-Mateo, A. & Manzanera, M. (2008). Salinity tolerance of the Mediterranean seagrass *Posidonia oceanica*: recommendations to minimize the impact of brine discharges from desalination plants. *Desalination (ELSEVIER), vol 221,* pp. 602-607.

Shao, D & Wing-Keung Lao. (2010). "Mixing and boundary interactions of 30º and 45º inclined dense jets". *Environmental Fluid Mechanics, vol 10, nº5,* pp. 521-553.

Terrados, J & Ros, J.D. (1992). Growth and primary production of *Cymodocea nodosa* (Ucria) Ascherson in a Mediterranean coastal lagoon: the Mar Menor (SE Spain). *Aquatic Botany, vol 43,* pp 63-74.

Tsihrintzis, V.A. and Alavian, V. (1986). Mathematical modeling of boundary attached gravity plumes. *Proceedings Intern. Symposium on Buoyant Flows, G. Noutsopoulos (ed), Athens, Greece,* pp.289-300.

Turner, J.S. (1966). Jets and plumes with negative or reversing buoyancy. *Journal of Fluid Mechanics,* vol. 26, pp 779-792.

Turner, J.S (1986). Turbulent entraintment: the development of the entraintment assumption, and its application to geophysical flows. *Journal of Fluid Mechanics, vol 173,* pp.431-471

VISJET: Innovative Modeling and Visualization Technology for Environmental Impact Assessment. http://www.aoe-water.hku.hk/visjet/visjet.htm

Zeitoun, M.A & McIlhenny, W.F. (1970). Conceptual designs of outfall systems for desalination plants. *Research and Development Progress Rept. No 550. Office of Saline Water, U.S. Dept, of Interior.*

14

Optimization of Hybrid Desalination Processes Including Multi Stage Flash and Reverse Osmosis Systems

Marian G. Marcovecchio[1,2,3], Sergio F. Mussati[1,4],
Nicolás J. Scenna[1,4] and Pío A. Aguirre[1,2]
[1]*INGAR/CONICET – Instituto de Desarrollo y Diseño, Avellaneda 3657 S3002GJC, Santa Fe,*
[2]*UNL – Universidad Nacional del Litoral, Santa Fe,*
[3]*UMOSE/LNEG-Und. de Modelação e Optimização de Sist. Energéticos, Lisboa,*
[4]*UTN/FRRo – Universidad Tecnológica Nacional, Rosario,*
[1,2,4]*Argentina*
[3]*Portugal*

1. Introduction

Distillation and reverse osmosis are the two most common processes to obtain fresh water from seawater or brackish water.

A leading distillation method is the Multi Stage Flash process (MSF). For this method, fresh water is obtained by applying thermal energy to seawater feed in multiple stages creating a distillate stream for fresh water uses, and a concentrated (brine) stream that is returned to the sea.

In Reverse Osmosis processes (RO), the seawater feed is pumped at high pressure to special membranes, forcing fresh water to flow through the membranes. The concentrate (brine) remains on the upstream side of the membranes, and generally, this stream is passed through a mechanical energy recovery device before being discharged back to the sea.

Desalination plants require significant amounts of energy as heat or electricity form and significant amounts of equipments. Reverse osmosis plants typically require less energy than thermal distillation plants. However, the membrane replacement and the high-pressure pumps increase the RO production cost significantly. Furthermore, even the salt concentration of permeated stream is low; this stream is not free of salt, as the distillate stream produced by a MSF system.

Therefore, hybrid system combining thermal and membrane processes are being studied as promising options. Hybrid plants have potential advantages of a low power demand and improved water quality; meanwhile the recovery factor can be improved resulting in a lower operative cost as compared to stand alone RO or MSF plants.

Several models have already been described in the literature to find an efficient relationship between both desalination processes (Helal et al., 2003; Agashichev, 2004; Cardona & Piacentino, 2004; Marcovecchio et al., 2005). However, these works analyse only specific fixed configurations for the RO-MSF hybridization.

In this chapter, all the possible configurations for hybrid RO-MSF plants are analyzed in an integrated way. A super-structure model for the synthesis and optimization of these structures is presented. The objective is to determine the optimal plant designs and operating conditions in order to minimize the cost per m³ of fresh water satisfying a given demand. Specifically, the work (Marcovecchio et al., 2009) is properly extended, in order to study the effect of different seawater concentrations on the process configuration. This will allow finding optimal relationships between both processes at different conditions, for a given fresh water demand.

2. Super-structure description

The modelled superstructure addresses the problem of the synthesis and optimization of hybrid desalination plants, including the Multi Stage Flash process: MSF and the Reverse Osmosis process: RO. The total layout includes one MSF and two RO systems, in order to allow the possibility of choosing a process of reverse osmosis with two stages. Many of the existing RO plants adopt the two stages RO configurations, since in some cases it is the cheapest and most efficient option.
Figure 1 illustrates the modelled superstructure. All the possible alternative configurations and interconnections between the three systems are embedded. The seawater feed passes through a Sea Water Intake and Pre-treatment system (SWIP) where is chemically treated, according to MSF and RO requirements. As Figure 1 shows, the feed stream of each process is not restricted to seawater; instead, different streams can be blended to feed each system. Then, part of the rejected stream leaving a system may enter into another one, even itself, resulting in a recycle. The permeated streams of both RO systems and the distillate stream from MSF are blended to produce the product stream, whose salinity is restricted to not exceed a maximum allowed salt concentration. Furthermore, a maximum salt concentration is imposed for the blended stream which is discharged back to the sea, in order to prevent negative ecological effects.

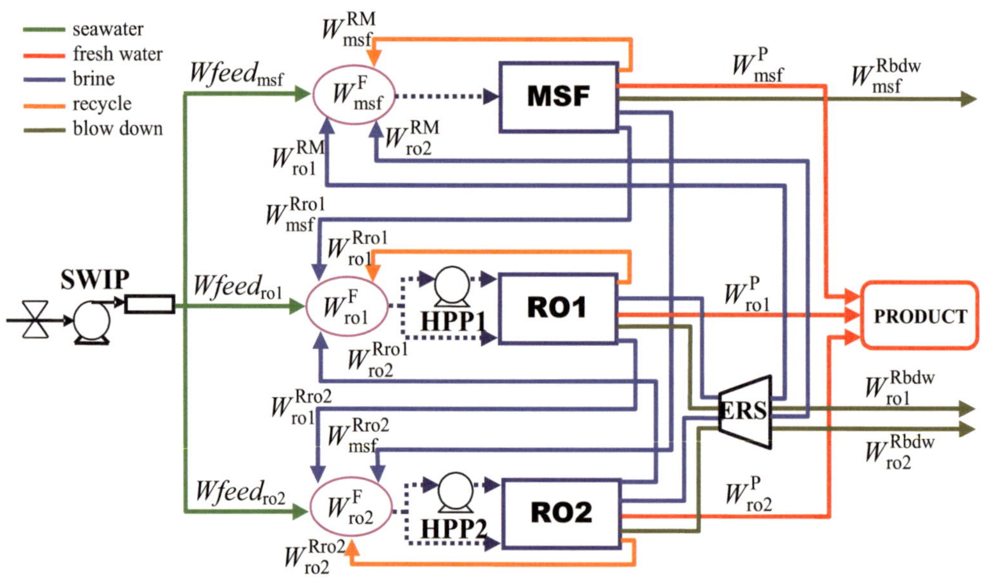

Fig. 1. Layout of the modelled superstructure

Seawater characteristics: salt concentration and temperature are given data, as well as the demand to be satisfied: total production and its maximum allowed salt concentration. On the contrary, the flow rate of the seawater streams fed to each system are optimization variables, as well as the flow rate and salt concentration of the product, blow down and inner streams.
The operating pressures for each RO system are also optimization variables. If the pressure of the stream entering to a RO system is high enough, the corresponding high pressure pumps are eliminated. Moreover, the number of modules operating in parallel at each RO system is also determined by the optimization procedure. The remainder rejected flow rate of both RO systems, if they do exist, will pass through an energy recovery system, before being discharged back to the sea or fed into the MSF system.
For the MSF system, the geometrical design of the evaporator, the number of tubes in the pre-heater, the number of flash stages, and others are considered as optimization variables.
The complete mathematical model is composed by four major parts: The Multi Stage Flash model, The Reverse Osmosis model, network equations and cost equations. The following section focuses on each of these four parts of the model.

3. Mathematical model

3.1 Multi Stage Flash model

The model representing the MSF system is based on the work (Mussati et al., 2004). A brief description of the model is presented here.
The evaporator is divided into stages. Each stage has a seawater pheheater, a brine flashing chamber, a demister and a distillate collector. Figure 2 shows a flashing stage.

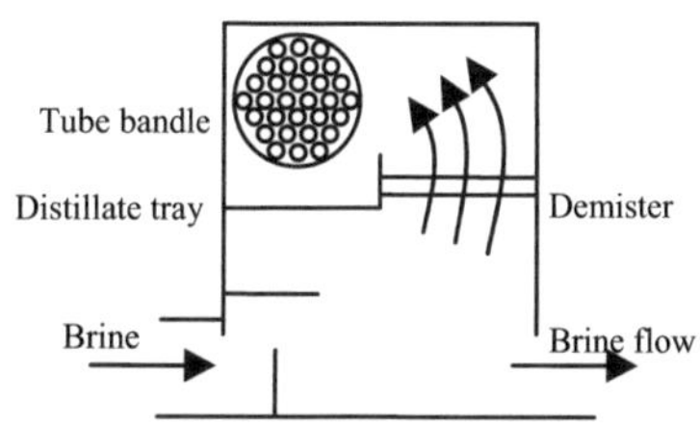

Fig. 2. Scheme of flashing stage

In a MSF system, feed stream passes through heating stages and is heated further in the heat recovery sections of each subsequent stage. Then, feed is heated even more using externally suplied steam. After that, the feedwater passes through various stages where flashing takes place. The vapor pressure at each stage is controlled in such way that the heated brine enters each chamber at the proper temperature and pressure to cause flahs operation. The flash vapor is drawn to the cooler tube bundle surfaces where it is condensed and collected as distillate and paseses on from stage to stage parallelly to the brine. The distillate stream is also flash-boiled, so it can be cooled and the surplus heat recovered for preheating the feed. Figure 3 shows an scheme of a MSF system with NS stages.
Often, part of the brine leaving the last stage is mixed with the incoming feedwater because it reduces the chemical pre-treatment cost. According to the interconections and recirculations considered in the modeled superstructure, two typical MSF operating modes are included: MSF-OT (without recycle) and MSF-BR (with recycle). However, more complex configurations are also included, since different streams can be blended (at different proportions) to feed the MSF system.

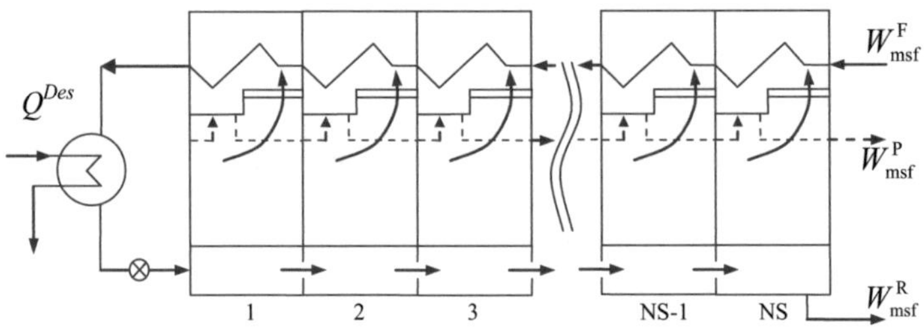

Fig. 3. MFS system

The MSF model considers all the most important aspects of the process.
The heat consumption is calculated by:

$$Q^{Des} = W^{F}_{msf}\ Cp_{msf}\ \Delta t\ 10^{-6}\ \rho^{b} \tag{1}$$

$$\Delta t = \Delta t_f + \Delta t_e + BPE \tag{2}$$

Total heat transfer area and number of flash stages are calculated as:

$$A_t = \frac{W^{F}_{msf}\ 10^{3}\ Cp_{msf}}{U}\ \ln\left(\frac{\Delta t - BPE}{\Delta t_e}\right)^{(T_{max} - \Delta t - T^{F}_{msf})/\Delta t_f} \tag{3}$$

$$NS = \left(T_{max} - \Delta t - T^{F}_{msf}\right)/\Delta t_f \tag{4}$$

The total production of distillate is evaluated by:

$$W^{P}_{msf} = W^{F}_{msf}\left[1 - \left(1 - \frac{Cp_{msf}\Delta t_f}{\lambda}\right)^{NS}\right] \tag{5}$$

The following equation establishes a relation between heat transfer area, number of tubes and chamber width:

$$A_t = \pi\ TD\ B_{msf}\ N_t\ NS \tag{6}$$

The stage height can be approximated by:

$$Hs = 2\ Lb + Ds \tag{7}$$

The number of rows of tubes in the vertical direction is related to the number of tubes in the following way:

$$N_{rt} = 0.481\ TD\sqrt{N_t} \tag{8}$$

The following equation relates the shell diameter to the number of rows of tubes and Pitch:

$$Ds = N_{rt}\ P_t\ \sqrt{2} \tag{9}$$

The length of the desaltor is constrained by the following two equations:

$$L_d = \frac{W^{\mathrm{P}}_{\mathrm{msf}}\ 10^{-3}}{B_{\mathrm{msf}}\ V_{vap}\ \rho_{vap}} \tag{10}$$

$$L_d = Ds\ NS \tag{11}$$

The total stage surface area is calculated by:

$$A_S = 2\ L_d\ B_{\mathrm{msf}} + 2\ Hs\ L_d\ + Hs\ B_{\mathrm{msf}}\ NS \tag{12}$$

Finally, the temperature of last flashing stage of the MSF system is calculated as:

$$T^{\mathrm{R}}_{\mathrm{msf}} = T_{\max} - NS\ \Delta t_f = T^{\mathrm{F}}_{\mathrm{msf}} + \Delta t \tag{13}$$

Despite the simplifying hypothesis assumed in the model, the MSF process is well represented and the solutions of this model are accurately enough to establish conclusions for the hybrid plant.

3.2 Reverse osmosis model

The model representing the RO system is based on the work (Marcovecchio et al., 2005). A brief description of the equations is presented here.

Each RO system is composed by permeators operating in parallel mode and under identical conditions. Particularly, data for DuPont B10 hollow fiber modules were adopted here. However, the model represents the permeation process for general hollow fiber modules and any other permeator could be considered providen the particular module parameters.

Figure 4 represents the RO system modeled for the hybrid plant.

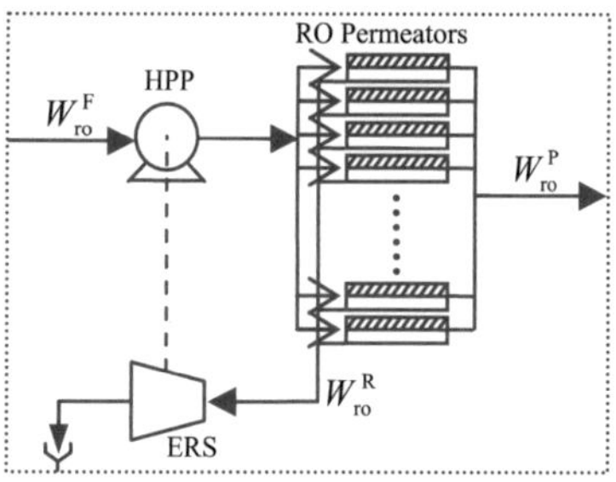

Fig. 4. RO system

Initially, pressure of inlet stream is raised by the High Pressure Pumps (HPP). Then, the pressurized stream passes through membrane modules, where permeation takes place. Part of the rejected stream could pass through the energy recovery system, before being discharged back to the sea or fed into the MSF system. Therefore, part of the power required for the whole plant is supplied by the energy recovery system, and the rest will be provided by an external source.

Equations (14) to (30) describe the permeation process taking place at one module of each system.

The transport phenomena of solute and water through the membrane are modeled by the Kimura-Sourirajan model (Kimura & Sourirajan, 1967):

$$J_s^w = 3600\, A \left(\overline{P}_s^b - \overline{P}_s^p - \frac{iRT_s\, \rho^b \left(C_s^m - C_s^P \right)}{10^6\, Ms\, 101325} \right) \qquad s=ro1,\, ro2 \tag{14}$$

$$J_s^S = \frac{3600\, B \left(C_s^m - C_s^P \right) \rho^b}{10^6} \qquad s=ro1,\, ro2 \tag{15}$$

The velocity of flow is:

$$V_s^w = \frac{\left(J_s^w + J_s^S \right)}{\rho^P} \qquad s=ro1,\, ro2 \tag{16}$$

The following equation gives the salt concentration of the permeate stream:

$$C_s^P = \frac{J_s^S 10^6}{V_s^W \rho^P} \qquad s=ro1,\, ro2 \tag{17}$$

Permeate flow rate is calculated as the product between the permeation velocity and the membrane area:

$$Q_s^P = V_s^w A_m \qquad s=ro1,\, ro2 \tag{18}$$

The total material balance for each permeator is:

$$Q_s^f = Q_s^b + Q_s^P \qquad s=ro1,\, ro2 \tag{19}$$

The salt balance in each permeator is given by:

$$Q_s^f C_s^F = Q_s^b C_s^R + Q_s^P C_s^P \qquad s=ro1,\, ro2 \tag{20}$$

The phenomenon of concentration polarization must be considered. The principal negative consequence of this phenomenon is a reduction in the fresh water flow. The approach widely used to model the influence of the concentration polarization is the film theory. The Sherwood, Reynolds and Schmidt numbers are combined in an empirical relation: equation (24) to calculate the mass transfer coefficient:

$$Sh_s = \frac{k_s\, 2\, r_0}{D} \qquad s=ro1,\, ro2 \tag{21}$$

$$\mathrm{Re}_s = \frac{2\, r_0\, U_s^S\, \rho^b}{\mu^b} \qquad s=ro1,\, ro2 \tag{22}$$

$$Sc_s = \frac{\mu^b}{\rho^b\, D} \qquad s=ro1,\, ro2 \tag{23}$$

$$Sh_s = 2.725\left(\mathrm{Re}_s\right)^{1/3}\left(Sc_s\right)^{1/3} \qquad \text{s=ro1, ro2} \tag{24}$$

The concentration polarization phenomenon is modeled by:

$$\frac{C_s^m - C_s^P}{C_s^R - C_s^P} = \exp\left(\frac{V_s^w}{3600\,k_s}\right) \qquad \text{s=ro1, ro2} \tag{25}$$

In order to estimate the average pressure drop in the fiber bore and the average pressure drop on the shell side of the fiber bundle, it is necessary to calculate the superficial velocity in the radial direction. According to (Al-Bastaki & Abbas, 1999), the superficial velocity can be approximated as the log mean average of the superficial velocity at the inner and outer radius of the fiber bundle:

$$U_s^{si} = \frac{Q_s^f}{3600\;2\,\pi\,R_i\;L} \qquad \text{s=ro1, ro2} \tag{26}$$

$$U_s^{so} = \frac{Q_s^f - V_s^w A_m}{3600\;2\,\pi\,R_o\;L} \qquad \text{s=ro1, ro2} \tag{27}$$

$$U_s^S = \frac{U_s^{si} - U_s^{so}}{\log\left(U_s^{si}/U_s^{so}\right)} \qquad \text{s=ro1, ro2} \tag{28}$$

The approximation for the pressure drop in the fiber bore is based on Hagen-Poiseuille's equation:

$$\overline{P}_s^P = 1 + \frac{1}{2}\frac{16\,\mu^P\,r_o\;V_s^w\;L^2}{3600\;r_i^4\;101325} \qquad \text{s=ro1, ro2} \tag{29}$$

Similarly, the pressure drop on the shell side of the fiber bundle is estimated by Ergun's equation:

$$\overline{P}_s^b = P_s^f - \frac{1}{2}\frac{150(1-\varepsilon)^2\,\mu^b U_s^S}{\varepsilon^3 d_p{}^2 101325}\left(R_o - R_i\right) - \frac{1}{2}\frac{1.75(1-\varepsilon)\rho^b\left(U_s^S\right)^2}{\varepsilon^3 d_p 101325}\left(R_o - R_i\right) \qquad \text{s=ro1, ro2} \tag{30}$$

Finally, the total flow rates of feed and permeate for each system are given by:

$$W_s^F = NM_s Q_s^f \qquad \text{s=ro1, ro2} \tag{31}$$

$$W_s^P = NM_s Q_s^P \qquad \text{s=ro1, ro2} \tag{32}$$

The chosen model considers all the most important aspects affecting the permeation process. Even thought, differential equations involved in the modeling are estimated without any discretization, the whole model is able to predict the flow of fresh water and salt trough the membrane in an accuracy way.

3.3 Network equations

The overall superstructure is modelled in such way that all the interconnections between the three systems are allowed, as it shown in Figure 1.

In effect, part of the rejected stream of each system can enter into another system, even itself. The fractions of rejected streams of RO systems that will enter into MSF system or that will be discharged back to the sea, will pass through the ERS. On the contrary, the fractions of rejected streams of RO systems that will enter into a RO system again, will not pass through the ERS, because the plant could benefit from these high pressurized streams. In fact, when all the streams entering to a RO system flow at a high enough pressure, the corresponding HPPs can be avoided. That RO system would correspond to a second stage of reverse osmosis. In that case, the pressure of all the inlet streams will be levelled to the lowest one, by using appropriated valves. However, if at least one of the RO inlet streams is coming from MSF system or from sea, the pressure of all the inlet streams will be lowered to atmospheric pressure, and before entering membrane modules, HPPs will be required. The network and cost equations are formulated is such way that the optimization procedure can decide the existence or not of HPPs and this decision is correctly reflected in the cost functions. When the whole model is optimized, the absence of a particular stream is indicated by the corresponding flow rate being zero. Furthermore, the optimization procedure could decide the complete elimination of one system for the optimal design. The energy and material balances guarantee the correct definition of each stream.

The total fresh water demand is 2000 m^3/h and is the result of blending the product stream of each system:

$$W_{\mathrm{msf}}^{\mathrm{P}} + W_{\mathrm{ro1}}^{\mathrm{P}} + W_{\mathrm{ro2}}^{\mathrm{P}} = prodc \tag{33}$$

The fresh water stream must not exceed a maximum allowed salt concentration. This requirement is imposed by the following constraint, taking into account that distillate stream is free of salt, but permeate RO streams are not.

$$c_{\max} \geq \left(NM_{\mathrm{ro1}} Q_{\mathrm{ro1}}^{\mathrm{P}} C_{\mathrm{ro1}}^{\mathrm{P}} + NM_{\mathrm{ro2}} Q_{\mathrm{ro2}}^{\mathrm{P}} C_{\mathrm{ro2}}^{\mathrm{P}} \right) \Big/ prodc \tag{34}$$

For ecological reasons, the salinity of the blended stream which is discharged back to the sea must not be excessively high. An acceptable maximum value for this salinity is 67000 ppm:

$$C_{\mathrm{msf}}^{\mathrm{R}} W_{\mathrm{msf}}^{\mathrm{Rbdw}} + C_{\mathrm{ro1}}^{\mathrm{R}} W_{\mathrm{ro1}}^{\mathrm{Rbdw}} + C_{\mathrm{ro2}}^{\mathrm{R}} W_{\mathrm{ro2}}^{\mathrm{Rbdw}} \leq 67000 \left(W_{\mathrm{msf}}^{\mathrm{Rbdw}} + W_{\mathrm{ro1}}^{\mathrm{Rbdw}} + W_{\mathrm{ro2}}^{\mathrm{Rbdw}} \right) \tag{35}$$

By considering all the possible streams that can feed MSF system, the following equations give the flow rate of MSF feed stream:

$$W_{\mathrm{msf}}^{\mathrm{F}} = Wfeed_{\mathrm{msf}} + W_{\mathrm{msf}}^{\mathrm{RM}} + W_{\mathrm{ro1}}^{\mathrm{RM}} + W_{\mathrm{ro2}}^{\mathrm{RM}} \tag{36}$$

Consequently, salt and energy balances for MSF feed are:

$$C_{\mathrm{msf}}^{\mathrm{F}} W_{\mathrm{msf}}^{\mathrm{F}} = Cfeed\ Wfeed_{\mathrm{msf}} + C_{\mathrm{msf}}^{\mathrm{R}} W_{\mathrm{msf}}^{\mathrm{RM}} + C_{\mathrm{ro1}}^{\mathrm{R}} W_{\mathrm{ro1}}^{\mathrm{RM}} + C_{\mathrm{ro2}}^{\mathrm{R}} W_{\mathrm{ro2}}^{\mathrm{RM}} \tag{37}$$

$$T_{\mathrm{msf}}^{\mathrm{F}} W_{\mathrm{msf}}^{\mathrm{F}} = Tfeed\ Wfeed_{\mathrm{msf}} + T_{\mathrm{msf}}^{\mathrm{R}} W_{\mathrm{msf}}^{\mathrm{RM}} + T_{\mathrm{ro1}} W_{\mathrm{ro1}}^{\mathrm{RM}} + T_{\mathrm{ro2}} W_{\mathrm{ro2}}^{\mathrm{RM}} \tag{38}$$

The overall mass and salt balances for MSF system are given by:

$$W_{msf}^{F} = W_{msf}^{P} + W_{msf}^{RM} + W_{msf}^{Rro1} + W_{msf}^{Rro2} + W_{msf}^{Rbdw} \tag{39}$$

$$C_{msf}^{F} W_{msf}^{F} = C_{msf}^{R} \left(W_{msf}^{RM} + W_{msf}^{Rro1} + W_{msf}^{Rro2} + W_{msf}^{Rbdw} \right) \tag{40}$$

Similarly to equation (36), the following equations give the flow rate of RO feed streams:

$$W_{ro1}^{F} = Wfeed_{ro1} + W_{msf}^{Rro1} + W_{ro1}^{Rro1} + W_{ro2}^{Rro1} \tag{41}$$

$$W_{ro2}^{F} = Wfeed_{ro2} + W_{msf}^{Rro2} + W_{ro1}^{Rro2} + W_{ro2}^{Rro2} \tag{42}$$

Equations (43) and (44) establish the division of the total rejected stream leaving each RO system in the different assignations:

$$NM_{ro1} Q_{ro1}^{b} = W_{ro1}^{RM} + W_{ro1}^{Rro1} + W_{ro1}^{Rro2} + W_{ro1}^{Rbdw} \tag{43}$$

$$NM_{ro2} Q_{ro2}^{b} = W_{ro2}^{RM} + W_{ro2}^{Rro1} + W_{ro2}^{Rro2} + W_{ro2}^{Rbdw} \tag{44}$$

The salt balances for RO system feeds are:

$$C_{ro1}^{F} W_{ro1}^{F} = Cfeed\ Wfeed_{ro1} + C_{msf}^{R} W_{msf}^{Rro1} + C_{ro1}^{R} W_{ro1}^{Rro1} + C_{ro2}^{R} W_{ro2}^{Rro1} \tag{45}$$

$$C_{ro2}^{F} W_{ro2}^{F} = Cfeed\ Wfeed_{ro2} + C_{msf}^{R} W_{msf}^{Rro2} + C_{ro1}^{R} W_{ro1}^{Rro2} + C_{ro2}^{R} W_{ro2}^{Rro2} \tag{46}$$

Meanwhile, energy balances for RO systems feeds are given by:

$$T_{ro1} W_{ro1}^{F} = Tfeed\ Wfeed_{ro1} + T_{msf}^{R} W_{msf}^{Rro1} + T_{ro1} W_{ro1}^{Rro1} + T_{ro2} W_{ro2}^{Rro1} \tag{47}$$

$$T_{ro2} W_{ro2}^{F} = Tfeed\ Wfeed_{ro2} + T_{msf}^{R} W_{msf}^{Rro2} + T_{ro1} W_{ro1}^{Rro2} + T_{ro2} W_{ro2}^{Rro2} \tag{48}$$

The overall mass balances for RO systems are:

$$W_{ro1}^{F} = W_{ro1}^{P} + W_{ro1}^{RM} + W_{ro1}^{Rro1} + W_{ro1}^{Rro2} + W_{ro1}^{Rbdw} \tag{49}$$

$$W_{ro2}^{F} = W_{ro2}^{P} + W_{ro2}^{RM} + W_{ro2}^{Rro1} + W_{ro2}^{Rro2} + W_{ro2}^{Rbdw} \tag{50}$$

The following equations establish the overall salt balances for RO systems:

$$C_{ro1}^{F} W_{ro1}^{F} = C_{ro1}^{P} W_{ro1}^{P} + C_{ro1}^{R} \left(W_{ro1}^{RM} + W_{ro1}^{Rro1} + W_{ro1}^{Rro2} + W_{ro1}^{Rbdw} \right) \tag{51}$$

$$C_{ro2}^{F} W_{ro2}^{F} = C_{ro2}^{P} W_{ro2}^{P} + C_{ro2}^{R} \left(W_{ro2}^{RM} + W_{ro2}^{Rro1} + W_{ro2}^{Rro2} + W_{ro2}^{Rbdw} \right) \tag{52}$$

Equations (53) to (60) assign to the variables $P_{ro1}{}^{in}$ and $P_{ro2}{}^{in}$ the minimal pressure over all the flows entering to the corresponding RO system. This assignation will allow the model to decide whether the HPPs before each RO system are necessary or not. In fact, if the minimal

pressure of the inlet streams: P^{in} is equal or greater than the pressure needed to pass through the membrane modules: P^{f}, then the corresponding HPPs are not necessary. On the other hand, if the value of P^{in} does not reach the operating pressure P^{f}, then the corresponding HPPs cannot be avoided. In the following section, this decision will be modelled by the cost functions.
If the stream feeding the RO1 system includes part of brine stream leaving the MSF system, equation (53) imposes that the corresponding variable $P_{ro1}{}^{in}$ be lower or equal than atmospheric pressure. On the contrary, if no stream coming from MSF system is feeding the RO1 system (i.e. $W_{msf}{}^{Rro1}=0$), then constraint (53) does not affect variable $P_{ro1}{}^{in}$ at all. Equation (56) performs the same imposition by evaluating the existence or not of stream coming from the sea in the RO1 feed.
Equations (54) and (55) evaluate the existence of streams coming from an RO system and feeding RO1 system. If any of these streams does exist (i.e. $W_{ro1}{}^{Rro1}>0$ or $W_{ro2}{}^{Rro1}>0$), the variable $P_{ro1}{}^{in}$ is imposed to be lower than the pressure of the corresponding stream.

$$W_{msf}^{Rro1}\left(P_{ro1}^{in}-1\right)\leq 0 \tag{53}$$

$$W_{ro1}^{Rro1}\left(P_{ro1}^{in}-(2P_{ro1}^{b}-P_{ro1}^{f})\right)\leq 0 \tag{54}$$

$$W_{ro2}^{Rro1}\left(P_{ro1}^{in}-(2P_{ro2}^{b}-P_{ro2}^{f})\right)\leq 0 \tag{55}$$

$$Wfeed_{ro1}\left(P_{ro1}^{in}-1\right)\leq 0 \tag{56}$$

Equations (57) to (60) act in analogous way to the four previous ones for the system RO2.

$$W_{msf}^{Rro2}\left(P_{ro2}^{in}-1\right)\leq 0 \tag{57}$$

$$W_{ro1}^{Rro2}\left(P_{ro2}^{in}-(2P_{ro1}^{b}-P_{ro1}^{f})\right)\leq 0 \tag{58}$$

$$W_{ro2}^{Rro2}\left(P_{ro2}^{in}-(2P_{ro2}^{b}-P_{ro2}^{f})\right)\leq 0 \tag{59}$$

$$Wfeed_{ro2}\left(P_{ro2}^{in}-1\right)\leq 0 \tag{60}$$

When the HPPs before an RO system are avoided, it is not convenient that the corresponding system operates at pressure lower than the available one. The following equations guarantee that, and also ensure the correct definition of associated cost functions.

$$P_{ro1}^{f}\geq P_{ro1}^{in} \tag{61}$$

$$P_{ro2}^{f}\geq P_{ro2}^{in} \tag{62}$$

Most of the constraints presented in this section are complementary to the cost functions described in the following section.

3.4 Cost equations

This section describes the cost equations of the total plant. The objective function to be minimized is the cost per m^3 of produced fresh water. Capital and operating costs are calculated. The cost equations were formulated in such way that they can correctly reflect the presence or absence of equipments, streams or systems.

Capital costs are calculated by equations (63) to (67), while equations (69) to (76) estimate the operating ones.

Cost function reported by (Malek et al., 1996) was adopted in order to estimate capital cost for the SWIP:

$$cc_{\mathrm{swip}} = 996\left((Wfeed_{\mathrm{msf}} + Wfeed_{\mathrm{ro1}} + Wfeed_{\mathrm{ro2}})\,24\right)^{0.8} \tag{63}$$

Capital cost of HPP is defined in the same way. As it was explained at section 3.3, the variables P^{in} assume the minimal pressure over all the streams feeding a RO system, while P^{f} is the operating pressure of the system. Equations (64) and (65) along with the optimization procedure, will make the variables cc_{hpp} to assume the capital cost of the HPP only when $P^{f} > P^{in}$, otherwise cc_{hpp} will assume value null.

$$\left(cc_{\mathrm{hpp1}} - \frac{W^{\mathrm{F}}_{\mathrm{ro1}}}{450}\left(393000 + 10710\, P^{\mathrm{f}}_{\mathrm{ro1}}\, 1.01325\right)\right)\cdot\left(P^{\mathrm{f}}_{\mathrm{ro1}} - P^{\mathrm{in}}_{\mathrm{ro1}}\right) \ge 0 \tag{64}$$

$$\left(cc_{\mathrm{hpp2}} - \frac{W^{\mathrm{F}}_{\mathrm{ro2}}}{450}\left(393000 + 10710\, P^{\mathrm{f}}_{\mathrm{ro2}}\, 1.01325\right)\right)\cdot\left(P^{\mathrm{f}}_{\mathrm{ro2}} - P^{\mathrm{in}}_{\mathrm{ro2}}\right) \ge 0 \tag{65}$$

Capital cost of the ERS is similar to the HPP one, since it consists of a reverse running centrifugal pump. Taking into account flow rate and pressure of the streams passing through the ERS, the capital cost is given by:

$$\begin{aligned} cc_{\mathrm{ers}} = {} & \frac{(W^{\mathrm{Rbdw}}_{\mathrm{ro1}} + W^{\mathrm{RM}}_{\mathrm{ro1}})}{450}\left(393000 + 10710\,(2P^{\mathrm{b}}_{\mathrm{ro1}} - P^{\mathrm{f}}_{\mathrm{ro1}})\,1.01325\right) + \\ & \frac{(W^{\mathrm{Rbdw}}_{\mathrm{ro2}} + W^{\mathrm{RM}}_{\mathrm{ro2}})}{450}\left(393000 + 10710\,(2P^{\mathrm{b}}_{\mathrm{ro2}} - P^{\mathrm{f}}_{\mathrm{ro2}})\,1.01325\right) \end{aligned} \tag{66}$$

The capital cost considered for the MSF system is the one due to the heat transfer area. According to (Mussati et al., 2006) this cost can be estimated as:

$$cc_{\mathrm{area}} = (A_t + A_S\, 25)\, 50 \tag{67}$$

Therefore, the plant equipment cost is: $cc_{\mathrm{eq}} = cc_{\mathrm{swip}} + cc_{\mathrm{hpp1}} + cc_{\mathrm{hpp2}} + cc_{\mathrm{area}}$. Civil work cost is estimated as a 10% of cc_{eq} (Wade, 2001). Indirect cost is estimated in the same way (Helal et al., 2003). Then, the Total Capital Cost (TCC) is given by:

$$TCC = cc_{\mathrm{eq}} + cc_{\mathrm{cw}} + cc_{\mathrm{i}} = 1.2\, cc_{\mathrm{eq}} = 1.2\,(cc_{\mathrm{swip}} + cc_{\mathrm{hpp1}} + cc_{\mathrm{hpp2}} + cc_{\mathrm{ers}} + cc_{\mathrm{area}}) \tag{68}$$

Capital charge cost is estimated as a 8% of the total capital cost (Malek et al., 1996):

$$co_{\mathrm{c}} = 0.08\, TCC \tag{69}$$

The cost due to permeators is included as operative cost, by calculating their annualized installation cost and considering the replacement of 20% of permeators per year. According to (Wade, 2001) this sum can be estimated as $397.65 per module per year.

$$co_{rp} = (NM_{ro1} + NM_{ro2})\ 397.65 \tag{70}$$

Energy cost is calculated by using the cost function given in (Malek et al., 1996) and the power cost reported in (Wade, 2001). The energy required by the SWIP and the HPP; and the energy provided by the ERS must be taken into account:

$$\begin{aligned} co_e = 0.03\ f_c \Bigg(& \frac{P_{swip}(Wfeed_{msf} + Wfeed_{ro1} + Wfeed_{ro2})\ 24}{eff_{swip}} \\ & + \frac{(P^f_{ro1} - P^{in}_{ro1})\ 1.01325\ W^F_{ro1}\ 24}{eff_{hpp}} + \frac{(P^f_{ro2} - P^{in}_{ro2})\ 1.01325\ W^F_{ro2}\ 24}{eff_{hpp}} \\ & - eff_{ers}\ 1.01325\ (2P^b_{ro1} - P^f_{ro1})\ 24\ (W^{Rbdw}_{ro1} + W^{RM}_{ro1}) - eff_{ers}\ 1.01325\ (2P^b_{ro2} - P^f_{ro2})\ 24\ (W^{Rbdw}_{ro2} + W^{RM}_{ro2}) \Bigg) \end{aligned} \tag{71}$$

Spares costs are calculated by using the estimated values reported by (Wade, 2001):

$$co_s = 24\ \ 365\ (W^P_{ro1} + W^P_{ro2})\, f_c\ 0.033 + 24\ \ 365\ \ W^P_{msf}\ f_c\ 0.082 \tag{72}$$

Chemical treatment costs is calculated using the cost per m³ of feed reported in (Helal et al., 2003):

$$\begin{aligned} co_{ch} = {} & 24\ \ 365\ (Wfeed_{ro1} + W^{Rro1}_{msf} + Wfeed_{ro2} + W^{Rro2}_{msf})\ f_c\ 0.018 \\ & + 24\ \ 365\ (Wfeed_{msf} + W^{RM}_{ro1} + W^{RM}_{ro2})\ f_c\ 0.024 \end{aligned} \tag{73}$$

General operation and maintenance cost is calculated according to the value per m³ of produced water reported in (Wade, 2001):

$$co_{om} = 24\ \ 365\ (W^P_{msf} + W^P_{ro1} + W^P_{ro2})\, f_c\ 0.126 \tag{74}$$

Similarly, power cost for MSF system is evaluated according to (Wade, 2001):

$$co_{pw} = 24\ \ 365\ W^P_{msf}\ f_c\ 0.109 \tag{75}$$

The cost of the heat consumed by MSF system is calculated by using the function proposed by (Helal et al., 2003):

$$co_{ht} = 24\ 365\ f_c\ (Q^{Des}\ 10^6 / \lambda)\ (T_{max} - 323)\ 0.00415\ /85 \tag{76}$$

Finally, the Annual Operating Cost (AOC) is given by:

$$AOC = co_c + co_{rp} + co_e + co_s + co_{ch} + co_{om} + co_{pw} + co_{ht} \tag{77}$$

By considering a plant life of 25 years (n) and a discount rate of 8% (i), capital recovery factor can be calculated, giving: $crf = ((i+1)^n - 1)/(i(i+1)^n)$. Finally, fresh water cost per m³ is given by:

$$\cos t = \frac{TCC/crf + AOC}{prodc\ 24\ 365} \tag{78}$$

Equations (1) to (78) define the model for the design and operation of a hybrid desalination plant, including MSF and RO systems.
In the following section, this model will be optimized for different seawater salt concentrations, and the obtained solutions will be analysed.

4. Results: Optimal plant designs and operating conditions

In this section optimized results are presented and discussed.
The proposed optimization problem P is defined as follows:

P: minimize *cost*

s. t. Equations (1) to (78)

while all the variables have appropriated bounds.
The optimization procedure will look for the optimal layout and operating conditions in order to minimize the cost per m³ of produced fresh water.
It is important to note that almost all discrete decisions were modelled exploiting the actual value of flow rates and pressures. Thus, no binary decision variables were included into the model. Only four integer variables are involved: the number of flash stages and the number of tubes in the pre-heater at the MSF system; and the number of permeators operating in parallel at each RO system.
Tables 1 and 2 list the parameter values used for the RO and MSF systems, respectively.

Parameters for RO systems	
i, number of ions for ionized solutes	2
R, ideal gas constant, N m / kgmole K	8315
Ms, solute molecular weight	58.8
T, seawater temperature, °C	25
ρ^b, brine density, kg/m³	1060
ρ^p, pure water density, kg/m³	1000
μ^p, permeated stream viscosity, kg/m s	0.9x10^{-3}
μ^b, brine viscosity, kg/m s	1.09x10^{-3}
D, diffusivity coefficient, m²/s	1x10^{-9}
P_{swip}, SWIP outled pressure, bar	5
eff_{swip}, intake pump efficiency	0.74
eff_{hpp}, high pressure pumps efficiency	0.74
eff_{ers}, energy recovery system efficiency	0.80
f_c, load factor	0.90

Table 1. Parameters for RO systems

Parameters and operating ranges of the particular hollow fiber permeator were taken from (Al-Bastaki &Abbas, 1999; Voros et al., 1997). These specifications constitute constants and bounds for some variables of the model.

Parameters for MSF system	
T_{max}, K	385
Cp_{msf}, Kcal/(kg K)	1
TD, m	0.030
Pitch: P_t	1.15
BPE, K	1.9
U, Kcal/m^2/K/h	2000
λ, Kcal/Kg	550

Table 2. Parameters for MSF system

The optimization model was implemented in General Algebraic Modeling System: GAMS (Brooke et al., 1997) at a Pentium 4 of 3.00 GHz. At first, the MINLP solver DICOPT was implemented to solve the problem. Unfortunately, the solver failed to find even a feasible solution for most case studies. Then, other resolution strategy was carried out in order to tackle the problem and obtain the optimal solutions.

Since it involves only 4 integer variables, the problem was solved in 2 steps. Firstly, the relaxed NLP problem was solved, i.e., the integer variables were relaxed to continuous ones. Departing from the optimal solution of the relaxed problem, the MINLP was solved by fixing the integer variables at the nearest integer values and optimizing the remaining variables. Since the MINLP problem presents a lot of non-convexities, a global search strategy was also implemented. In fact, for each study case, the previous 2 steps were repeated starting the optimization search from different initial points, and then, the best local optimal solution was selected. The generalized reduced gradient algorithm CONOPT was used as NLP solver. This resolution procedure was successful, providing optimal solutions in all case studies. The total CPU time required to solve all the cases was 1.87s, what proves that the proposed procedure is highly efficient and the model is mathematically good conditioned.

11 case studies were solved for seawater salt concentration going from 35000 ppm up to 45000 ppm. The total production was fixed at 2000m^3/h with a maximum allowed salt concentration of 570 ppm.

Table 3 shows the values of the main interconnection variables for the optimal solutions: feed flow rates, product and internal streams, as well as their salt concentrations.

Table 4 reports design variables and operating conditions for each process for the optimal solutions.

For seawater salt concentrations between 35000 and 38000 ppm, the optimal solutions do not include the MSF system. In fact, for these salinities, the optimal hybrid plant designs consist on a typical two stage RO plant. However, if the seawater salinity is greater than 38000 ppm, both desalination processes are present in the optimal design of the plant; that is: including MSF system is profitable.

Figure 5 shows a scheme of the optimal design of the plant obtained for seawater salinities between 35000 and 38000 ppm.

Optimal solutions for the hybrid plant: MSF-RO. Total Production: 2000m³/h. Maximum allowed salt concentration: 570pm												
Seawater salinity *Cfeed*,ppm		35000	36000	37000	38000	39000	40000	41000	42000	43000	44000	45000
Production flow rates for MSF and RO processes												
W_{msf}^{P}, m³/h		-	-	-	-	132.4	210.6	285.9	358.5	428.8	496.4	561.9
($W_{ro1}^{P}+W_{ro2}^{P}$), m³/h		2000	2000	2000	2000	1867.6	1789.4	1714.1	1641.5	1571.2	1503.6	1438.1
Seawater feed: *Wfeed*, m³/h												
MSF		-	-	-	-	304.0	543.9	794.6	1057.7	1336.0	1631.7	1947.5
RO1		4241.3	4373.0	4498.0	4790.1	4441.0	4376.8	4315.4	4256.7	4199.9	4144.8	4091.6
RO2		-	-	-	-	-	-	-	-	-	-	-
Flow rates of the interconnection streams: *W*, m³/h												
MSF	F	-	-	-	-	1146.4	1808.6	2444.1	3044.2	3574.3	4073.0	4546.9
MSF	P	-	-	-	-	132.4	210.6	285.9	358.5	428.8	496.4	561.9
MSF	RM	-	-	-	-	725.4	1169.2	1599.3	1986.5	2238.3	2441.3	2599.4
MSF	Rro1	-	-	-	-	-	-	-	-	-	-	-
MSF	Rro2	-	-	-	-	-	-	-	-	-	-	-
MSF	Rbdw	-	-	-	-	288.6	428.7	558.8	699.2	907.2	1135.3	1385.6
RO1	F	4241.3	4373.0	4498.0	4790.1	4441.0	4376.8	4315.4	4256.7	4199.9	4144.8	4091.6
RO1	P	1499.6	1495.1	1492.3	1540.7	1385.2	1322.5	1262.4	1204.8	1149.1	1095.8	1044.6
RO1	RM	-	-	-	-	-	-	-	-	-	-	-
RO1	Rro1	-	-	-	-	-	-	-	-	-	-	-
RO1	Rro2	2741.7	2877.9	3005.7	2808.8	3055.8	3054.3	3053.0	3051.8	3050.8	3049.0	3047.2
RO1	Rbdw	-	-	-	440.6	-	-	-	-	-	-	-
RO2	F	2741.7	2877.9	3005.7	2808.8	3055.8	3054.3	3053.0	3051.8	3050.8	3049.0	3047.2
RO2	P	500.4	504.9	507.7	459.3	482.4	466.9	451.7	436.8	422.1	407.8	393.7
RO2	RM	-	-	-	-	117.1	95.4	50.2	-	-	-	-
RO2	Rro1	-	-	-	-	-	-	-	-	-	-	-
RO2	Rro2	-	-	-	-	-	-	-	-	-	-	-
RO2	Rbdw	2241.3	2373.0	2498.0	2349.5	2456.3	2492.0	2551.2	2615.0	2628.6	2641.2	2653.4
Salt concentration of the interconnection streams: *C*, ppm												
MSF	F	-	-	-	-	60317.6	58059.1	56833.3	56050.2	55726.6	55530.6	55431.6
MSF	R	-	-	-	-	68194.8	65712.0	64363.0	63531.0	63322.7	63237.4	63247.0
RO1	F	35000	36000	37000	38000	39000	40000	41000	42000	43000	44000	45000
RO1	P	383.8	402.4	419.8	437.8	455.4	475.3	496.2	518.2	541.5	566.1	592.1
RO1	R	53934.2	54493.6	55160.9	55810.8	56472.6	57113.9	57747.9	58375.5	58992.1	59610.0	60220.3
RO2	F	53934.2	54493.6	55160.9	55810.8	56472.6	57113.9	57747.9	58375.5	58992.1	59610.0	60220.3
RO2	P	940.5	951.6	978.9	1013.6	1055.3	1095.3	1137.1	1180.6	1226.4	1274.4	1324.8
RO2	R	65764.6	65884.8	66174.0	66522.5	66861.1	67221.6	67577.7	67928.1	68269.1	68617.4	68959.8
Cost of fresh water, \$/m³		0.6784	0.6952	0.7121	0.7410	0.7717	0.7985	0.8240	0.8486	0.8726	0.8968	0.9198

Table 3. Optimal solutions for the hybrid plant: interconnection variables

Optimal solutions for the hybrid plant: MSF-RO. Design variables and operating conditions.											
Seawater salinity: *Cfeed*, ppm	35000	36000	37000	38000	39000	40000	41000	42000	43000	44000	45000
MSF											
Q^{Des}, Gcal/h	-	-	-	-	8.80	12.93	16.78	20.46	23.89	27.10	30.14
NS	-	-	-	-	19	23	26	28	30	32	33
A_S, m^2	-	-	-	-	828.8	1103.5	1332.8	1515.5	1679.8	1836.0	1957.8
A_t, m^2	-	-	-	-	11858.9	19552.9	27100.3	34308.4	40568.8	46406.2	52594.8
Δt, K	-	-	-	-	7.24	6.75	6.48	6.34	6.30	6.28	6.25
Δt_f, K	-	-	-	-	3.54	2.95	2.63	2.46	2.34	2.23	2.19
Δt_e, K	-	-	-	-	1.80	1.89	1.95	1.99	2.07	2.15	2.16
N_t	-	-	-	-	368	501	615	723	798	855	940
L_d, m	-	-	-	-	8.55	12.09	15.13	17.66	19.88	21.96	23.74
Hs, m	-	-	-	-	1.45	1.53	1.58	1.63	1.66	1.67	1.72
D_S, m	-	-	-	-	0.45	0.53	0.58	0.63	0.66	0.67	0.72
T_{msf}^F, K	-	-	-	-	310.5	310.3	310.3	309.9	308.6	307.4	306.3
T_{msf}^R, K	-	-	-	-	317.7	317.1	316.7	316.2	314.9	313.7	312.6
RO1											
NM_1	4633	4777	4915	5232	4843	4773	4706	4642	4580	4520	4462
P^f_1, atm	67.900	67.900	67.900	67.899	67.900	67.900	67.899	67.900	67.893	67.898	67.900
HPP1	yes	yes	yes	yes	yes	yes	yes	yes	yes	yes	yes
T_{ro1}, K	298.0	298.0	298.0	298.0	298.0	298.0	298.0	298.0	298.0	298.0	298.0
RO2											
NM_2	3129	3189	3286	3063	3332	3331	3330	3328	3327	3325	3323
P^f_2, atm	67.868	67.868	67.868	67.866	67.867	67.867	67.867	67.867	67.860	67.865	67.866
HPP2	no	no	no	no	no	no	no	no	no	no	no
T_{ro2}, K	298.0	298.0	298.0	298.0	298.0	298.0	298.0	298.0	298.0	298.0	298.0

Table 4. Optimal solutions for the hybrid plant: design variables and operating conditions

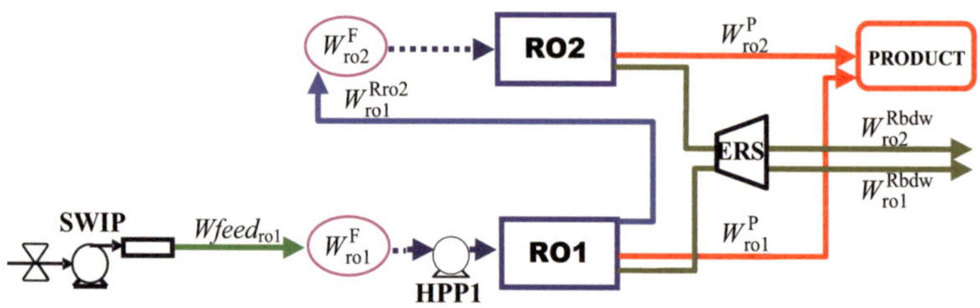

Fig. 5. Scheme of the optimal design for seawater salinities between 35000 and 38000 ppm

The stream with flow rate W_{ro1}^{Rbdw} is only present for 38000 ppm of seawater salinity. For salinities lower than 38000 ppm, the totality of the stream rejected from the first RO stage: system RO1, enters into the second RO stage: system RO2. Then, the stream entering into the system RO2 is sufficiently pressurized. Therefore, the high pressure pumps before system RO2 are avoided in the optimal solutions. This decision is properly made by the optimization procedure, and it is correctly reflected in the cost function.

Figure 6 shows a scheme of the optimal solutions obtained for seawater salt concentrations between 39000 and 45000 ppm.

For these case studies, both desalination processes are present at the optimal hybrid plant design. The RO systems work as a two-stage RO plant, i.e.: system RO1 is fed directly from sea, while its rejected stream is the fed stream for system RO2. No other streams are blended to feed RO systems.

Regarding MSF system, it operates with an important recycle. This re-circulated stream reduces the chemical pre-treatment and raises the feed stream temperature with the consequent reduction of external heat consumption. Both factors straightly impact on the final cost.

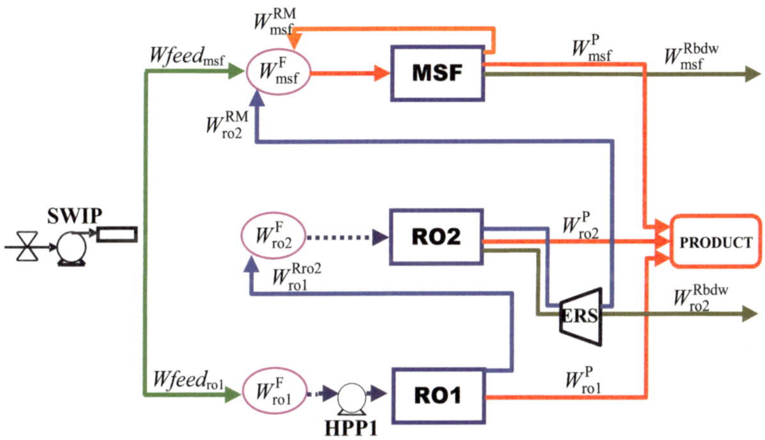

Fig. 6. Scheme of the optimal design for seawater salinities between 39000 and 45000 ppm

As it is also shown in Table 3, the three first cases presented in Figure 6 include the stream with flow rate $W_{ro2}{}^{RM}$. However, for seawater salt concentrations higher than 41000 ppm this flow rate is null and the stream does not exist. Then, for the last four case studies, even though the two desalination processes are selected for the optimal plant design, they operate in independent way. In fact: there is no stream connecting the MSF and RO processes. However, both processes share the intake and pre-treatment system. Furthermore, the salinity of the product stream satisfies the maximum allowed salt concentration requirement because the three product streams are blended. As it can be seen at Table 3, if only the permeate streams coming from RO systems are blended, then the salt concentration of the resulting stream will be far above the maximum allowed salt concentration.

Again, the stream feeding system RO2 is composed only by the stream rejected from system RO1 and it is high pressurized. Thus, the high pressure pumps before system RO2 are unnecessary and consequently, they are avoided at the optimal design.

Figure 7 shows the fresh water produced by each desalination process for all the case studies.

As it was mentioned, for seawater salt concentrations below 38000 ppm, MSF system is not present, thus the demand is totally satisfied by RO systems. On the other hand, for seawater salinities higher than 38000 ppm, both processes contribute to satisfy the demand. Although the RO systems produce more fresh water than MSF system, the MSF production increases according to the seawater salinity rise.

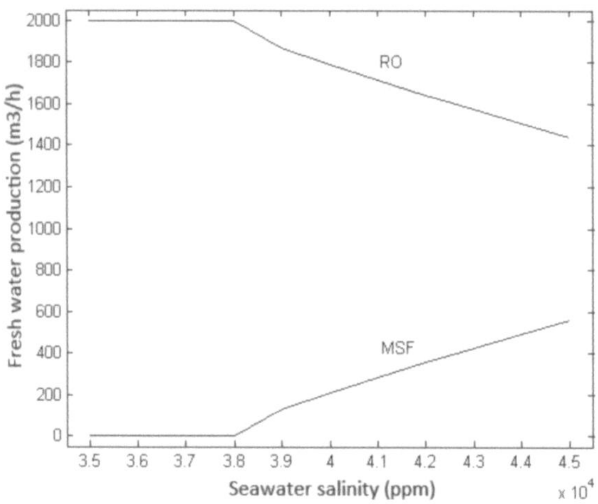

Fig. 7. Fresh water production

If the MSF system would not be considered, for the optimal design of a two stage RO plant the capacity of the second stage will decrease when the seawater salinity increases (Marcovecchio et al., 2005). In fact, even though the stream rejected from the first RO stage is high pressurized and it could enter into a second stage with no need of high pressure pumps; in the optimal design, part of this stream is discharged back to the sea. The second stage capacity will continue decreasing until only one RO stage is the optimal design for high feed salt concentration. The reason why it is no longer profitable to use the stream rejected from the first RO stage is that its salinity is too high. Thus, the salt concentration of the potential permeate will be lower but also high and then, it is not possible to satisfy the maximum allowed salt concentration even by blending with the first stage permeate. Therefore, the fresh water produced by the first stage must be higher in order to satisfy the demand. Consequently, the flow rate of seawater is increased. As a consequence, the cost per m^3 of fresh water increases, since many costs are directly affected.

Contrary, in a hybrid plant where the MSF process is available, in that break point where the optimal design of a RO plant changes, it begins to be profitable to complement the RO production with distillated from MSF system.

Then, for feed salinities higher than 38000 ppm, the growth of MSF system production is approximately linear. With this plant design, there is no stream rejected from the first RO stage being discharged back to the sea, i.e. the totality of that stream enters into the second RO stage. And the total production of the plant reaches the requirement of maximum allowed salt concentration by blending the slightly concentrated permeate from RO systems with the free of salt distillate of MSF process.

Finally, Figure 8 compares the cost per m^3 of fresh water produced with the optimal configurations obtained for the hybrid plant with those obtained for the RO stand alone plant.

As it is shown in Figure 8, the cost reduction reached with the hybrid RO-MSF plant is considerable. For feed salinities between 39000 and 44000 ppm, the cost function has an

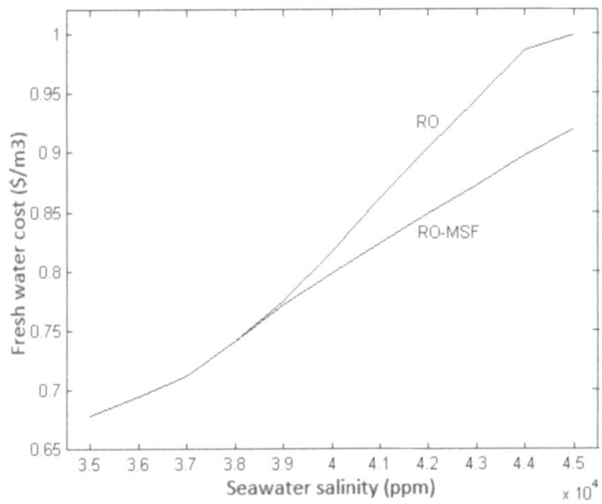

Fig. 8. Fresh water cost for hybrid RO-MSF plants and RO stand alone plants

almost linear growth with respect to the seawater salinity, for both: RO and hybrid plant. However, the growth rate associated to the hybrid plant cost is far lower.

For comparative purposes, optimal designs for the MSF stand alone plant were calculated. That is, it was calculated the cost per m^3 of fresh water produced by the MSF-once through process satisfying the same demand: 2000 m^3/h. In the implemented model, this cost is not affected by the feed salinity. The cost obtained for the case studies was $1.1683. Also, the designs obtained are the same for all the case studies, since the only constraint that could affect the solution is the one requiring that the concentration of the stream discharged back to sea be lower than 67000 ppm, but this limit is not reached in any case.

5. Conclusion

In this work, a MINLP mathematical model for the optimal synthesis and design of hybrid desalination plants, including the two conversion processes: reverse osmosis and multiple stage flash evaporation, was presented.

The MSF model is based on a previous work presented by (Mussati et al., 2004). It involves real-physical constraints for the evaporation process and is derived on energy, mass and momentum balances. In addition, geometric dimensions of stages including chambers and pre-heaters are considered as optimization variables. Heat exchange areas of condensers are also design variables to be determined.

The RO model with hollow fiber permeators is based on the work (Marcovecchio et al., 2005). For this model, the transport phenomena of solute and water through the membrane are modelled by Kimura-Sourirajan model. The concentration polarization phenomenon is taken into account. The Hagen-Poiseuille and Ergun equations are employed to calculate the pressure drops. In the RO model, the number of permeators operating in parallel, the operating pressure and flow rates are the main optimization variables.

The modelled hybrid plant includes two RO and one MSF systems. The proposed superstructure allows optimizing not only operating conditions but also process configurations simultaneously. Thus, the model includes network constraints which are related to all potential interconnections between the three systems.

Network constraints ensure the correct definition of flow rates, salt concentrations and temperatures for each stream.
Cost equations take into account all the factors affecting the cost of each process. Certainly, capital investment and operating cost of all process equipments were considered.
Optimal solutions for eleven case studies were obtained, for different seawater salinities. Then optimal designs and operating conditions were determined by minimizing the cost per m^3 of produced fresh water. Cost equations are able to reflect accurately the presence or absence of certain equipment, stream or even a whole system.
From the optimal solutions, it can be concluded that the RO stand alone plant is the best option for feed salinities between 35000 and 38000 ppm. In fact, the optimal design obtained for these cases consist on a two-stage RO plant while the MSF process was completely eliminated.
However, when the seawater salinity rises, it is profitable to integrate the MSF system in a hybrid plant. Actually, for feed salinities higher than 38000 ppm both desalination processes are present at the optimal plant design. In these cases, the integration of MSF process allows a better use of the rejected streams leaving the first and second RO stages. As a consequence, the final fresh water cost is reduced. It is important to note that although the RO production is higher than the MSF one, the MSF capacity increases according to the seawater salinity rise.
Then, important conclusions about the relationship between membrane and thermal desalination processes can be established from the optimal solutions presented in this work. In fact, the optimal hybrid plants were described for different seawater conditions, in order to minimize the cost of producing fresh water.
In future works, more detailed models for each process will be included in the superstructure problem, in order to improve the model presented here. In addition, other interconnections between the two studied processes will be considered, as the incorporation of streams coming from RO system in different stages of the MSF evaporator. Then, the interaction between both desalination processes will be more flexible and may lead to reduce the total cost of the process.

6. Acknowledgements

The authors acknowledge financial support from 'Agencia Nacional de Promoción Científica y Tecnológica' (ANCyT), and 'Consejo de Investigaciones Científicas y Técnicas' (CONICET), Argentina.

7. Nomenclature

Subscripts

msf	Multi Stage Flash System
ro1	Reverse Osmosis System 1
ro2	Reverse Osmosis System 2

Superscripts

F	Feed
P	Permeate - Product

R	Rejected – Concentrated brine
Rbdw	Rejected to be blown down
RM	Rejected brine entering into MSF system
Rro1	Rejected brine entering into RO1 system
Rro2	Rejected brine entering into RO1 system

ρ^b	brine density - RO, kg / m^3
ρ^p	pure water density - RO, kg / m^3
μ^b	brine viscosity - RO, kg / (m s)
μ^p	permeated stream viscosity - RO, kg / (m s)
λ	latent heat evaporation – MSF, Kcal / kg
Q^f	feed flow rate per membrane module - RO, m^3/h
Q^p	permeate flow rate per membrane module - RO, m^3/h
Q^b	brine flow rate inside the shell per membrane module - RO, m^3/h
C^m	salt concentration at the membrane wall - RO, ppm
ρ_{vap}	vapor density – MSF, kg / m^3
Δt	temperature drop – MSF, K
J^w	water flux - RO, kg/m^2.h
J^S	solute flux - RO, kg/m^2.h
ε	void fraction - RO
Δt_e	effective driving force for the heat transfer operation – MSF, K
Δt_f	temperature drop for the flashing operation – MSF, K
P^f	feed stream pressure - RO, atm
$\overline{P}^p$	Average pressure in the fiber bore - RO, atm
$\overline{P}^b$	Average pressure on the shell side of the fiber bundle - RO, atm
V^w	velocity of permeation flow - RO, m/h
U^{so}	Superficial velocity at the outer radius of the fiber bundle - RO, m/s
U^{si}	Superficial velocity at the inner radius of the fiber bundle - RO, m/s
U^S	superficial velocity in the radial direction of the bulk stream - RO, m/s
A	pure water permeability constant - RO, kg/m^2.s.atm
A_m	module membrane area - RO, m^2
AOC	Annual operating cost, $/y
A_S	Total stage surface area – MSF, m^2
A_t	total heat transfer area – MSF, m^2
B	solute permeability constant - RO, m/s
B_{msf}	chamber width – MSF, m
BPE	boiling point elevation, K
C	salt concentration, ppm
cc_{area}	capital cost of heat transfer area of MSF system, $
cc_{cw}	capital cost of civil work, $
cc_{eq}	total equipment cost, $
cc_{ers}	capital cost for the Energy Recovery System, $
cc_{hpp1}	capital cost of High Pressure Pumps for system RO1, $
cc_{hpp2}	capital cost of High Pressure Pumps for system RO2, $

cc_i	indirect capital cost, $
cc_{swip}	capital cost for the Seawater Intake and Pre-treatment system, $
$Cfeed$	feed salt concentration, ppm
c_{max}	maximum salt concentration allowed for the product stream, ppm
co_c	capital charge cost, $/year
co_{ch}	chemical treatment cost, $/year
co_e	energy cost, $/year
co_{ht}	cost of the heat consumed by system MSF, $/year
co_{om}	general operation and maintenance cost, $/year
co_{pw}	power cost for system MSF, $/year
co_{rp}	cost of permeator replacement, $/year
co_s	spares cost, $/year
$cost$	cost per m^3 of produced fresh water, $/$m^3$
Cp_{msf}	heat capacity – MSF, Kcal / (kg K)
crf	capital recovery factor
D	diffusivity coefficient - RO, m^2 / s
d_p	specific surface diameter - RO, m
Ds	shell diameter – MSF, m
eff_{ers}	energy recovery system efficiency
eff_{hpp}	high pressure pumps efficiency
eff_{swip}	intake pump efficiency
f_c	load factor
Hs	chamber height – MSF, m
i	number of ions for ionized solutes - RO
k	mass transfer coefficient - RO, m/s
L	length of fiber bundle - RO, m
Lb	level of brine in the flashing chamber – MSF, m
L_d	length of desaltor – MSF, m
Ms	solute molecular weight – RO
NM	number of membrane module operating in parallel mode in each RO system
N_{rt}	number of rows in the vertical direction – MSF
NS	number of flashing stages – MSF
N_t	number of tubes – MSF
p^{in}	pressure of the stream entering into each RO system – RO, atm
$prodc$	total plant production, m^3/h
P_{swip}	seawater intake system outlet pressure, bar
P_t	Pitch – MSF
Q^{Des}	external heat consumption – MSF, Gcal/h
R	ideal gas constant - RO, N m / (kgmol K)
Re	Reynolds number $(2.r_0.U_s^S.\,\rho^b / \mu^b)$ - RO
R_i	inner radius of the fiber bundle - RO, m
r_i	inner fiber radius - RO, m
R_o	outer radius of the fiber bundle - RO, m
r_o	outer fiber radius - RO, m

Sc	Schmidt number ($\mu^b / \rho^b . D$) - RO
Sh	Sherwood number ($2.k.r_o/D$) - RO
T	temperature, K
TCC	total capital cost, $
TD	tube diameter - MSF, m
Tfeed	seawater temperature, K
T_{max}	maximum brine temperature - MSF, K
U	overall heat transfer coefficient - MSF, Kcal / (K m^2 h)
V_{vap}	vapor velocity - MSF, m / s
W	flow rate, m^3/h
Wfeed	seawater feed flow, m^3 / h

8. References

Agashichev, S.P. (2004). Analysis of integrated co-generative schemes including MSF, RO and power generating systems (present value of expenses and "levelised" cost of water). *Desalination*, 164 (3): 281-302, ISSN: 0011-9164.

Al-Bastaki, N.M. & Abbas, A. (1999). Modeling an industrial reverse osmosis unit. *Desalination* 126 (1-3): 33-39, ISSN: 0011-9164.

Brooke, A.; Kendrick, D.; Meeraus, A. & Raman, R. (1997). GAMS Language Guide, Release 2.25, Version 92. GAMS Development Corporation.

Cardona, E. & Piacentino, A. (2004). Optimal design of cogeneration plants for seawater desalination. *Desalination* 166: 411-426, ISSN: 0011-9164.

Helal, A.M.; El-Nashar, A.M.; Al-Katheeri, E. & Al-Malek, S. (2003). Optimal design of hybrid RO/MSF desalination plants. Part I: Modeling and algorithms. *Desalination* 154 (1): 43-66, ISSN: 0011-9164.

Kimura, S. & Sourirajan, S. (1967). Analysis of data in reverse osmosis with porous cellulose acetate membranes. *AIChE Journal* 13 (3): 497-503, ISSN: 1547-5905.

Malek, A.; Hawlader, M.N.A. & Ho, J.C. (1996). Design and economics of RO seawater desalination. *Desalination* 105 (3): 245-261, ISSN: 0011-9164.

Marcovecchio, M.G.; Aguirre, P.A. & Scenna, N.J. (2005). Global optimal design of reverse osmosis networks for seawater desalination: modeling and algorithm. *Desalination* 184 (1-3): 259-271, ISSN: 0011-9164.

Marcovecchio, M.G.; Mussati, S.F.; Aguirre, P.A. & Scenna, N.J. (2005). Optimization of hybrid desalination processes including multi stage flash and reverse osmosis systems. *Desalination* 182 (1-3): 111-122, ISSN: 0011-9164.

Marcovecchio, M.G.; Mussati, S.F.; Scenna, N.J. & Aguirre, P.A. (2009). Global optimal synthesis of integrated hybrid desalination plants, *Computer Aided Chemical Engineering*, Vol 26: 573-578, ISBN-13: 978-0-444-53433-0. Elsevier Science B.V. 19th European Symposium on Computer Aided Process Engineering (ESCAPE19), 14-17 June 2009. Cracow, Poland.

Mussati, S.F. ; Aguirre, P.A. & Scenna, N.J. (2006). Superstructure of alternative Configurations of the multistage flash desalination process. *Industrial & Engineering Chemistry Research* 45 (21): 7190-7203, ISSN: 0888-5885.

Mussati, S.F.; Marcovecchio, M.G.; Aguirre, P.A. & Scenna, N.J. (2004). A new global optimization algorithm for process design: its application to thermal desalination processes. *Desalination* 166: 129-140, ISSN: 0011-9164.

Voros, N.G. ; Maroulis, Z.B. & Marinos-Kouris, D. (1997). Short-cut structural design of reverse osmosis desalination plants. *Journal of Membrane Science* 127 (1): 47-68, ISSN: 0376-7388.

Wade, N.M. (2001). Distillation plant development and cost update. *Desalination* 136 (1-3): 3-12, ISSN: 0011-9164.